Springer Climate

Springer Climate is an interdisciplinary book series dedicated to climate research. This includes climatology, climate change impacts, climate change management, climate change policy, regional climate studies, climate monitoring and modeling, palaeoclimatology etc. The series publishes high quality research for scientists, researchers, students and policy makers. An author/editor questionnaire, instructions for authors and a book proposal form can be obtained from the Publishing Editor.

Now indexed in Scopus® !

Jaap M. de Heer · Giasuddin Ahmed Choudhury

Strategies for Adaptation to Climate Change in a Transformative Approach

Building on the Bangladesh Experience

Springer

Jaap M. de Heer
Groot-Ammers, The Netherlands

Giasuddin Ahmed Choudhury
Dhaka, Bangladesh

ISSN 2352-0698 ISSN 2352-0701 (electronic)
Springer Climate
ISBN 978-3-031-88883-0 ISBN 978-3-031-88884-7 (eBook)
https://doi.org/10.1007/978-3-031-88884-7

This Springer imprint is published by the registered company Springer Nature Switzerland AG
The registered company address is: Gewerbestrasse 11, 6330 Cham, Switzerland

To Ada and Rezina,
for understanding and patience
during numerous busy days and absence!

Foreword

We are pleased to prepare this book in which the making and content of the Bangladesh Delta Plan 2100 (BDP 2100) are highlighted and expanded for a broad audience. Moreover, this publication provides an introductory and concluding part to place the Delta Plan in a broader scientific as well as practical context. Adaptation to climate change in a transformative approach is based on the understanding that adaptation is not a routine or well-known activity but that transformation of institutional, organizational as well as of physical systems is required in an uncertain context. In this respect, Bangladesh Delta Plan 2100 is a unique adaptive plan following its own approach of 'adaptation by development' for which a number of professional practices have been applied. It has been prepared to jointly address adaptation to climate change and natural hazards, food security and socio-economic development. Bangladesh has been experiencing rapid socio-economic development in recent times. However, making this development sustainable is challenging due to vulnerability to natural calamities and climate change impacts. We expect that BDP 2100 serves the country, ensuring long-term water and food security, economic growth and environmental sustainability while reducing vulnerability to natural disasters and building resilience to climate change and other delta-related challenges. It would be great if building on the Bangladesh experience and expanding it inspires others to plan and work for achieving an adapted and safe future.

Groot-Ammers, The Netherlands — Dr. Jaap M. de Heer
Dhaka, Bangladesh — Giasuddin Ahmed Choudhury, M.Eng.

Preface

Many people are aware of the discourse about climate change, and the way public debates developed last years. Institutions and different media are frequently showing climate-related studies and reports of scientists, taken seriously by activists in a variety of actions and protests and increasingly by the larger public and a growing number of companies. The warnings are interpreted in completely different ways from support to denial by political and business leaders and the broader public. The regular COP meetings draw frequent attention, showing the difficulty of making progress in worldwide decision making on financial arrangements and measures to limit greenhouse emissions to stop further temperature rise. Reduction of carbon dioxide and methane emissions, energy transition and also financing of climate adaptation measures show progress but still prove to be controversial at the global level. Some political leaders and parts of companies and the public tend to be neutral or even take positions of denial, refusing to take or accept action. Others are requesting for more robust decisions on reduction of the greenhouse gases as well as on financing of climate change adaptation and damage and loss due to impacts of the changes already taking place. The Green Climate Fund is, e.g. in terms of financial resources, more or less complete but is at the same time too bureaucratic to provide funds effectively for covering the costs of adaptation and of damage and loss in developing countries. For the latter, it's positive that an additional fund has been generated under COP28 to finance damage and loss costs related to climate change impacts. Also here, the challenge is to get necessary funds timely available in an effective probably more decentralized way.

With this publication, we clearly take part in the discourse. Data from IPCC and other sources combined with real-life observations increasingly show that climate change is happening and, e.g., temperature is increasing, probably to an unwanted or even dangerous level. Urgency of policies and adaptive decision making on robust strategies and measures to mitigate climate change, especially the rise of temperature and greenhouse gases is increasingly based on scientific climate data, observations and analysis, but many worldwide conferences show that it is hard to develop consistent joint action. Besides mitigation strategies and measures, adaptation on the ground in terms of protection and water safety and security is gradually achieving more attention, especially in countries threatened or already harmed by the impacts of climate change. Unexpected speed and seriousness of developments like extreme weather conditions, severe drought situations and fire incidents, high seawater temperature, extremely heavy rains, floodings, and disasters as well as accelerated melting processes are impressive and calling for action. Our consideration is clear, take mitigation actions, be prepared for possible serious consequences and adapt to (expected) impacts on the ground while there is some lead time. The focus of this book is adapting to expected impacts of climate change, learning from examples, especially from one of the largest and most vulnerable deltas in the world, the Bangladesh delta. This means that adaptation to climate change is here primarily related to the issues and challenges of low-lying delta areas. However, the overall approach and applied practices or a selection of them can be used in a variety of climate change adaptation circumstances.

'Adaptation' is however an abstract concept, and it needs to be seen in a specific context, where impact of climate change can already be observed or where it can be expected. Adaptation may also be confusing in the sense that it may not be enough to just make some adjustments to new circumstances because climate change requires most probably also renewal of systems, innovative solutions, and strategic thinking. Moreover, adaptation needs to be related to in that context living people, their cities, houses, infrastructure, environment, economy, physical systems as well as to the existing policies, rules, plans to overcome impacts of climate change by measures with respect to, e.g., water resources or to heat stress. Adaptation to climate change means in other words more than technical adjustment of systems, it should be seen and treated as a process of change which touches on the many aspects of water and environment and thus sectors of human life, society, and ecology. When the concept of adaptation is used in this book, it is meant in this broader meaning.

Adaptation over time can be interpreted as a transition process, the process or a period of changing from one state or condition to another. The concept of transition may apply on an abstract, overarching level, where it means moving from a situation facing climate change without adequate measures to a state with flourishing living conditions. However, climate change touches in the course of time upon so many aspects of society, economy, culture, physical and ecological systems, it is time bounded, multifaceted, and consequently multi-sectoral by nature. Because of this, adaptation is complicated with respect to strategizing, planning, and taking actions because (1) climate change implies multi-sectoral involvement with various transformations over time and (2) the end situation is uncertain and unclear, making it hard to generate ways to deal with the ongoing changes of the climate and its impacts. Uncertainty and ambiguity about the way and the extent to which the climate will change and about the regional and local implications leads to time-related open-ended situations with many contingencies. The concepts of transition and transition management fall short here as they suppose a clear start and transition process to a defined end state. Climate change and its impacts are too broad, complicated, and open-ended as it relates to societal, political, administrative, organizational, technical, and physical processes. So defining the beginning and end of a transition is difficult while the suggestion that such transition can be managed in one go may not be realistic.

What deserves more attention with respect to climate change adaptation is an action-oriented strategic management approach which makes it possible for government institutions, private sector, and stakeholders to generate actions with respect to contingencies, uncertainties, and preparing for the future. To find an adequate approach of adaptation to climate change, the idea of transformation with dedicated changes and practices has been introduced, building on literature and practices of strategic management. This book further conceptualizes adaptation to climate change in this way. Adaptation literature often focuses on impact-related technical content. Interesting is also how to approach and organize adaptation, find useful practices and interventions on the ground. This publication builds on insights from strategic management and organization sciences, blending them with technology and related disciplines. The book shows how adaptation has been applied in this way in a number of countries vulnerable for climate change impacts, especially focusing on Bangladesh, and briefly on the Netherlands and Vietnam. These countries follow adaptive approaches with some similarities but also significant variations. The experiences with the Bangladesh Delta Plan 2100 provide the main practical input. BDP 2100 sounds as a single and final plan but is in fact the start of an adaptive process with an inbuilt pattern of frequent updating, adjustment, or renewal in relation to actual or expected circumstances and developments.

Solving Bangladesh's existing water problems and coping with still uncertain climate change-induced problems requires longer-term and consistent strategic thinking and planning with probably frequent adjustments and transformations over time. The BDP 2100 aims to address this objective by introducing an adaptive and as far as possible evidence-based delta plan. For formulating such delta plan, the focus on content plays an important role, defining the state of the delta systems with their issues and challenges and search for alternative solutions and interventions. For developing strategies and measures under uncertain climatic conditions and in terms of interpretation of such ambiguous situations, formulating a vision, aspirations and goals, the processes of strategic management and adaptive decision making come forward. Besides, it is essential to look at the broader context with requirements of climate change, societal dynamics, and economy. Closely related is the governance to steer adaptation and transformation processes through an adequate institutional framework, institutions at different levels coupled with the different stakeholders.

Worldwide, Bangladesh is considered to be one of the most vulnerable countries concerning climate change but is also seen as a resilient nation, overcoming many disasters and setbacks. Since its independence in 1971, the country has faced many natural hazards and is well aware of climate change and possible impacts. Observations show, that already a combination of sea-level rise, (possible) changes in the monsoon as well as erratic heavy rainfall and temperature rise with more extreme events have created large-scale social, environmental, and economic impacts and pressure on millions of vulnerable people living in the Bangladesh delta. In the future, climate change is likely to increase flood and drought risks, increase monsoon river floods, and saltwater intrusion in the country. Under the characteristic uncertainties, the government considered the climate change challenges as highly risky and initiated a long-term, holistic planning process, and serious institutional transformation.

The title of this book refers to how adaptation to climate change is viewed as transformative approach in relation to a wide palette of factors and practices and a broad array of relevant trends and developments. As said, adaptation to climate change is an uncertain endeavor. So, it is useful to utilize appropriate practices to design the strategy process and to apply them effectively within that process. Adaptation to climate change cannot be seen as a routine planning practice because it shows new challenges and uncertainties, but it is evolving based on useful practices, generating insights, strategies, actions, and transformations. One of them is imagining the future in different ways and forming narratives of possible futures, considering a set of uncertain but dominant developments. Under different scenarios, the strategies and actions may differ, providing the possibility to fit in the unfolding reality. Impacts of climate change not only call for responses but also for critical evaluation and reflection leading to proactive action. However, these bring about interventions and change, e.g. institutionally, in organization and management setups and working processes as well as in the physical systems. This whole set of interrelated changes has been indicated as a transformative approach. Transformative means causing a marked change or reform in a relevant situation, structure, process, system or area, like in institutional arrangements, organizations, systems for water safety and security, agriculture, navigation, urban areas, etc. Transformation takes time, happens in a process of becoming, and requires continuous reality construction. Elements of this are creative thinking, considerations of stakeholders, an essential governmental role, mobilization of the private sector and managerial (inter)action. A crucial lever of change here is sensemaking in close interaction (sometimes called 'management of meaning') between involved parties and communication in various forms and with different purposes. Important interactions are seen in the broader discourse on climate change and adaptation. Sensemaking and sensegiving in interactions lead to understanding and finding each other, e.g. in a joint narrative on adaptation in a dedicated network organization and in cross-cultural collaboration on adaptation. Another important lever of change is forming strategies, based on issues and challenges, utilizing technology as the various relevant sectors of society have their own technology including policies, regulations, specific information, knowledge, projects, techniques, and practices. A third lever of change is adaptive decision making and related power and influence aspects. In the government environment where adaptation processes mostly develop, approval processes, decision making, negotiation on investments, and related power games shape the way adaptation is finally formed. Hence, these levers of change and their closely related factors constitute what has been called a transformative approach. The three levers of change play a role, e.g. in performing diagnostic search in the different sectors, diagnosing the issues and challenges of climate change, and finding the ways to elaborate on them in strategies and decide on measures. Translating measures into feasible proposals focuses especially on transformative technology management.

The main elements of the book are based on strategic management and insights of water governance and technology, combined with organizational dynamics and specific ideas on adaptation to climate change. The experience with Bangladesh Delta Plan 2100 plays an important role to highlight the ins and outs of the developed transformative approach. The main messages of the book show how a country can organize and prepare itself to deal with climate change impacts in addition to the usual challenges. The following items give an impression of such a transformative process in Bangladesh:

- Introduction of the BDP 2100 start-up in 2014: Setting the scene: a vulnerable delta confronted with climate change challenges, food security issues while seeking options for socio-economic development. The conceptual framework and design of the adaptive strategy making process and approach are developed here.
- Preparing the ground for strategic delta plan formulation: Baseline study phase focused on 26 different delta themes, to diagnose the delta situation from a holistic perspective, trying to understand delta dynamics as a whole while considering climate change. Issues and challenges have been characterized by identifying six Hotspot areas, each with for that area typical problems and phenomena.
- Dealing with contingencies and uncertainty about what could happen to the delta. How to meet uncertainties and their impacts in scenarios formulated as narratives of different possible futures with main drivers involved, autonomous drivers selected; external, policy-free scenarios formulated.
- People's participation and consultation during a roadshow through the country and in many other sessions for necessary interaction with local people from various backgrounds in the identified Hotspot areas. Interaction took place in the so-called Delta Ateliers which are informative, diagnostic workshops added by an important strategy design and measures focused part. This social design type of approach makes use of focused interaction, local experience and interests, creative thinking, supported by use of Touch Tables and experts to lead the sessions and others to elaborate and make sense of it.
- Strategy formulation based on findings of the Baseline Studies and input from the Delta Ateliers. Elaboration in a large amount of potential National, Hotspot and thematic strategies. Expert judgment to funnel all potential strategies, first into a smaller set of realistic strategies and then further narrowing down the sets of strategies into preferred strategies anchored in extensive consultation and in feedback from Steering Committee, Inter-Governmental Committee Bangladesh-The Netherlands and the National Advisory Committee (Cabinet level).
- Starting the approval process of the Delta Plan document in 2017, followed by further consultation and refinements. Decision preparation and decision making of BDP 2100 on September 4, 2018, by the National Economic Council.
- Start of realization preparations and funding in 2018–2023; working on readiness to implement and face climate change impacts; improvement of five necessary implementation capacities while preparing proposals for projects and executing projects and measures.
- Indicating the next moves and way forward for continued adaptation.

The Bangladesh experience shows that adaptation to climate change starts with sensemaking of a certain, unclear but potentially threatening development and interaction about it by involved officials and national and international actors. Their communication, activities, choices, and decisions need to be seen in conjunction with existing weather patterns, natural hazards, and the manifestation of climate change with impacts on the natural resources systems. Moreover, adaptation to climate change has next to the technical content a socio-economic dimension in relation to the many societal and economic aspects of water and climate change. Bangladesh formed its BDP 2100 with its own characteristics and needs in terms of water safety and security, food security and socio-economic development. The philosophy is 'adaptation by development', by comprehensive strategy making and realization, which are embedded in the countries' policy traditions, administrative routines, societal and political dynamics, technologies, governance, and funding. This means that adaptation strategy making and implementation, while focusing on the physical system, climatic conditions and use of technologies, touches on deeper cultural processes, with a struggle for meaning, sensemaking and sensegiving to define the issues and challenges and formulate strategies and actions. Besides, the power-related aspects of negotiation and adaptive decision making including budget allocation for investments are evident as well. The Bangladesh Delta Plan 2100 is in other words an expression of the applied practices and the politics of climate change and culture.

Groot-Ammers, The Netherlands Dr. Jaap M. de Heer
Dhaka, Bangladesh Giasuddin Ahmed Choudhury, M.Eng.

Acknowledgements We are grateful to a large number of people for contributions to this book in various ways. The focus of the book is on adaptation and transformation in the face of climate change and other developments. We built our insights for a considerable part on the strategy making and implementation experiences in Bangladesh. So, there are in the first place the people of Bangladesh who contributed in numerous sessions all over the country including many officials and representatives from the civic society and private sector. Whether it was in the Delta Ateliers, the stakeholder workshops, consultation sessions, or in meetings of the guiding committees at different levels on the development of the BDP 2100 and its implementation, they all contributed to the challenge of finding ways for adaptation to climate change combined with socio-economic development.

In addition, we thank the officials from both governments from Bangladesh and the Netherlands, especially also the Embassy of the Kingdom of the Netherlands, for establishing the BDP 2100 projects and enabling our leading positions to contribute to the formulation of the Bangladesh Delta Plan and to assist to the implementation process later on. The collaboration on these activities with the various institutions of the Government of Bangladesh, especially the General Economics Division, Planning Commission of the Ministry of Planning and the Bangladesh Water Development Board of the Ministry of Water Resources was remarkable and of great help. We also thank the many external reviewers and experts who contributed to understanding and insights.

Many thanks of course to our friends and colleagues of the BDP 2100 and SIBDP teams for closely working together in these projects and for the meaningful sharing, exchange and generation of knowledge and insights as well as many precious team experiences. We are especially grateful for the collaboration with William Oliemans MSc. of Deltares and Dr. Hero Heering for the many contributions and pleasant interactions. We are happy to have been in the position to lead the consortia on formulation of the BDP 2100 and on its implementation. The first consortium, Bangladesh-Dutch Delta Advisory Services (BanDuDeltAS), consisted of eleven international companies, which was a challenge in itself to manage and produce the required outcomes. The second consortium, (SIBDP) consisted of six companies. Finding promising strategies and pathways for adaptation to and transformation for climate change, even under different scenarios, is not an easy task in the complicated delta area of Bangladesh. We are grateful for the professional contributions of the team members and for the fruitful cooperation. We thank the lead partner of both consortia, TwynstraGudde Holding, management consultancy firm in the Netherlands, for trust and great support during the project periods.

Special thanks to Dr. Fahim Faisal, North South University, Bangladesh and Ir. Richard Jorissen, Rijkswaterstaat (Department of Public Works and Water Management, the Netherlands)/ Delft University of Technology (TU Delft), for their valuable contributions, discussions on texts and improvements of earlier versions of the manuscript.

Special thanks also to Martijn de Heer for his suggestions on this publication and for the preparation of a number of figures. Besides, Amirul Hossain, Sabbir Ahmed, as well as Joost Meijer and Ricardo van Breemen delivered valuable insights, comments or other contributions.

The support or contributions of you all in the development of this book or in other ways was of great help and is highly appreciated!

Competing Interests The authors have no competing interests to declare that are relevant to the content of this manuscript.

Contents

Part II Situational Characteristics of Adaptation

Part III Adaptation and Transformation Practices

Part V Adaptation and Transformation Experiences

About the Authors

Dr. Jaap M. de Heer was born (1957) on an island in Zeeland, south-western corner of the Netherlands. He studied civil engineering (1979) and worked as student seven months in Iran as site engineer on a harbor project at Bandar Abbas. This was an amazing experience just before the revolution in Iran. His experiences stimulated interest in tension, uncertainty, and major change which he focused on strategic change and transformation in his next study, Organization Sciences and Business Administration, at Erasmus University Rotterdam (1985). He worked at the Department of Public Works and Water Management called 'Rijkswaterstaat' for six years, where he was involved in planning and change processes, e.g., of a river institute. After that, he became researcher and associate professor at Leiden University where he got his Ph.D. in 1991, on the subject of 'strategy and change in organizations under pressure'. The study elaborates on and showcases strategy processes encompassing austerity, decline, and revitalization. He became policy and strategy consultant in 1995 at TwynstraGudde management consultants, where he was appointed as partner in 2000 and became managing partner a couple of years later. He was professor with Special Appointment at VU University Amsterdam from 2007 to 2014 with the Chair 'Strategy and Cultural Change' which focused on the importance of sensemaking, power games and interaction in strategy formulation, decision making and change. His inaugural address concerns 'strategy and change as expressions of the politics of culture'. He (co)authored books and articles. He worked part time in Bangladesh since 2001 to 2012 as strategy and institutional specialist in water governance, together with representatives of Rijkswaterstaat. He was furthermore involved in a number of assignments like decision making on seaward expansion of the Port of Rotterdam, policy review of the Dutch Coastal Zone Policy, and the preparation of the Dutch Delta Program. He started in 2014 as leader of the international BanDuDeltAS Consortium consisting of eleven companies to prepare a long-term delta plan focused on water safety and security, climate change adaptation, food security, and socio-economic development in Bangladesh. Besides, he was involved in projects in Indonesia and Vietnam on water governance also in relation to agricultural transformation. In 2018, he was again appointed as leader of the international SIBDP Consortium, now to prepare implementation of the BDP 2100. In

2024 he started jhrStratis, his own consultancy business concerned with adaptation and transformation related strategy development and implementation.

Giasuddin Ahmed Choudhury M.Eng. obtained his Bachelor of Science in Civil Engineering degree from BUET, Dhaka, in 1968 and his master's degree in Engineering from IHE, Delft, the Netherlands, in 1997. He joined the Bangladesh Water Development Board in 1970 as an Assistant Engineer and retired as the Additional Director General in 2005. During this period, he served as Director General of the Bangladesh Haor and Wetland Development Board and as Director General of the Water Resources Planning Organization. On his retirement from BWDB, he joined CEGIS as its Executive Director in 2005 and retired in 2012. During this period, he served as Member of the Joint Rivers Commission (Bangladesh-India). He was involved with international committees e.g. on irrigation and drainage (ICID), on large dams (ICOLD), and hydrological programmes (IHP). He also served as an Advisor to CEGIS. He joined Mott MacDonald Company in 2014–2018 as the Deputy Team Leader of the Bangladesh Delta Plan 2100 Formulation Project. From 2018 to 2024, he served as Deputy Team Leader, Support to Implementation BDP 2100 Project. He has more than fifty years of experience in the field of water resources, environment and natural resources management and planning, with proven ability in providing leadership in the formulation of the BDP 2100, National Water Policy, National Water Management Plan, Master Plan for Haor Areas as well as many feasibility studies of water sector projects. He has longtime experience in developing and leading participatory water management. He has extensive experience in process, program and project management, leadership experience with solid expertise in adaptive delta management, vision and scenario development, adaptive pathways, strategic planning, investment planning as well as implementation of large-scale and complex projects. He is (co)author on several books related to water resources management. He has represented Bangladesh on key national and global platforms, including the Global Water Partnership-South Asia, the World Water Forum, and the United Nations Framework Convention on Climate Change (UNFCCC). He is fellow of the Institution of Engineers, Bangladesh. His enduring contributions continue to shape sustainable water policies and practices in Bangladesh and beyond.

Abbreviations

ADB	Asian Development Bank
ADM	Adaptive Delta Management
AF	Adaptation Fund
ATP	Agricultural Transformation Process
BADC	Bangladesh Agricultural Development Corporation
BanDuDeltAS	Bangladesh-Dutch Delta Advisory Services
BBS	Bangladesh Bureau of Statistics
BD	Bangladesh
BDP 2100	Bangladesh Delta Plan 2100
BDT	Bangladesh Taka
BEZA	Bangladesh Economic Zones Authority
BIWTA	Bangladesh Inland Water Transport Authority
BMD	Bangladesh Meteorological Department
BMDA	Barind Multipurpose Development Authority
BUET	Bangladesh University of Engineering and Technology
BWDB	Bangladesh Water Development Board
CBA	Community-Based Adaptation
CC	Climate Change
CDSP	Char Development and Settlement Project
CEGIS	Center for Environmental and Geographic Information Services
CHT	Chattogram Hill Tracts
CIDA	Canadian International Development Agency
COP	Conference of the Parties
CSIRO	Commonwealth Scientific and Industrial Research Organization
DAE	Department of Agriculture Extension
DGC	Delta Governance Council
DLS	Department of Livestock
DoE	Department of Environment
DoF	Department of Fisheries
DP	Drought Prone
DPHE	Department of Public Health Engineering

DPP	Development Project Proforma
DPT	Delta Preparatory Team
DRF	Delta Results Framework
DTW	Deep Tube Well
DW	Delta Wing
ECNEC	Executive Committee of the National Economic Council
EEP	Environment and Ecosystems Program
EKN	Embassy of the Kingdom of the Netherlands
EZ	Economic Zone
FAO	Food and Agriculture Organization of the UN
FAP	Flood Action Plan
FCD(I)	Flood Control, Drainage (and Irrigation)
FEWS	Flood Early Warning System
FFWC	Flood Forecasting and Warning Center
FRM	Flood Risk Management
FYP	Five-Year Plan
GBM	Ganges-Brahmaputra-Meghna
GCF	Green Climate Fund
GDP	Gross Domestic Product
GED	General Economics Division
GHG	Greenhouse Gas
GMIS	Groundwater Management Information System
GoB	Government of Bangladesh
GoN	Government of the Netherlands
HIES	Household Income and Expenditure Survey
IDRP	Integrated Development and Restoration Program
IFI	International Financial Institution
IMED	Implementation, Monitoring, and Evaluation Division
IPCC	Intergovernmental Panel on Climate Change
IUCN	International Union for Conservation of Nature
IWM	Institute of Water Modeling
IWMFM	Institute of Water Modeling and Flood Management
IWRM	Integrated Water Resources Management
IWT	Inland Water Transport
JRC	Joint Rivers Commission
LGED	Local Government Engineering Department
M&E	Monitoring and Evaluation
MDG	Millennial Development Goal
MoEFCC	Ministry of Environment, Forest, and Climate Change
MoL	Ministry of Land
MoP	Ministry of Planning
MOU	Memorandum Of Understanding
MoWR	Ministry of Water Resources
MSL	Mean Sea Level
NEC	National Economic Council

NGO	Non-Governmental Organization
NWMP	National Water Management Plan
NWPo	National Water Policy
O&M	Operation and Maintenance
OECD	Organization for Economic Cooperation and Development
PIC	Project Implementation Committee
PIU	Project Implementation Unit
PMU	Program Management Unit
PPP	Public-Private Partnership
PPSC	Projects/Programs Selection Committee
PWD	Public Works Department
PWM	Participatory Water Management
RBE	Right bank Embankment
RBM&E	Result-Based Monitoring and Evaluation
RDF	Results Development Framework
RHD	Roads and Highways Department
SDG	Sustainable Development Goal
SIBDP	Support to Implementation of Bangladesh Delta Plan
SLR	Sea-Level Rise
SSWRDP	Small-Scale Water Resources Development Program
STW	Shallow Tube Well
TA	Technical Assistance
UN	United Nations
UNDP	United Nations Development Programme
UNDRR	United Nations Disaster Risk Reduction
UNFCCC	United Nations Framework Convention on Climate Change
WARPO	Water Resources Planning Organization
WASA	Water Supply and Sewerage Authority
WB	World Bank Group
WMO	Water Management Organization
WSS	Water Supply and Sanitation

List of Figures

List of Tables

Part I
Adaptation to Climate Change and to Other Issues

In Search of Sensitizing Concepts

Chapter 1
Introduction and Overview

Abstract This chapter presents an introduction to climate change adaptation in general and in the way a threatened country, Bangladesh in particular, made up its mind about how to deal with it. The main messages of the book show how a country can organize and prepare itself to deal with climate change impacts in addition to the usual challenges. The chapter also provides an overview of the book. The impression is sometimes that the focus on climate change mitigation and reduction of greenhouse gases, although indeed really urgent, comes at the cost of timely adaptation to climate change impacts on the ground, which is also urgently required at the short term with a clear orientation on possible future interventions. Compared to the massive public and decision-maker's attention for mitigation of climate change, adaptation to impacts on the ground seems to be underestimated. The focus in this publication is in line with the mentioned urgency: full focus on adaptation to climate change impacts while there is still time, ultimately in order to take measures, start adjustments or renewal which also takes a lot of time. Unexpected impacts on changing often threatening water and land conditions may ask less for routine and more for innovative methods to cope with upcoming situations in related sectors of society and in ecosystems. Ideas about adaptation can be found in the body of literature of Strategic and Adaptation Management. Characteristic for this publication is the attention for the idea of adaptation and transformation, how one can organize governance and other systems and actors roles in sensemaking, strategy formulation and choice of interventions in the course of time.

Keywords Climate change · Adaptation · Mitigation · Transformation · Transformative approach · Delta · Strategy · Strategic planning · Practices · Delta Plan · Uncertainty · Inertia

J. M. de Heer and G. A. Choudhury, *Strategies for Adaptation to Climate Change in a Transformative Approach*, Springer Climate,
https://doi.org/10.1007/978-3-031-88884-7_1

1.1 Adaptation in a Transformative Approach

Since ancient times, many people are situated in areas where they like to live, build their futures, and try to realize their dreams. These areas are most probably experiencing change of seasons or changes of weather, in some cases more drastically and even causing dangerous situations than in others. An interesting type of area is a delta area where many settlements arose and have been developed into cities, nowadays frequently major cities, centers of trade and logistics where land meets the sea.

A delta area is an area where rivers discharge into the sea. Characteristic is that a delta is a low-lying area where river flows enter the large volume of seawater. Here, river water loses its velocity, and consequently, the sediment loads carried by the river settle down somewhere in the coastal zone, contributing to dynamic delta development with erosion and accretion processes and formation of new land. Normally, delta areas are rich of biodiversity and lively ecosystems. As said, since long people like to live and work in these fertile delta areas because of connections with other people or countries, exchange of goods and trade opportunities, development of ports and businesses as well as building cities offering attractive places to live. The location near the sea also means that people are dealing with tidal dynamics with frequent high spring tides. Besides, there are, dependent of the location, frequent confrontations with natural hazards as storms, heavy rains, and storm surges. The salt seawater is an asset in terms of shipping, fisheries, and as source of seafood and salt but less so for agriculture due to salt intrusion, saline ground and groundwater. The relatively recent change of climatic conditions due to increasing volumes of greenhouse gases and rising temperature confronts delta areas with extra threats. Clearly, impacts of climate change as rising sea level, increasing frequency of storms (hurricanes, cyclones, typhoons) and erratic heavy rainfall endangers an increasing number of countries and especially low-lying delta areas.[1]

Change of climate is still a highly complicated issue with many uncertainties how this change will develop over time, whether or not extreme changes will occur and what possible impacts will be. For the subject of our conversation here, climate change as such is considered to be a fact and a major issue in recent global developments. This doesn't imply that there are no uncertainties, on the contrary, there are many contingencies in relation to, e.g., the pace of climate change with the behavior of the drivers behind it, the global ability to sufficiently reduce emission of carbon dioxide, methane, and other gases to avoid too high temperatures, behavior of warmer seawater, melting of ice caps and glaciers, etc. The point of departure is acceptance of the fact that conditions are changing with serious impacts on the ground and consequently that strategies and measures are required to significantly reduce emission of greenhouse gases to slow down and stop temperature rise while at the same time investing in adaptation to impacts of climate change since there is some lead time ahead. Investing in water safety, security, and disaster management has already proven to be effective in coastal areas or in riverine countries. Building on these experiences or where needed introducing effective water governance and management

[1] See for backgrounds also: Ahmed et al. (2023).

is, as many recent examples show, needed already on the short term. But not only that, spatial planning and food security are issues as well to be addressed in time. The impression is sometimes that the focus on climate change mitigation and reduction of greenhouse gases, although indeed really urgent, comes at the cost of timely adaptation which is urgently required at the short term! The focus of this publication is in line with the last mentioned urgency: adaptation and transformation to climate change are needed while there is time for it, ultimately in order to take measures to adjust or to find new ways in relation to changing water and land conditions, to cope with upcoming situations in impacted sectors of society and ecosystems. Transformation is viewed here as a process of meaningful transforming, a significant, dramatic, and marked change in form, nature, or appearance often implying improvements of a situation for the persons involved. Transformation management incorporates making sense of the way an organization or other entity is performing in relation to opportunities or challenges, developing a vision of how it could be, and defining direction or proactive adjustments and the subsequent management of the moves to achieve the new vision. Literature on strategic management and change offers several schools of thought and approaches accordingly.

Ideas about adaptation can be found in the body of literature of Strategic Management and Adaptation Management. The focus is here on adaptation of natural systems by seeking flexible responses with respect to natural resources management while uncertainty exists about the real developments and their drivers and also about the responses defined in terms of appropriateness and effectiveness in relation to these developments and drivers. Characteristic for this literature is 'learning' as metaphor for adaptation. A learning model, often shown as cycle, consists of several stages over time: reflection on the current status and performance, strategy making and planning in the face of contingencies and uncertainty, deciding what to do to be safer or perform better, implementing measures and projects, generating feedback on results and outcomes in relation to the initial issues and challenges, defining action again, etc.

In an earlier and slightly different view, the concept of adaptation is well known in biological sciences where it refers to the process by which a species becomes fitted to its environment. Adaptation is the trait (also seen as quality or capacity) and process of adjusting features, structures, physiology or behavior of an organism to environmental changes to become more suited to exist and survive in that environment. Adaptation incorporates a, in most cases, gradual process of change by which an organism or population tries to fit better in its environment or seek another environment. This biological view can be used as metaphor for organizations which are dealing with the usual challenges but in addition try to cope with pressing changing environmental conditions, e.g., due to climate change. Closely related to this is the view that environments or even the universe, are impermanent and constantly changing. Morgan's (1997) metaphor of Flux and Transformation emphasizes such continuous change with reference to Heraclites' saying: 'one can never step twice in the same river' because the stream of water is continuously different although the riverbed may seem stable. The range of metaphors or images is useful for perceiving our living environment in various, complementary ways which is important for right interpretations and full understanding required for meaningful adaptation. Such transformative view means that it is important to understand the forces that generate organizations or countries and the sources or drivers of change to find ways of dealing effectively with it in a transformation process, utilizing suitable transformative technologies over time. Although climate change is a continuously ongoing process, recently indeed showing flux and transformation, many involved institutions show(ed) inertia to effectively understand and deal with it. Inertia refers according to Naslund and Pemer (2012) to unwillingness or inability to accept certain information and to alter norms, values and residing stories and restricts change. This suggests that by interpreting the essence of storytelling, it is possible to understand the ways how dominant narratives can lead to inertia to act. Inertia to take adequate mitigation decisions in global conferences, inertia to contribute to agreed funds, inertia to take local decisions or change populations and individual behavior. This inertia refers to resistance to change for many different reasons, based on interests, differing perceptions and interpretations in extrema resulting in climate change denial. Here interesting metaphors are institutions viewed as 'political arenas', 'instruments of domination', or even 'psychic prisons'. The stories that there is no such thing as climate change or the idea that it is impossible to stop climate change and temperature rise make that people turn away, become indifferent or depressed or become extreme climate activists. To turn the pessimistic views and information about climate change and inertia in a more positive direction, Ritchie (2024) argues that, although climate change is a serious issue which needs to be urgently addressed, it is not the end of the world. The pessimistic dominant narrative about inescapable serious climate change leads in her view to indifference or inertia. The dominant narrative needs additional elements recognizing the many positive developments as rising awareness and willingness to act, less use of fossil fuel, huge increase in clean energy, more positive decisions at the global level.

Adaptation to the impacts of climate change is clearly not a routine or automatic response or some kind of acclimatization. A specific adaptive trait and transformation process are required in many cases to fit in the changing environment and that requires specific action with some weight and determination on the institutional level. 'Taking measures in time' raises questions: when to act, how to act, what can be done and by whom? Leary et al. (2007) concluded already in 2008 in their study on adaptation to climate change on general lessons: (1) adapt now; (2) create conditions to enable adaptation; (3) integrate adaptation with development; (4) increase awareness and knowledge; (5) strengthen institutions; (6) protect natural resources; (7) provide financial assistance; (8) involve those at risk; and (9) use place specific strategies. These are wise lessons but the question is of course who is willing and able to do all these things. These lessons show at least, that adaptation has to be interpreted as urgent and that a fundamental process of change and interventions is required, involving necessary decision making and taking appropriate measures. See also Burton et al. (2008) on better understanding of climate change and adaptation.

What can or should be done depends in the first place of the perception of policy makers at the highest level of the climatic conditions concerning their country of possible negative impacts on society, economy or politically. In this perception, the interpretation of the existing state of the physical system and the quality of built infrastructure including protection works play certainly a role. Besides, the willingness to act relates to the expected threatening developments, in such a way that they result in more and more problems in the usual way of living in a certain area and for doing things in every's day life one is used to do. Important is to see, that climate change impact adds to or combines with already existing weather and physical characteristics and their issues and challenges. Climate change may in that combination lead to higher temperatures, too much water in certain areas offering threats of frequent or even continuous flooding causing dangerous situations and bad water quality. Climate change may also result in too little water, leading to water scarcity, deteriorating water quality and ecological systems, severe drought. Due to sea-level rise, saline water will gradually become a serious problem, e.g. with shortages of drinking water and reduction of food production. These examples contribute to an environment where it is becoming harder to live and build a prosperous future. Adapting to such a changing environment depends highly on the features of the natural habitat and the dynamics of the natural system. This is also dependent of community-based adaptation with a tradition of dealing with such challenges and of already taken measures in the existing situation. Furthermore, adequate larger-scale adaptation has to be addressed at the national level and is highly dependent of willingness, capability, and the availability of resources to find possibilities to modify the natural system, to behave differently, e.g., with evacuation arrangements or ultimately to move to other places.

Who should take action on climate change impacts is in the first place a matter of politics and administrative system in which opinions and values compete with respect to seriousness of climate change as such and to fundamental responsibilities in terms of individual or governmental responsibility and to acting in a reactive or proactive way. Small-scale individual action can add up to significant results. On the larger scale, government institutions have mandate and means to take up necessary action and stimulate societal involvement of adaptation to impacts in climate change. Also, community-based adaptation depends on government's allocation of budget to finance the actions. So, meaningful political choices are important in the first place to take action and followed by government institutions with policies, plans funding and programs or projects. Providing feedback on the status and progress of actions in relation to ongoing climate developments is another important principle of an adequate governmental framework for adaptation to climate change. In many countries, the political and administrative dynamics cause problems and tension with respect to choices on climate change adaptation measures. Besides national troubles, a related area which needs to be included in political considerations concern national safety and may sometimes be of geo-political importance as has been argued by retired general Middendorp (2022) in his publication 'Building Resilience' tensions in or between countries can cause instability and tensions which may result in serious confrontations.

Impacts of climate change touch on interests of many institutions and stakeholders which emphasizes the need for an open mind towards each other, collaboration and if possible joint planning. Differences in opinion and even conflicts will be there but may find solutions with diplomacy and real stakeholder involvement. Insights in how interaction and sensemaking as well as sensegiving develops in the related policy, planning and implementation processes are of crucial importance for developing the required administrative and social construction of adaptation to climate change. In this sense, adaptation as used in many reports on climate change, is a too abstract and ambiguous concept which needs to be unraveled in a systematic way in terms of (1) what are manifest and expected impacts with their issues and challenges of climate change in a certain area; (2) what is in relation to (1) a diagnostic characterization of that area and upcoming problems, indicating which specific impact may be expected on a national scale and area or hotspot wise; (3) what can be done considering the factor time, on short, medium, and long terms to deal with impacts of climate change?

1.2 Strategic Planning Practices: The Bangladesh Experience

Considering the above-mentioned items, adaptation to climate change and related decision making will most probably be confronted with uncertainty and high priority changes, effecting overall political choices and administrative and societal sectors. This brings forward a process of gradual change of awareness raising and perhaps some sort of planning while interventions are prepared and executed possibly alternated with major interventions as response to incidents. This leads altogether to a status of 'accommodating climate change impacts' or more reactive, a wait and see what happens response. Others take, being aware of possible threats and vulnerabilities, a more proactive stance and prepare and take action. Bangladesh is e.g. already dealing with river floods and natural hazards to which climate change is added. The government is aware of vulnerabilities and wants to avoid setbacks of economic development and threats to its fast-growing population and rapid industrialization and urbanization. It is also provided with potential opportunities to be captured as education, innovation in various sectors and overall socio-economic development including poverty reduction.

Acting on the perceived opportunities and threats, the Government of Bangladesh (GoB) wanted to realize a more sustainable growth and development path by reducing the frequent socio-economic setbacks due to natural hazards and damage and loss. A high ambition was and still is to develop and realize a long-term vision to be a safe and prosperous nation, to be better protected and prepared for existing natural hazards and to be prepared with respect to the upcoming climate change impacts. GoB, being aware of the long-term delta planning in the Netherlands, has requested the Government of the Netherlands (GoN) in 2010 to provide assistance for the formulation of a Bangladesh Delta Plan 2100. To explore the usefulness and specific requirements of the Bangladeshi situation, a Bangladesh Delta Plan Preparatory Team (DPT) under Bangladeshi leadership and combining Bangladeshi and Dutch expertise was formed in 2011. The DPT assessed the needs, thematic topics and requirements for a delta plan as long-term integrated plan for sustainable development of the Bangladesh

Delta.[2] This preparatory study forms the basis for the final decisions by GoB to prepare a Bangladesh Delta Plan and for a specific Memorandum of Understanding between the two countries in which long-term collaboration and support have been arranged.[3] The preparatory study and MOU formed building blocks for the joint formulation of the project in 2012 and tender documents, including the Terms of Reference for the formulation of the Delta Plan of Bangladesh. The tender of 2013 with main Project Document argues about usefulness and necessity of a delta plan and shows the way GoB and GoN have been discussing and collaborating for the preparation of the Delta Plan formulation project (EKN, 2013). GoB owns and leads the formulation process via the Ministry of Planning, General Economics Division of the Planning Commission. Its Project Management Unit was assisted by a team of consultants from both countries to assist and support the background studies and delta plan formulation process.

GoB decided to develop a Bangladesh Delta Plan 2100, including a strategic and practical approach to realize this plan, inspired by Dutch ideas. The GoB formulated a number of requirements in order to develop a delta plan according to the requirements and characteristics of the Bangladeshi situation, standards and needs. The delta plan should: 1. be a long-term plan with short-term investment plan; 2. facilitate socio-economic development to become middle-income country; 3. cover the delta and relevant adjoining areas; 4. develop a comprehensive holistic and multi-sectoral approach in relation to water safety and security, food security and socio-economic development with specific attention for ecology; 5. consider ideas of Adaptive Delta Management as already practiced in the Netherlands and Vietnam; 6. build on available studies and reports, no additional fundamental research; 7. develop a specific institutional framework to make implementation and update of BDP 2100 possible. By formulating these requirements, the government made clear it wanted a delta plan described as 'strategic long-term, holistic, water-centric and techno-economic plan' for the future development of Bangladesh.[4] In doing so, GoB indicated it was inspired by the Dutch delta program approach but wanted to make its own plan, fitting in the Bangladeshi setting, administrative culture, and situations.[5]

The delta plan formulation started with studies on the state of BD delta, followed by diagnostic analysis of issues and challenges, leading to tentative conclusions about which challenges needed to be addressed by potential strategies to reach the formulated vision and mission. After extensive consultation through the whole country and expert meetings, the preferred strategies were formulated: 1. national-level strategies; 2. Hotspot strategies; 3. thematic strategies. These strategies have been elaborated into about 300 project concept notes as basis for the investment plan. The investment plan was jointly formed with additional assistance of a World Bank team. After extensive multi-stakeholder consultations on and expert reviews of the draft delta plan, the GoB approved BDP 2100 with its Investment Plan of $37 bln. in the meeting of the

[2] See: DPT, Preparation for the Bangladesh Delta Plan (2012).

[3] MOU between Bangladesh and the Netherlands, 2012.

[4] Frequently mentioned by GoB/Planning Commission officials.

[5] See for broader political and economic background, e.g., Jahan (2005).

National Economic Council (NEC) on September 4, 2018. The BDP 2100 was then mainstreamed at national level and accordingly in the National 8th Five-Year Plan. The plan realization started directly after approval, with support of a renewed project of Bangladesh and the Netherlands, called Support to Implementation of Bangladesh Delta Plan 2100 (SIBDP 2100). SIBDP prepared the changes of institutional framework and facilitated the decision making about it and worked on readiness of required implementation capacities. SIBDP elaborated the national BDP 2100 into six more specific regional or basin-wise implementation programs. Moreover, an additional Investment Plan has been provided on the basis of these programs. During SIBDP, project preparation and execution gradually took shape while proposals for tapping into the Green Climate Fund were prepared. At the end of SIBDP in December 2023, GoB and GoN discussed ideas for follow up, e.g., on further project execution and for updating of BDP 2100 given the perceived worsening climatic conditions. Moreover, the COVID-19-related economic and financial problems need government attention with possible implications on investments. Herewith, the adaptive nature of BDP 2100 becomes visible in a new situation, with the challenge of updating and (re)considering priorities.

1.3 Overview of the Book

This publication intends to provide an overview and understanding of the necessity and relevance of a transformative approach of adaption to climate change impacts in relation to already existing climatic, natural system, and socio-economic conditions. The main messages are as follows:

- Adaptation to climate change has by the direct involved actors and stakeholders been focused on multi-sectoral challenges concerning water safety and security, contribution to food security next to utilization of opportunities for socio-economic development.
- In addition, adaptation to climate change concerns managing the contingencies and uncertainties in a strategy process and in related transformational changes and meaningful interventions in the physical systems. These have to be embedded in a countries' governance, policies, planning, and programming including a financing arrangement and budget allocation. To achieve this, understanding and attention for administrative culture is required. Besides, design of strategies and measures has to fit in a well understood natural resources system, while where possible revitalizing these.
- Hence, adaptation to climate change is, although to a large extent techno-economic in nature, a cultural process of perception and interpretation of developments, of sensemaking and sensegiving in interactions on uncertain climatic conditions with possible serious impacts, on opportunities and threats for socio-economic development and in decision making on what can be done on risks and challenges given limited funding and capacities. Besides, it is a power game during and after

the strategy process, starting, e.g., with the process design, role of stakeholders, results of Baseline Studies, proposals on strategies and measures, use of adaptive pathways, selection of interventions, setting priorities, funding, multi-sectoral discussions and decision making.

- In the chosen holistic approach, many involved aspects (or disciplines) and sectors, the knowledge and practices of these aspects and sectors can be utilized for adaptation and transformation and make this so-called transformative technology management of crucial importance.
- Methodology on adaptation to climate change impacts can be fruitfully derived from theories and practices of strategic management and change as well as from decision-making theory and practices, of course in conjunction with the various scientific disciplines on climate change, civil engineering and a range of related technical and social aspects.

In the course of time from the start early 2014 till 2024, episodes with their own characteristic focus and activities have been identified and indicated with the main applied practices in relation to: strategy process agreement and design; diagnostic search: identifying issues and challenges; strategy formulation BDP 2100: adaptation by development; developing a dedicated institutional framework and tools; investment planning and regional or basin-wise adaptive programming; realization of measures and interventions, including adaptive design; funding from various sources; adaptation is ongoing transformation with reflective evaluation, updating and adaptive adjustment.

For an overview of the book, the structure of the storyline can be used. It has been structured by dividing the book into five parts: Part I starts broadly on adaptation; II focuses on the Bangladesh delta and on physical characteristics, issues, and challenges as well as on climatic conditions; III highlights strategy development under uncertainty; IV shows aspects and aspirations of implementation and realization; V presents lessons learned, broadens the scope to other examples of adaptation, leading to concluding remarks and suggestions. To go into more detail:

Part I presents the introduction to the central theme, with societal relevance and discourse. Besides, adaptation to climate change has been conceptualized with theoretical orientations. Adaptation is indicated as temporal and culturally embedded concept, dominated by stakeholders views and interests in decision making while being shaped into administrative and technical approaches and strategies and measures for action.

Part II dives into characteristics of the Bangladesh delta in general, the interesting river systems, the variety of water resources and the challenging coastal zone. Besides, socio-economic characteristics have briefly been presented. Moreover, the climatic conditions have been elaborated.

Part III is showing the Bangladesh experience with adaptation and scenario-based strategy making. This covers the whole adaptation strategy process and practices including tracking of already followed strategies and plans, analysis of delta issues and challenges, long-range scenario-based strategy development. Three types of strategies are indicated: national level, hotspot level, and thematic strategies.

Part IV deals with realization and implementation aspects such as strategies and practices regarding the institutional framework, participation and consultation, and investment planning as well as building realization capacities and progress feedback based on reflective evaluation with indications for updating.

Part V broadens the scope to experiences with adaptation in the Netherlands and Vietnam. Lessons learned will be evaluated against the background of all experiences while a way forward also will be considered.

References

Ahmed, M., Choudhury, G. A., & De Heer, J. M. (2023). Bangladesh Bodwip: Ekush Shotoker Bybosthapona. *Concise book in Bangla on the Bangladesh Delta and Delta Plan 2100*. Prothoma Prokashoni.

Burton, I., Dickinson, T., & Howard, Y. (2008). Upscaling adaptation studies to inform policy at the global level. *The Integrated Assessment Journal Bridging Sciences & Policy, 8*(2), 25–37.

Choudhury, G. A., Terwisscha van Scheltinga, C., Van den Bergh, D., Chowdhury, F., De Heer, J. M., Hossain, M., & Karim, Z. (2012). *Preparations for the Bangladesh Delta Plan*. Alterra Report 2300. ISSN 1566-7197.

Embassy of the Kingdom of the Netherlands. (2013). *Project Document. The formulation of the Bangladesh Delta Plan*.

Jahan, R. (2005). *Bangladesh Politics. Problems and Issues*. UPL.

Leary, N., Adejuwon, J., Barros, V., Burton, I., & Lasco, R. (2007). *Climate change and adaptation*. Earthscan.

Middendorp, T. (2022). *De klimaat generaal. Bouwen aan weerbaarheid*. Podium.

Morgan, G. (1997). *Images of Organization* (2nd Ed.). Sage.

Naslund, L., & Pemer, F. (2012). The appropriated language: Dominant stories as a source of organizational inertia. *Human Relations, 65*(1), 89–110.

Ritchie, H. (2024). *Not the End of the World. Surprising facts, dangerous myths and hopeful solutions for our future on planet Earth*.

Chapter 2
Unraveling the Concept of Adaptation

Abstract The discourse about climate change and adaptation is briefly described in order to indicate main subjects of concern, like ongoing or upcoming changes over longer-term periods. Thinking strategically about these changes and their uncertainties is important, considering the differences in perceptions and interpretations by the different parties in the discourse. Learning to adapt has been introduced, meaning that awareness and adaptive capacity but also the willingness to act in society may be increased. To do that, a receptive and adequate government role is important, taking a proactive stance and facilitating adaptation. Adaptation is in texts so widely used that it comprehends and varies much in current climate change discourse. It all boils down to in the discourse involved actors and their framing of different phenomena and situations with acceptance of different realities. Trying to understand of what is going on and of what we are actually talking about is a challenge in itself, let alone preparing actions on the ground to what are seen as the real challenges. To become more precise for the purpose of this book, definition of problems and challenges related to climate change impact is of crucial importance, in addition to already existing societal and physical issues and proposed policy solutions. Adaptation to a phenomenon as climate change relates to many aspects of the physical environment of society and economy and because of that, isn't an easy game to deal with. The climate change phenomenon and adaptation to it need careful analysis and interpretation, seeking the logics of change and thereby the levers of change in institutional, societal, economic, and physical processes and systems which are potentially influenced by changing conditions. This chapter presents some relevant concepts to approach adaptation to climate change impacts in order to better understand it and to utilize insights in practices of strategizing, decision making, and implementation.

Keywords Adaptation · Adaptive decision making · Strategic choice · Mitigation · Climate change · Discourse · Narrative · Stakeholders · Holistic · Sensemaking · Sensegiving · Power · Decision making · Policy window · Mixed scanning · Strategy · Strategic choice · Long term · Uncertainty · Adaptive pathway · Temporal · Spatial · Institutional · Institutional framework · Multi-level · Governance · Organizational · Practices · Praxis · Practitioners ·

J. M. de Heer and G. A. Choudhury, *Strategies for Adaptation to Climate Change in a Transformative Approach*, Springer Climate,
https://doi.org/10.1007/978-3-031-88884-7_2

Leadership · Socio-economic · Technology · Technology management · Physical system · Intervention · Measure · Conceptual model

2.1 Identified Concepts

Adaptation to climate change should, as argued before, not be treated as abstract phenomenon. The challenge is instead to operationalize it as it happens in practice, in such a way that it opens up in a number of meaningful concepts. We approach adaptation as creative development, as response or anticipation to major changes which cannot be ignored. Initial findings on the possibility of a changing climate were followed by more and more proof of and debate about that 'it' actually is happening: it clearly generated a broad and still lively discourse over time. We see that potential impacted countries with their representing governments gradually realize that action is or may be needed. They form a narrative expressing their interpretation about possible impacts for vulnerable areas. This is often a dominant narrative although also substitutes or counternarratives are formed by political parties or other groups in society. The role of institutes like IPCC, universities, and knowledge institutions is important here as they generate data, prepare analyses and reports, and their narratives. However, differences in disciplines, in research findings, and opinions lead to different narratives on what should be done. With respect to a delta area, this can be, e.g., traditional or innovative protection, moving along with developments, building with nature, seaward approaches, spatial planning, and migration or a mixture of these. Once a government is initiating action on plan preparation or interventions, decision making is required to set a course for action and way forward, mostly including budget reservations. Adaptation to climate change takes place without all information about what will happen with temperature rise, sea-level rise, etc. There is a general need to reduce uncertainty or manage uncertainty in a strategy process which forms various scenarios of possible futures. A transformative approach to climate change adaptation should contain those concepts which characterize the most involved phenomena and open the floor for a manageable overall process leading to the expected outcomes over time. Managing uncertainty means in this respect that no final answers or maximum solutions need to be provided but a long-term adaptive strategy including a dedicated institutional framework, financial perspective and social and technological measures with (no-regret) physical adjustments or renewal.

Before talking about responses, the effects of climate change necessitate critical analysis and introspection that results in understanding and possible measures. These, however, mostly will result in interventions and change, for example, in the physical systems, organizational and management performance and planning procedures. This entire collection of connected adjustments has been described as a transformative approach. Transformative refers to bringing about a significant change or reform in a situation, structure, process, system, or region. Examples of this include institutional arrangements, organizations trying to deal with deep uncertainty, new measures on water safety and security systems, agricultural systems, navigation systems, urban areas, etc. Transformation shows and stimulates ongoing reality construction, as a process of becoming. Such transformation takes time, because it includes innovative thinking, stakeholder considerations, a crucial role of the government, private sector mobilization, and managerial (inter)action.

Interesting is to find useful levers of change to generate and develop transformation and utilize its opportunities and advantages. Adaptation to climate change is in the first place the work of human beings, as policies, strategies, plans, and measures as well as decisions are all man made. They come forward in and through organizations with their creative thinking, interaction, technology, etc. A crucial lever of change here is sensemaking in close interaction (sometimes called 'management of meaning') between involved parties and communication in various forms and with different purposes. Important interactions are seen in the broader discussions and knowledge exchange as well as in the discourse on climate change and adaptation. Sensemaking and sensegiving in interactions lead to understanding between parties, possibly finding each other, e.g. in a joint narrative on adaptation in a dedicated network organization and in cross-cultural collaboration on adaptation. Another important lever of change is forming and implementing strategies, based on issues and challenges, utilizing technology as various relevant societal and economic sectors use their own technology including policies, regulations, data, knowledge, projects, techniques, and practices. Transformative technology management is a key focus when formulating strategies and measures and for converting measures into workable plans and projects to realize adaptation on the ground. A third lever of change is adaptive decision making and related power and influence aspects. In the government environment where adaptation processes mostly develop, approval processes, decision making, negotiation on investments, and related power games shape, the way adaptation is finally formed. Hence, these levers of change and their closely related factors constitute what has been called a transformative approach. The three levers of change play a role, e.g. in performing diagnostic search in the different sectors, diagnosing the issues and challenges of climate change, and finding the ways to elaborate on them in strategies and decide on measures.

Based on earlier scientific research and longtime consultancy experiences on strategy and decision making on infrastructure development in an administrative setting and culture, the following conceptual model has been prepared in which the briefly mentioned concepts are summarized; see Fig. 2.1. The concepts provide theoretical orientations, are seen as sensitizing concepts, and will be further explored.

2.2 Climate Change Discourse

Analysis of the climate change discourse shows the dynamics of the upcoming climate issue in the course of time. Taylor (2013) argues that climate change has been represented in a variety of ways and that these have enacted their own discursive patterns, which people discuss and act upon at local, national, and global scales.

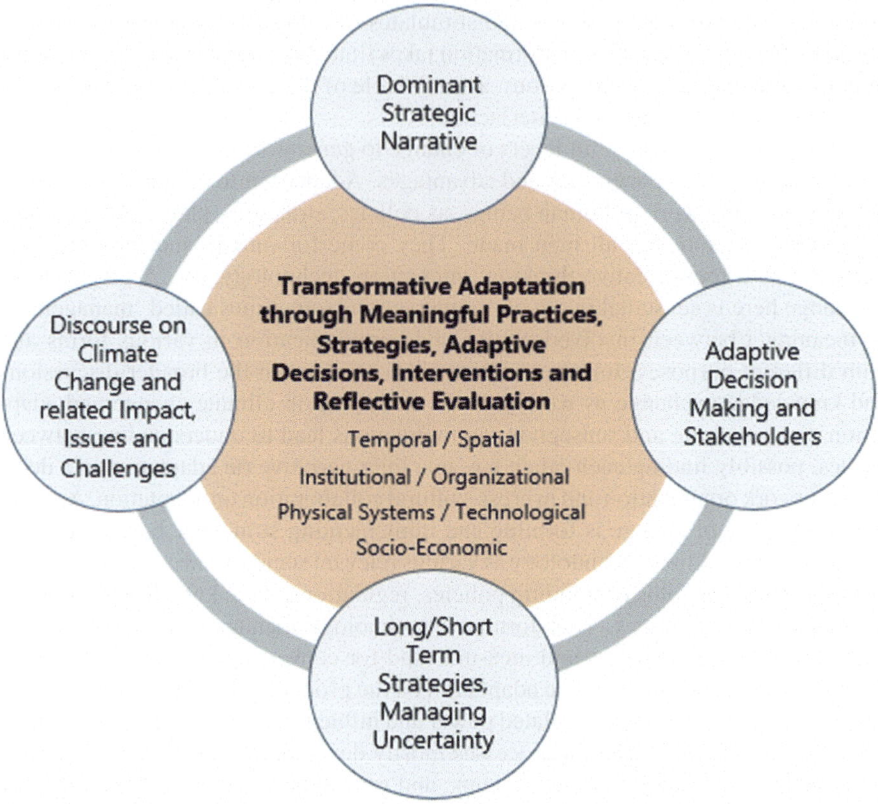

Fig. 2.1 Tentative conceptual model of transformative adaptation

Characteristics of the climate change discourse are in the first place the strong scientific roots, even domination, with a lot of scientific research and discussions, trying to make sense of studies, data, computer-based modeling, professional reporting. Amidst the scientific discussions and uncertainties, institutional framing took place in 1994 when UN member states decided to constitute the United Nations Framework Convention on Climate Change (UNFCCC). This was a remarkable step at the time, based on the principle (Montreal Protocol, 1987), that member states should act in the interest of human safety even in the face of scientific uncertainty. Scientific evidence however proved to be not easy to review and understand (UNFCCC, 1987). Many studies, measurements, modeling an projections resulted in a sequence of reports especially provided by the IPCC with an important synthesis report (IPCC, 2023) showing the dynamics and increasingly problematic development of swelling quantities of greenhouse gasses in relation to temperature rise with a range of consequences. This highly technical discourse together with unpopular messages and according to some doom scenarios showed initially a gap with mindset of and understanding by policy makers and the larger public. Environmentalists, frequently with strong ties to climate and nature-related scientific disciplines, became more and more involved and worried. They played an important role in mobilizing public and governmental awareness. Against this background, the climate change discourse shows further institutionalization, especially with the Kyoto Protocol (UNFCCC, 1997). This agreement operationalizes the UNFCCC-agreement by committing industrialized countries and upcoming economies to reduce greenhouse gasses.

Indications about possible human-induced climate change are known for quite some time but were by most people not interpreted as real, as sufficiently proven let alone seen as threatening. Till today political leaders react in different ways from acceptance and action, to playing it down and stretching response policies over 30–50 years, to blunt denial. Full heartedly recognizing the climate change issue with willingness for acting upon it is only seen in the last couple of years. Examples as ‘it’s a hoax’, ‘science can’t really proof it’, versus spirited debates and strong protests by scientists, global or local movements who don’t shy away from civil disobedience are characteristic for the exciting discourse.

Well known are the many large-scale COP and other conferences with various success, mainly focused on mitigation of climate change by attempts to reach agreements on reduction of emission of carbon dioxide and other greenhouse gases to avoid further temperature rise. The Paris Agreement (UNFCCC, 2015) incorporates three pillars of climate action: (a) mitigation, which involves the reduction of greenhouse gas emissions through such activities as swapping fossil fuel-based power generation to solar and wind systems and reducing agricultural and industrial activities that are highly polluting. Mitigation is seen as the first line of defense against future impacts and the best way to avoid or avert loss and damage from climate change. (b) Adaptation concerns preparing strategies and measures, communities and infrastructure for climate impacts. This can include measures such as protecting communities from sea-level rise by building sea defenses, preparing for extreme events like cyclones by investing in early warning systems and cyclone shelters, or switching to drought- or flood-resistant crops. Adaptation helps reduce or minimize loss and damage from climate change. Enhancing adaptive capacities and ensuring support are important to this. (c) The third pillar is loss and damage, and this addresses the impacts of climate change which were not covered by other mitigation or adaptation measures. Adaptation has limits, so the IPCC refers to ‘hard’ and ‘soft’ adaptation limits. Hard limits refer to situations where adaptation is not possible, while soft limits are there where options to adapt may exist but are not available. Addressing loss and damage may involve rebuilding infrastructure such as roads and bridges, hospitals and schools, the reconstruction of homes, support to establish new livelihoods, mental and physical health care, programs to revitalize culture, and acts of memorial and remembrance. The COP28 started in 2023 with the promising establishment of a dedicated fund for covering costs of damage and loss due to climate change impact in poor countries. A day later, the scientific evidence for the relevance of fossil fuel reduction and mitigation of climate change was denied in the same conference (UNFCCC – COP28, 2023). The variation in opinions and interests behind them becomes apparent, especially in the COP meetings. Mitigation therefore still dominates in public and political debates, focused on influencing the triggers of climate change. The relation with adaptation to impacts of climate change is clearly important because substantial mitigation contributions may lead to less climate change and by consequence to less investments on adaptation on the ground. There are still many ongoing political debates on contributing to climate change mitigation, e.g., in relation to developing countries as well as to adaptation to impacts, e.g., with respect to promises on contributing to the Green Climate Fund. Remarkable is the limited action on real adaptation on the ground including prevention. Increasingly, however, people are confronted with dramatic events and impacts of heavy storms, extreme rainfall, large-scale flooding, extreme heat stress, increasing water scarcity. In any case, the need of more action on climate change impacts becomes visible. Some countries take a leading role, apparently depending on their physical situation and expected impact of climate change and how this gained political momentum. The global and national discourse on climate change forms a dynamic context in which adaptation takes place. Against the backdrop of these real experiences, the needs for action are articulated into demands to politicians and administration. Here, officials experience pressure to act in the interest of the people, ecology, and businesses.

2.3 Thinking and Interacting Strategically in Narratives of Change

2.3.1 Formulation of Narratives

The discourse on climate change as briefly discussed indicates ongoing conversations, based on countless interactions and attempts to make sense[1] of research findings, proposed solutions, decisions on problematic situations, etc. Depending among others on position in the discourse and knowledge and relationships, narratives play an important role in the ongoing conversations. Narration is basically an interactive process between narrator and listener(s) which is part of a larger frame of interaction and activities (Gulich & Quasthoff, 1986). A narrative discourse refers to the use of language to convey stories, describe events, and express ideas. In the ways these are expressed, the narrative shows intentions of the narrator, values and meaning, they possibly express ways of behaving and organizing. Morgan's (1997) images or metaphors not only offer ways to look at and understand organizational or societal dynamics but also offer meaningful narratives about how organizations and behavior can be viewed and interpreted. Against this background, a narrative can be formed about adaptation with recognition and interpretation of significant developments in the climate, societal views, and ideas about physical environment and some form of variation or modification to cope with the developments in order to survive or satisfy certain needs. Adapting to this can best be narrated as process of change

[1] See on sensemaking e.g. Weick (1995); Gioia and Chittipeddi (1991). See also De Heer et al. (2009a).

while moving from a relative stable situation to a state of flux and transformation in many areas, with uncertainty and ambiguity of unknowns and unstable situations.

An important challenge for a country is how to use the narrative(s), e.g., to give meaning to the climate situation and how government intends to address it. Such narratives or substitutes or even counternarratives can provide meaning about how to behave or organize oneself in the face of climate change and intrinsic societal, economic, and physical uncertainties and tensions. Most manifestations of climate change relate to a higher temperature with probably changing weather patterns and water related issues. A country may face serious climate change impacts with respect to water in various ways, e.g., too much water causing devastating floods with related damage and loss as well as diseases or too little water causing drought and scarcity of drinking water. In many cases, the expected or immediate impacts of these challenges trigger a response by involved governments. This may be in terms of risk reduction, disaster management (e.g. evacuations in India and Bangladesh concerning cyclones), and damage control or with a comprehensive design process as reaction to the huge damages and losses and societal disruption. These reactive approaches address serious issues when a natural disaster occurs or took place. Increasingly, governments are more interested in proactive, preventive approaches related to the gradual climate change and involved processes instead of a reactive approach to incidents.

2.3.2 Managing Selected Practices to Adapt as Main Message

Adaptation is in this publication primarily based on trying to understand the logics of change with application and management of suitable practices by qualified practitioners. These actors form an important link in the chain of academia, government institutions, private sector and the public. The mixed assistance team for BDP 2100 as well as for SIBDP played this role as group of dedicated practitioners aiming to adapt to uncertain changing circumstances because of gradually or rapidly shifting climate conditions and the real impacts on the ground. The narratives of the assistance team come forward from their professional background and application of the practices and relate, e.g., to water safety and security as well as to temperature related challenges as agricultural drought. The adaptation challenges may show variations of location and physical conditions, related public and private sectors, and societal as well as political and governance processes. Choices have to be made on expected climate change developments, but also on narratives of possible futures with sets of required measures and investments. Here we see the importance of perception and sensemaking of the changing situation among others by leaders: Do they see climate change as a minor issue, a hoax, a threat to be technically controlled or a development opportunity seeking synergy of changes in multiple socio-economic sectors?

2.4 Adaptive Decision Making and Stakeholders' Influence

Another relevant body of knowledge concerns strategic choice and decision making. Perception of the logics of change and uncertainties associated with them focuses also attention on problem analysis and the search for a response repertoire. Considering choices and making decisions comprise an important part of a response to threats and the manifest impacts on human life and ecology, striving for survival or sustainable development. The assumption is sometimes that research and evidence-based well-defined studies and plans, originating from sound methods and techniques will convince professional decision-makers and will more or less automatically lead to expected adaptive action and outcomes. Interesting however is that administrative practices and culture as well as political realities prove differently. Literature on decision-making research and theories show in the course of time many approaches to explain and even prescribe a decision-making process. Still relevant are the findings of Simon (1957) and his reflections on decision making in and by organizations, characterized by bounded rationality, constraints or influences. Ideas about incremental decision making (Lindblom, 1959) or synoptic rational models (Dror, 1964) offer valuable insights in various approaches and practices. A for the purpose of this book interesting way of thinking about decision making has been provided by Etzioni (1967, 1986) in his view on adaptive decision-making, called the mixed scanning approach. This approach focuses on a mixture of broad, shallow and in depth search and analysis of data. This leads to (a) exploration and consideration of a broad array of themes, findings, and choices followed by and combined with (b) detailed study of a for the situation interesting focused subset of facts and choices. In doing so, mixed scanning involves two sets of judgments: the first are broad and fundamental choices about the basic policy and direction; the second are more incremental decisions that prepare the way for new, basic judgments and that implement them once they have been made. As said, so many items and sectors relate to climate change adaptation and necessary transformation, that the mixed scanning approach can indeed be promising. It may lead to frequently scanning for adaptation relevant developments and themes, selecting and focusing on the main issues and challenges which can be further elaborated into strategies and implementable measures.

Another important aspect has been highlighted by Kingdon (1995), who argues that decision making needs to be seen in the first place as subject of dynamic processes of agenda setting with respect to problematic issues. Moreover, he emphasizes that streams of problem, policy and politics proposals, interests and influences come forward and are successfully dealt with or not due to actions of certain officials, political groups, or unpredictable circumstances. Kingdon calls the momentum when problem, policy, and politics streams meet or intersect the policy window which opens only now and then, e.g., with formation of a new government, with a specific issue to be settled by ministries or other government institutions and during budget allocations. When problems, proposals, and policy needs or opportunities come together, a proposal can be picked up by policy makers while other issues and proposals are left out. The dynamics behind this are influenced[2] by changing public and or government values and preferences, information technology, input from policy entrepreneurs close to centers of decision making, rivalry between parties, pressure from certain communities or interest groups, global discourse, e.g. on climate change and felt urgency to take action. The already mentioned deep uncertainty concerning external developments is accompanied by decision process related ambiguity and power games which make decision making even more complicated.

Sometimes consultation or even training and stakeholder sessions are mentioned as ways to gain understanding and support for decision-making proposals. However, this as such doesn't lead to acceptance and final approval by those in charge. We're talking here about policy or decision-makers as human beings with their own perception, interpretation and sensemaking of a situation, processes anchored in their cultural background, societal network, and in interaction with colleagues, family, and friends. Moreover, these actors are mostly part of a political party most likely involved in trying to address societal issues based on their values and interests, making part of a dominant party or coalition in government or competing with other parties to become the dominant party or coalition. What a certain dominant party or coalition wants to achieve may be completely different from another one because of differences in perception, e.g., of climate change or of needs and priorities, e.g., in terms of water demand and cost of supply, safety levels, etc. Moreover, there most probably will be differences in values, beliefs, and interests with possible culture wars in a party's history. Experiences with BDP 2100 show that policy and decision making takes place by a limited number of key people in and surrounded by influential institutions experienced in preparing political decisions, in legal aspects, in financial allocation and in implementation. Their political and administrative rationality is as Kingdom shows related to competing issues and values, to many factors and circumstances and is quite different from the technical and mostly content-focused rationality of, e.g., scientists and technical experts. They often produce valuable analyses and, e.g., a set of alternative solutions even with decision support tools but it is uncertain what comes out of the administrative process. This makes decision making on climate change-related water problems or even water crises an important field of interest when talking about adaptation to climate change.

The OECD (2011) concluded that water crises are or increasingly become governance crises, not so much regarding what needs to be done on short notice but in terms of vision and perspective, decision making, also why and when in relation to other pressing issues; by which institutions on what level and who is going to finance it; how to achieve adequate results acceptable for the public especially with an eye on the constituency of politicians.[3] Uncertainties about climate change and impacts together with an increasing gap between water demand and supply make it even more complicated. The OECD suggests principles on water governance (OECD, 2015) which formed an important source of inspiration for reflection on decision

[2] See also: Pal (2014).

[3] See for Bangladesh, e.g., Khan et al. (2008).

making and preparing the BDP 2100 institutional framework, for multi-level considerations, and for dealing with stakeholder involvement, their interests and ownership and in participation and consultation processes. The principles have been applied as well as possible in the given situation and governance setup and administrative culture.[4]

Strategic change is full of choices at various involved levels dealing with the issues and opportunities as the many items and trade-offs with respect to adaptation suggest. This suggests that decision making is not only about responding to a certain situation but also anticipating proactively to it. As Child argued, strategic change and decision making are concerned with contingency factors and contingency choices (Child, 1972). Complexity with respect to adaptation involves decision-making-related factors and choices about:

- An increasing sense of urgency requires, given the subject and possible drastic interventions, answers on what to do, when, where, how, by whom. Under conditions of natural disasters or severe impacts of climate change, decision making may require crisis decision making.
- Stakeholder interests and demands. Broad public a stakeholder's participation will be needed for ownership and acceptance of execution of measures while it generates at the same time resistance by non-believers and 'not in my back yard' voices.
- Long-term considerations need to be linked with short-term decision making. Climate change develops over longer periods of time but impacts require short-term investment decisions, preparations, and timely interventions.
- Information and communication problems due to uncertainty about climate change and ambiguity among scientists with their multiple interpretations and opinions of what is going to happen and should be done.
- Importance of looking ahead to stories of what might happen in the future and what a vision narrative on a desired future would incorporate, requiring various scenarios or narratives of possible futures. Because people want to make sense of the situation, they are more or less aware of, they become worried when a vision is lacking for too long or when they feel it is dominated by an elite of experts.
- Issues and challenges to be seen in a holistic view in terms of multifaceted content of problems, e.g. in a delta area. This also regards governance, with many institutions and mandates, several multi-level, and sectoral interdependencies requiring coordinated decisions and action.
- Scenario-based strategic planning to be decided upon and translated into implementation programs, investment decisions, funding, and projects to execute.

For dealing with such issues in agenda setting and decision making, the importance is emphasized of policy makers' perception, values and interpretations for sense-making and sensegiving in interaction with the relevant institutions. What Weick (1995) argues with respect to organizations in terms of interests, powerplay and

[4] Van Alphen, De Heer and Minkman (2021) reviewed the application of the OECD principles on water governance with respect to the Dutch Delta Program and the Bangladesh Delta Plan 2100.

ambiguity in a struggle for meaning making or management of meaning, counts also for a network of organizations involved in decision making on adaptation to climate change impacts and maybe even stronger given the variety of opinions in the discourse. The strategy making and planning under these circumstances can be seen as expressions of politics of culture or the struggle for meaning making.[5] Important is the administrative culture showing a struggle over framing of problems and on definition of ideas for policy solutions[6] of the dominant government coalition, especially for decision making under conditions of adversity and stress. Research based on a narrative approach to decision making by Merkus (2014) clearly shows administrative and political coping mechanisms and performative action.

Decisions are taken when key actors not only are convinced of and expect benefits from proposed strategies and actions but especially when the power constellation is in favor of it and when they personally strive for it given the values and interests at stake. This requires attention for two additional concepts: (1) sensemaking and sensegiving in the cultural and political processes of governance and (2) power and influence in and around the involved government and societal institutions. Earlier, these policy and decision-making dynamics have been addressed under the phrase 'strategy making as expression of the politics of culture' (De Heer, 2009). The 'politics of culture' refer to differences in perception, framing and interpretation of a situation, differences in sensemaking in interaction between actors, leading to ambiguity between or power play by involved individuals.[7] A possible way out of this is combining the idea of a deeply uncertain future with various external scenarios which can form a basis for bridging gaps by formulating narratives of different possible or plausible futures. These scenarios and narratives are useful to strategic conversation and joint sensemaking between involved actors, to formulate strategies in line with each scenario or a robust strategy performing well under a number of scenarios.

Against the backdrop of the highlighted dynamics of strategy and decision making, the following elements are of importance here: (a) significant problematic situation, (b) dominated by uncertainty and ambiguity, (c) actors highlight their issue in agenda-setting actions, (d) actors express the vision and objectives of what they want to achieve over time, (e) actors make strategic choices under these conditions, (f) the desire to frequently assess impact of choices, and (g) a willingness to possibly adjust measures in order to achieve the vision and objectives. In the course of time, actors can also decide to revise the original vision. So in fact, the process of strategic choice and ultimately of adaptive decision making is aimed at uncertain moving targets.

A definition of adaptive decision making is based on the described aspects and mentioned elements. Adaptive decision making refers to a process of agenda setting and strategic choice, aimed at realizing longer-term vision and objectives while significant uncertainty and ambiguity dominate the choice behavior of actors who intend to frequently assess and generate feedback on the drivers of uncertainty and

[5] De Heer (2009b). See also PhD dissertation of Merkus (2014).

[6] Also: Fisher and Forrester (1993).

[7] See also Merkus et al. (2014).

efficacy of choices to enable adjustment of choices where needed based on critical evaluation, research and/or innovation.

To summarize, the main elements in this definition of adaptive decision making are as follows:

- A process of agenda setting.
- Preparing and making strategic choices.
- Considering realization of longer-term vision and objectives.
- Significant uncertainty and ambiguity among decision-makers and stakeholders.
- Competing values and interests influence the choice behavior of actors.
- Intention to frequently assess and generate feedback on the drivers of uncertainty and efficacy of choices where needed.
- Modification, renewed adaptation based on critical evaluation, research and/or innovation.

Adaptive decision making is viewed as of crucial importance as it not only plays a role in the initial strategy and decision-making process but also during implementation and during the frequent upcoming adaptations. This justifies specific governance qualities and possibly requires transformation.

2.5 Strategy Making in an Uncertain and Volatile Context

So far many relevant aspects of adaptation to climate change in combination with already existing natural hazards and challenges have been highlighted. They may be seen as sensitizing concepts which help to structure the way action can be generated and organized. Adaptation is in fact based on sensemaking and deciding to deal with major drivers of developments and change. Sometimes this may be in a robust strategic way, involving drastic choices. In many cases, adaptation takes place in gradual, searching way in a state of flux and transformation. Either way, the strategic change is often qualified as 'transition' or 'transition process', suggesting a movement from one situation into another. However, the other situation under conditions of climate change is unknown. We most probably know tendencies and may have a sense of direction towards a new situation under the changing climate but are not sure how moderate, bad, or extreme it will be or whether or not it will stabilize in a certain period. So the idea of managing an open-ended transition is difficult and unrealistic to handle in a complicated societal situation with many governmental sectors and costly interventions involved. In the case of adaptation to climate change, the body of knowledge on adaptive strategic management and transformative change is useful as it is dealing with long-term developments, with an important role of uncertainties as characteristics for future developments.

In this sense, Mintzberg and Waters (1985) observed strategy features in businesses which are highly relevant to adaptation: strategies appeared to be flexible and adaptive. In practice of studied companies, not the whole intended strategies turn out to be realized while the remaining deliberate strategies were supplemented by emergent strategies and jointly led to the realized strategies. Intended strategy is the strategy mostly elaborated in details and come forward as outcome of bargaining and compromise, which an organization expects to realize. However, in many cases parts of such intended strategy are not realized. Unrealized strategies are abandoned strategy parts, e.g., due to high costs, changing preferences, innovations, or dominant external developments. What remains of the intended strategy is the deliberate strategy process which aims at realizing the still important strategy elements. The deliberate strategy is often added by a spontaneous upcoming, emergent strategy, which is an unplanned strategy and arises in response to unexpected opportunities and challenges. In the course of time, reflective evaluation feedback can show the unrealized and also the emergent strategy elements and possibly changed circumstances which all may require adjustment of the strategy in order to realize adequate adaptation; see Fig. 2.2.

The dynamics of unrealized and emergent strategy show flexibility and adaptive capacity, needed in a volatile and uncertain situation which is characteristic for adaptation to climate change. Useful is here the idea of 'tracking of strategies', meaning that the already followed strategies are reviewed to see what has been done, what works well and what not.

Although the literature on strategic management is primarily focused on organizations and management, especially in businesses, parallels in thinking are evident. The importance is emphasized of drivers of change, making sense of a situation and giving

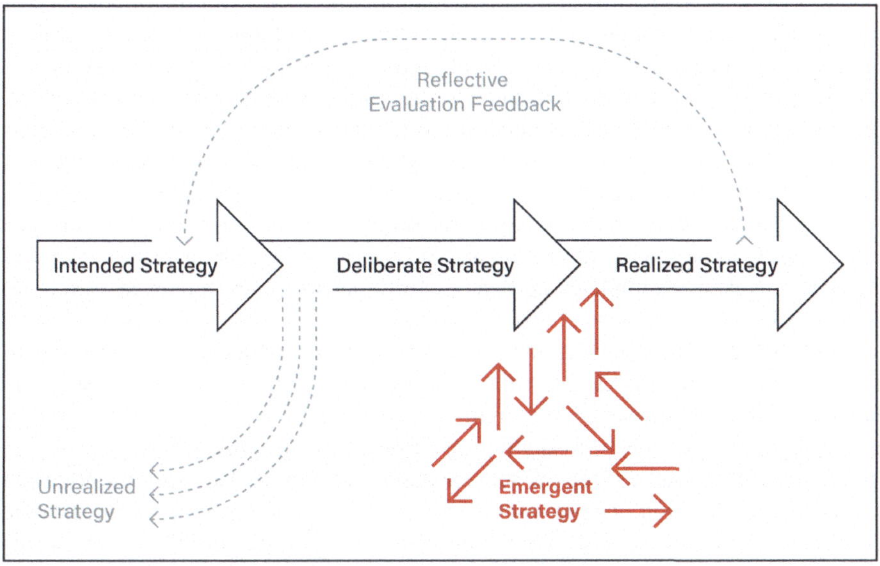

Fig. 2.2 Features of the strategy process: intended and emergent strategies for adaptation. Adapted from: Mintzberg and Waters (1985)

direction in terms of vision, mission, and goals which can be elaborated into meaningful action to realize a desired future state.[8] The traditional strategic management and planning approaches focused on a single future direction, which is understandable because they were developed before the 1970s in a far less complex and volatile world. Oil crises of the late 1970s and 1980s already led to scenario-based strategizing. The further globalization, the digital revolution, the 2008 financial crisis, increasing market volatility, geopolitical controversy, the COVID-19 pandemic, climate change, unrest and war generated pressures and strategic uncertainties. These implied new management and planning challenges, requiring strategic thinking and policy making incorporating these dynamics in relation to political, social, and economic choices. Adaptation to climate change can in this perspective easily shift to the background. However, new and sometimes unexpected extreme events with heat stress, heavy rains, and floodings occur unexpectedly. Adaptation than takes suddenly shape as revitalization process, consisting of managing hard times or even decline in certain countries or sectors, in conjunction with improvement, innovation in order to increase the adaptive capacity.[9] Strategic thinking and strategy making therefore need to be ahead of serious trouble and be able to (a) incorporate volatility and complexity, (b) recognize uncertainty, and (c) be more flexible and adaptive to changing environments. Attention for these requirements is found in decline and revitalization literature, crisis literature as well as in strategic management literature. This goes back on strategic management and change ideas with notions as contingency factors and contingency choices (Child, 1972), intended and emergent strategies (Mintzberg & Waters, 1985), strategic drift and transformational pressure with the importance of strategy making, adaptation, power, and the cultural web (Johnson & Scholes, 2002). The risk of strategic drift has been highlighted with respect to business strategies moving gradually away from relevance to the forces at work in their environment. This concept can be applied to adaptation to climate change: if procrastination takes too long, in fact ignoring environmental forces takes place, the drift becomes apparent and societal performances are likely be affected. Responses may enter a state of flux without clear direction with possibly even more negative impact on performances or qualities, leading to non-adaptation and or disaster. When adequate strategies and measures are taken and implemented successfully, negative impacts can be overcome; see Fig. 2.3.

Child (1972) and, e.g., Carter et al. (2008) discuss strategy literature and in their opinion missing aspects. Here the importance of contingency choices, strategic choice, and the role of power in and around organizations with respect to strategy and strategic choices has been highlighted as well as perception, interpretation of situations as important aspects of sensemaking and action. This focuses attention to strategy and decision making as being an expression, an outcome of the struggle for meaning and consequently for taking action.[10] Pettigrew (1992) points in this

---

[8] See, e.g., Johnson and Scholes (2002); De Wit and Meyer (2010); Quinn et al. (1988).

[9] De Heer, Strategy and Change in Organizations under Pressure (1991), with review of harsh conditions of strategy making in organizations in decline or revitalization.

[10] See, e.g., De Heer et al. (2015).

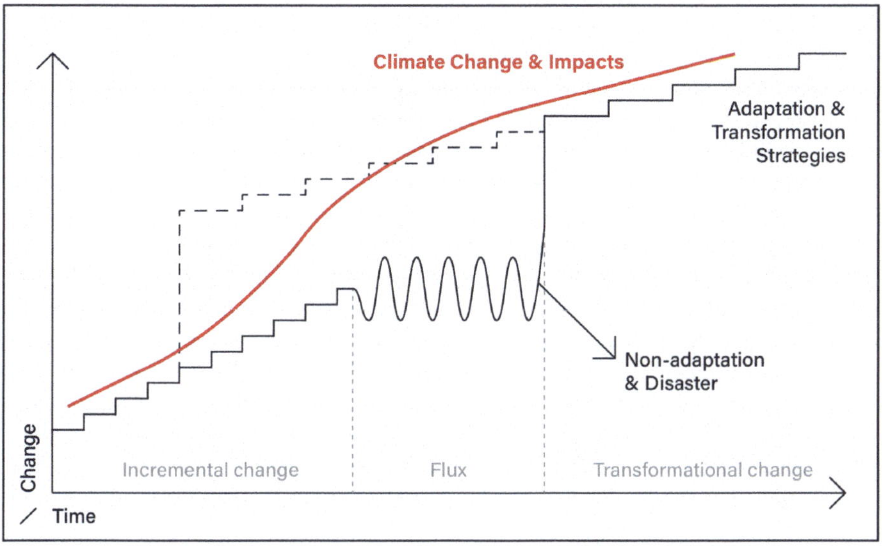

Fig. 2.3 Strategic drift and various adaptation responses. Adapted from: Johnson and Scholes (2002)

respect to the importance of temporal interconnectedness of change in past, present and future time. Besides, he also emphasizes the embeddedness of change, implying the need to understand it in the context of interconnected levels of analysis. Thirdly, Pettigrew explores context and action: creating reality by action while context is also a stimulus to action. On managing strategic change, Pettigrew (1992) elaborates on the importance of interdependent exploration of content (the 'what' of change), process (the 'how' and 'by whom'), and context (the 'why' and 'when' of change).

Under the current circumstances of climate change, delays in global decision making on mitigating climate change, trying but hardly succeeding to reduce further rise of temperature, the consequence is that adaptation to it becomes increasingly a matter of transformational change, hence significant leaps in time need to be made to adjust in time. The discussed theoretical insights offer the therefor required points for orientation and inspiration.

2.5.1 Managing a Process of Scenario-Based Strategy Making

Uncertainty about how climate change and related drivers of development will be in the near and further future leads to the requirement of thinking in possibilities which may occur in different futures with various conditions. So next to decision making in a mixed scanning approach, uncertainty about future climate change development needs to be addressed to improve strategic choices and decision making even further. Thinking about different possible futures has been practiced for quite some time but draw especially attention since the scenario development at oil company Shell.[11] Scenarios are in this view alternative descriptions or stories of environmental characteristics and based on certain drivers their possible future development. Scenario writing provides narratives of possible futures following different assumption and thereby providing different storylines. Although the scenario approach was broadly welcomed as it, other than the forecasting methods, included and accepted uncertainty about the future into strategizing, some criticism increased gradually. The first reason therefor is that scenarios were not easy applicable in actual management decision making. Secondly, scenarios were in most cases developed by a specialized team, leading to misunderstanding in organizations and opinions of 'nice things of top management'. To overcome this, e.g., Schwenker and Wulf (2013) further developed the scenario methodology with a six-step approach and related tools. A more or less similar improvement has been provided by De Ruijter and Alkema (2014) who introduces eight strategic topics for a process of strategic thinking and acting:

1. Mission, the reason of existence.
2. Trends, what is happening around us.
3. Scenarios: in which situation could we end up.
4. Options: is our current course futureproof and if not, what can we do.
5. Vision: where do we want to go, where do we want to be.
6. Roadmap: how do we get there, what will the strategy be.
7. Action; elaborating scenarios, strategy and execution aligned with planning and budgeting.
8. Monitoring: to stay on course.

These topics pay attention to activities to be carried out and structure the strategy process over time. The earlier problems with decision making and communication were addressed in this way. Content as well as stakeholders get full attention during the process. Schwenker and Wulf (2013) emphasized in their approach comparable attention for a well-structured and open process. They suggest: step 1 attention to definition of the scope, by preparing a framework of a. goal of the scenarios, b. strategic level of analysis, c. participants, d. definition of stakeholders end e. time horizon. They suggest as step 2 Perception Analysis, to make assumptions and mental models of management on the industry or other environmental developments explicit. This is confronted with a 3600 stakeholder feedback to learn more about existing perceptions and sensemaking. Step 3 concerns trends and uncertainties in the environment. It uses a weak-strong potential model and impact/low–high uncertainty matrix (following Van der Heijden) to conclude on critical impacts and uncertainties. Next step 4 indicates the formulation of scenarios, e.g., along two major drivers from the impact/uncertainty matrix, resulting in four different scenarios. This is followed by strategy definition and elaboration in various measures.

---

[11] See, e.g., Wack (1985); De Geus (1988), Duncan and Wack (1994), Van der Heijden (1996). On scenario-based planning also, e.g., Schwenker and Wulf (2013).

2.5.2 *Measures Suggested in Adaptation Pathways*

An interesting question is how effective certain measures will be in relation to the challenges in the course of time. Measures come normally forward from existing knowledge and already applied solutions and can be renewed gradually or may be replaced in case of innovation. The uncertainty about climate change and its impact means, that suitability and effectivity of measures is also uncertain because of the question whether or not challenges can be met after some time by the chosen measures. This can imply that a shift could be needed from a no longer effective or efficient measure to other ones that would perform better under the circumstances. Another point of attention is that measures best can be taken in such a way that they leave open the possibility for applying various follow-up interventions and by doing that, avoiding lock-in situations. This way of thinking has been elaborated under the concept of adaptive pathways (e.g. Haasnoot, 2013). Kwakkel et al. (2015)[12] which recognizes certain thresholds in the application of measures and shifts over time to other measures when required. Question to form an adaptation pathway is which initial fitting measures could be implemented, keeping lock-in situations in mind, and which possible alternatives can be found and applied when a shift of measures is needed to achieve a required situation, e.g. in terms of safety, social acceptance, investment costs, and financial consequences. Well-chosen indicators signal that measures are no longer effective. This can be related to possible tipping points in the main developments taking place as climate change, transboundary problems, but also in social acceptance, legal implications, or financial constraints. A shift to another scenario for strategy making and measures could become appropriate. Thinking in terms of adaptive pathways is valuable for strategy making under uncertain conditions as it implies the need to think over the longer term in alternative options and decide in time on the right measures to achieve the adaptation goals.

In many situations, no-regret measures can be sorted out which can be taken on short notice and are meaningful on the longer term. This type of measures needs often to be taken anyway, like substantial maintenance programs which can be designed in such a way that they serve adaptation to climate goals as well. In Bangladesh, these programs are called rationalization programs while in the Netherlands regular maintenance is added by 'Replacement and Renovation' programs. Also here the efficacy and lock-in phenomena should be kept in mind in the perspective of a certain adaptation pathway. A possible complication in applicability and use of adaptation pathways is not technical in nature but concerns the governance and political rationality as described earlier. Policy- and decision-makers are dealing with many societal and political issues in different ways of thinking, in taking decisions on competing investments, etc. Acting in this multi-sectoral and frequently multi-level political rationality may lead to different decisions than approval of a longer-term adaptive pathway possibly presumes.

---

[12] See Werners et al. (2021), for overview and evaluation of pathway approaches.

2.5.3 *The Importance of Suitable Practices in the Strategy Process*

Jarzabkowski argues that adaptation, being varying degrees of change from incremental adjustment to radical reorientation, may be explained using the theory of social becoming (Pettigrew, 1990; Sztompka, 1991): 'change is carried out within the internal context in interaction with the external context. There is thus an ongoing process of social becoming that is realized through a chain of social events, or practice'. (Jarzabkowski, 2004). In the so-called practice turn of strategy, attention is focused on what strategists or managers are actually doing in the strategy process and why they do it like that, against which knowledge and experience background and within a specific context the process takes place. Whittington (2006) separates three related aspects on this strategy as practice approach: (a) strategy praxis which is the part of reality in which the strategy process is embedded; (b) the practice or the set of activities carried out by (c) the practitioners, professionals carrying out the practices; see Fig. 2.4. In this figure, originally, Whittington distinguishes Praxis in one circle. In Fig. 2.4, praxis has been divided into two circles: the strategy praxis from professional practitioners formulating draft strategies and measures and the strategy praxis of officials involved in decision making about the strategies and measures. During strategy formulation, coming from initial ideas to a strategy ready to be discussed is a different process than scrutinizing the submitted documents, organizing external review and feedback as well as conducting consultation in society.

Figure 2.5 shows the relationship between the three elements of strategy as practice. The relevance is, that the earlier mentioned steps of strategy making and 'having a strategy' as such do not mean so much. It is important what practitioners write or discuss a mission, what they include or leave out and why or which trends they observe and analyze and which ones not, etc. Many interpretations and implicit choices are already taking place here. Practitioners take their practices from their professional background and institutes where professional practices are formed and applied, and where they become in many cases recursive in nature[13] while other practices may be adaptive, open for further development.

The strategy process can in this way be seen as a sequence of episodes in the relevant adaptation praxis where professional practitioners such as strategy project managers, consultants, researchers, engineers apply their practices. The results of those episodes have to be explained with the ins and outs of what has been done to responsible management so that they can make sense of it and make the relevant decisions. Strategy as performative practice has been highlighted by Kornberger and Clegg (2011) in the case of Sydney 2030.

Leadership of the involved practitioners is an interesting and very important aspect of strategy making. Designing a process with rational, logical steps, and ingredients is one thing but practitioners will do the actual work in which many (implicit) choices are involved. In a comprehensive strategy process, good leadership is required to steer

[13] See Jarzabkowski (2004): Strategy as practice: Recursiveness, adaptation and strategic practices-in-use.

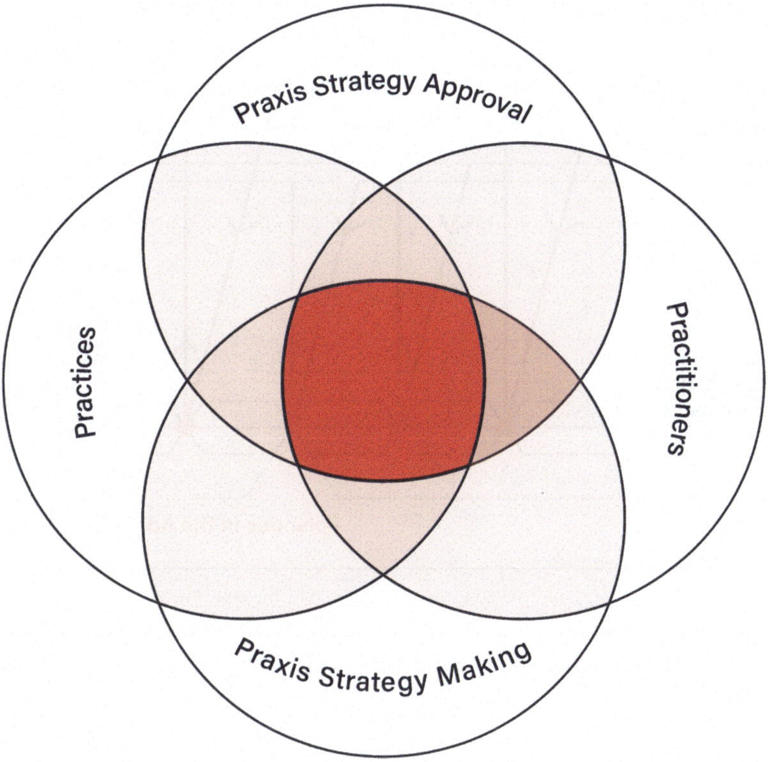

Fig. 2.4 Practices applied by practitioners in the praxis of strategy making and praxis of strategy approval. Adapted from: Whittington (2006)

the strategy process through a number of decision gates to end results. Leadership should know and if needed question what the practitioners actually do when they apply their practices, which choices have been made and how these influence strategy outcomes or alternatives. In the approval processes and final decision making in which investments play an important role, the assumptions and choices as well as the application of the practices need to be clarified.

2.6 Transformative Adaptation in the Core Domains

2.6.1 Temporal and Spatial

How can adaptation to climate change be seen in relation to required action, given (1) the magnitude of the changing climate as long range external driver, (2) major multi-level changes stretching over decades with global, societal, and organizational

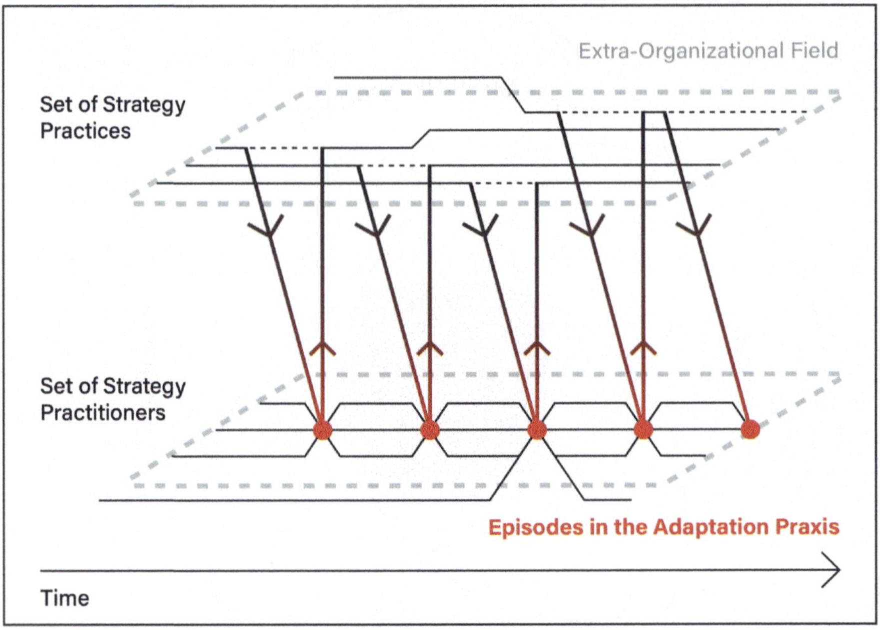

Fig. 2.5 Practices applied by practitioners within episodes of the strategy praxis. Adapted from: Whittington (2006)

impacts, and (3) fierce competition from other issues and challenges? That's a massif question we are confronted with in managing long-term uncertainties in an operational multi-sectoral and facetted setting. It may be useful thinking about it as a process of becoming, from hidden, unfolding to the manifest unfolded reality. It's there and for a lot of people it's becoming clear it's there because of the increasingly manifest logics of change: climate apart from normal variations shows fundamental changes in increase of atmospheric and seawater temperature, faster melting processes, sea-level rise, higher frequency of heavy storms, torrential and erratic rainfall and floods, extreme temperatures, and droughts. How far this will develop is unclear, hidden in the future. Intriguing is, that climate change is ongoing. This means that some impacts, as rising temperature and sea level with their drastic consequences, issues, and challenges are already manifest, are unfolded in terms that they happened and are still happening. Some impacts like melting of North and South Pole, of glaciers and rise of seawater temperature and, e.g., the legacy of sea-level rise are unfolding in a process of becoming, and some unknowns are hidden in the future to come. Whether there will be a gradual or erratic and unpredictable development of the unfolding climate change consequences and unknowns in line or not with the already manifest impacts, issues, and challenges is the big question. Hence the importance of introducing the concept of uncertainty and ambiguity in the approach of adaptation, climate data and sound scientific analysis and reporting are of crucial importance but can't give guarantees about the development of the climate change patterns or drivers behind it. Adaptive strategy making and implementation are important to deal with long-term goal achievement under uncertain conditions. The point of adaptivity over time is the inbuilt necessity to frequently look what is happening, reflecting on it and making a next move. This can take place in a new cycle of updating earlier plan and strategies or in type of interventions. Besides, an adaptive approach over time balances over- or under-investment and contributes to sustainable transformation of the socio-economic and natural systems including ecology.

The spatial dimension of climate change and its impacts is evident. Low-lying areas as deltas are confronted with sea-level rise and larger river discharges due to heavy rains. Urban areas are experiencing drainage problems and heat stress. Water scarcity threatens other areas with the problem of groundwater depletion and food insecurity. The relationship with the temporal dimension is that initially impacts will be low and may be ignored but gradually they increase to a level in certain areas; they cannot be ignored when people start suffering from unexpected weather, floods, heat which can become life threatening. There are great variations in impacts per region. In a not so large country as Bangladesh, six hotspot areas with specific and typical problems for each area were identified for developing specific strategies and measures. The spatial dimension points to differentiated characteristics of climate change and its impacts in global regions, so downscaling of global climate change models is an important challenge. Besides, each country shows per area differences of impacts, e.g. with respect to coastal zone, main rivers, urban areas, drought-prone areas, hilly or mountainous areas, areas with specific physical conditions. This means, that next to national level or generic strategies and measures, area or hotspot specific strategies and measures will be needed and that baseline studies need to be sufficiently focused on area-specific data and research.

2.6.2 Institutional and Organizational

The willingness and capacity to adapt depend to a large extend of the way the government is governed and equipped: by political choices of the dominant political coalition, by quality and preparedness of the administrative system including institutional setting of rules and regulations, existence of qualified organizations and staff, budget conditions, etc. Strategic planning for climate change adaptation should seek a balance between among others the climate, environmental and technological aspects of adaptation and the governance, institutional, organization and management and stakeholders aspects of adaptation processes. It is government organizations in which adaption needs are addressed and strategies, measures, and technical proposals are prepared and accepted or not, elaborated and financed. Climate change adaptation concerns mostly more than one policy sector and intervenes significantly in societal processes. For preparing such a major plan like a multi-sectoral delta plan for a whole country, designing a strategy process incorporating and even facilitating direct involvement and interaction with responsible policy makers and stakeholders is important. This type of adaptation focused strategy making should be close to the heart of political decision making, where mainstreaming of plans and allocation of budgets take place. Levers of change which can be utilized to shape strategies leading to transformation are related to the real issues and challenges in the impacted societal systems and structures. Governance and institutional development face however the challenge of incorporating climate change in the core governance and relevant policy sectors for preparing decision making and integrated plans and programs, for adjusting institutions, making new spending priorities on required funding and related taxation, cost recovery or loan arrangements. Doing all this requires a body of governance, institutional and public management technology, based on knowledge and experience of government performance, data and information of what needs to be changed and of methods and techniques to be applied for preparing and executing necessary actions. Each impacted sector and related challenge implies its own approach and technology to come to solutions. The technologies required to achieve necessary transformation is called transformational technology, by which adaptation to climate change can be realized. Here, only some aspects of government with respect to adaptation to climate change have been highlighted. Many more can of course be said about the sectors in the public sphere, the private sector, personal situations, etc. Seen in this way, adaptation with its broad scope refers to fundamental changes or transformation processes, impacting many aspects of society: of the governance and institutional as well as socio-economic and physical systems.

2.6.3 Socio-Economic Development

Societies and their economies are continuously on the move, sometimes in a stable episode of evolution, sometimes in situations of revolution and dramatic changes. Climate change relates to this in a serious way, as it touches many aspects of society and its people. Society and its institutions can be seen from different points of view. As briefly indicated before, Morgan (1997) introduced useful multiple images or metaphors to look at and interpreted different characteristics of organizational dynamics. Organizations, e.g., seen as interdependent 'Organisms', show various needs and required resources to keep the organism alive in their living environment. It also shows the tendency to compete about scarce resources as well as the capacity to learn and to adapt to changes in the ecosystem. Organizations can also be seen as 'Cultural Process', in which traditions, values, norms, and interests play an important role in perception and interpretation of reality on which an ongoing process of sensemaking and action is based. Differences in interpretation and in opinion can lead to ambiguous situations where actors strive for domination, rivalry or even culture wars. In connection to this, organizations may be seen as 'Political Arena' with attention for identity, interests or values to strive for. Influencing, participation, negotiation, and powerplay offer ways for dealing with power and decision making on as important regarded issues. Organizations seen as 'Machines' focus attention to a useful production process with division of work in well-structured command, execution, and control tasks in a process in which input is efficiently transferred into output, useful products. 'Flux and Transformation' is a metaphor for the flow of events in organizations and society, based on actors' perceptions, interpretations, choices and actions, with a capacity to self-organize, change and self-renew in line with a desire for sustained existence, development, and a certain identity. Society and its institutions can be viewed as a constant state of flux embodying features of both permanence and change.

Aspects of socio-economic development in relation to climate change and strategizing become evident when looking at the different aspects of the ongoing reality construction. Choices about the scope of e.g. a delta plan strategy will most likely be related to the needs, the values, the interests of stakeholders but also to the production processes and even the socio-economic development of a nation as a whole. This may concern a narrow scope, focusing on water safety or broader scope which includes not only water safety but also water security, health and sanitation, water for agriculture and food security, for navigation, water and urban development, etc. In this respect, looking at the broad scope of the Bangladesh Delta Plan 2100, 'adaptation by development' has been frequently mentioned in this publication.

2.6.4 Technological Aspects

The technology of involved organizations and disciplines can be utilized and managed as good as possible to realize climate change adaptation. The word 'technology' can be used to refer to a collection of knowledge, skills, techniques, mostly from several disciplines. In this context, it is the current state of knowledge of how to combine resources to produce desired products or outputs, to solve problems, to fulfill needs. This is the subject of transformative technology management, which utilizes existing bodies of technology and tries to innovate where needed. As said, climate change and adaptation have impact on many sectors of society which means that change in these sectors directly relates to the characteristics and modes of operations of these sectors in terms of contextual aspects and dynamics, content-wise issues and challenges and ongoing processes between societal organizations, government institutions, stakeholders, etc. Transformation of a sector means one is dealing with the body of knowledge or the technology concerning that sector, especially with respect to the issues and challenges of climate change and closely related processes concerning water safety and security, disaster management, spatial planning, food security and, e.g., renewable energy issues. Preparing a delta plan for a country like Bangladesh comprises all these aspects and means in fact the management of related technologies. Because of that, utilizing those technologies per sector is useful because this offers opportunities to define interventions including innovations as way out to adequate adaptation. This concerns also managing transformative technologies available for guiding strategy making, institutional change, financial opportunities, strategies, and measures for preparing no-regret projects with adaptive design and execution of measures on the ground. In this sense, transformative utilizing technology in a broad meaning for achieving desired goals and vision. This can be done by research of problem situations and baseline conditions, searching for and managing relevant data, knowledge development, instruments, and technologies. The aim is that transformative change can take place of as problematic defined situational circumstances, e.g., transforming the living conditions into more favorable ones.

2.6.5 Physical Systems

Climate change as such and its impact manifest themselves in the physical domain in various ways, like the global atmosphere, the warming of air and seawater, melting of ice caps or the Antarctic's or Himalayans, flooding and salt intrusion in low-lying delta areas, droughts, etc. The physical domain is clearly of utmost importance to understand climate change, how to limit it but also how to deal with impacts on the ground by developing adequate measures and projects. Implementation of these requires literally intervening in the spatial environment, in the natural resources, especially land, water and ecosystems.

Knowledge of the physical domain is certainly not complete as the uncertainties of climate change and related weather patterns show. Moreover, areas where interventions have been programmed need thorough study to gather necessary data and information about the state of those areas, possibilities for designing interventions, options for innovation and specific aspects of execution of the works. Local stakeholders are normally well informed about characteristics of their area and have detailed information on natural resources and ecology.

References

Carter, C., Clegg, S. R., & Kornberger, M. (2008). *A very short, fairly interesting and reasonably cheap book about studying strategy*. Sage Publications.

Child, J. (1972). Organizational structure, environment and performance: The role of strategic choice. *Sociology, 6*, 2–22.

De Geus, A. (1988, Mar–Apr 2–6). Planning as learning. *Harvard Business Review*.

De Heer, J. M. (1991). *Strategie en verandering in organisaties onder druk (Strategy and Change in Organizations Under Pressure)*. (PhD Dissertation). Leiden University. VUGA.

De Heer, J. M., Van Marrewijk, A., & Veenswijk, M. (2009a). Making sense of complexity in cross cultural collaboration. In *Unravelling organizational dynamics*. Prentice Hall/Pearson Education.

De Heer, J. M. (2009b). *Inaugural address for the chair 'Cultural Change and Strategy'*. VU University Amsterdam.

De Heer, J. M., Veenswijk, M., Merkus, S. (2015). *De bestuurlijke cultuur van besluitvorming in de Nederlandse infrastructurele sector*.

De Ruijter, P., & Alkema, H. (2014). *Scenario based strategy. Navigate the future*.

De Wit, B., & Meyer, R. (2010). *Strategy: Process, content*. Cengage Learning.

Dror, Y. (1964). Muddling through: Science or Inertia? In: Public Administration Review, Vol. 30.
Duncan, N. E., & Wack, P. (1994). Scenarios designed to improve decision making. *Planning Review, 22*(4).
Etzioni, A. (1967). Mixed-scanning: A "third" approach to decision-making. *Public Administration Review, 27*(5), 385–392.
Etzioni, A. (1986, Jan–Feb). Mixed scanning revisited. *Public Administration Review, 46*(1), 8–14.
Fisher, F., & Forrester, J. (Eds.). (1993). *The argumentative turn in policy analysis and planning.* Durham.
Gioia, D. A., & Chittipeddi, K. (1991). Sensemaking and Sensegiving in Strategic Change Initiation. *Strategic Management Journal, 12*(6), 433–448.
Gulich, E., & Quasthoff, U. M. (1986). Story-telling in conversation: cognitive and interactive aspects. *Poetics.*
Haasnoot, M. (2013). *Anticipating change: Sustainable water policy pathways for an uncertain future.*
IPCC. (2023). *Climate change 2023: Synthesis report. Contribution of working groups i, ii and iii to the sixth assessment report of the intergovernmental panel on climate change.* In Lee, H., & Romero, J. (eds.).
Jarzabkowski, P. (2004). Strategy as practice: Recursiveness, adaptation and strategic practices-in-use. *Organization Studies, 25*(4).
Johnson, G., & Scholes, K. (2002). *Exploring corporate strategy* (6th ed.).
Khan, S. I., Islam, A. A., & Haque, M. I. (2008). *Political culture, political parties and the democratic transition in Bangladesh.* UPL.
Kingdon, J. W. (1995). *Agendas, alternatives, and public policies* (2nd ed.).
Kornberger, M., & Clegg, S. R. (2011). Strategy as performative practice: The case of Sydney 2030. *Strategic Organization, 9*(2), 136–162.
Kwakkel, J. H., Haasnoot, M., & Walker, W. E. (2015). Developing dynamic adaptive policy pathways: a computer-assisted approach for developing adaptive strategies for a deeply uncertain world. *Climatic Change, 132*, 373–386.
Lindblom, C. E. (1959). The science of muddling through. In: Public Administration, Vol. 19, pp. 79–88.
Merkus, S., De Heer, J. M., & Veenswijk M. (2014). Decision-making as performative struggle: Strategic political-executive practices influencing the actualization of an infrastructural development. *Journal of Organizational Ethnography.* ISSN: 2046-6749
Merkus, S. (2014). *Struggling for meaning in the Dutch infrastructure sector.* VU University.
Mintzberg, H., & Waters, J. A. (1985, July–Sept). Of strategies, deliberate and emergent. *Strategic Management Journal, 6*(3).
Morgan, G. (1997). Images of organization, 2nd Edition.
OECD. (2011). *Water governance in OECD countries: A multi-level approach.* OECD Publishing.
OECD. (2015). *Centre for entrepreneurship, SMEs.* OECD Publishing.
Pal, L. A. (2014). *Beyond policy analysis—Public issue management in turbulent times* (5th ed.).
Pettigrew, A.M. (1990). Longitudinal field research on change: Theory and practice. In: Organization Studies, 1, pp. 267–292.
Pettigrew, A. M. (1992). The character and significance of strategy process research. *Strategic Management Review, 13*(2).
Quinn, J. B., Mintzberg, H., & James R. M. (1988). *The strategy process. Concepts, contexts, and cases.*
Schwenker, B., & Wulf, T. (2013). *Scenario-Based strategic planning.*
Simon, H. (1957). A behavioral model of rational choice. In: Models of Man, Social and Rational.
Sztompka, P. (1991). The theory of social becoming: An outline of the conception. In: The Polish Sociological Bulletin, No. 96, pp. 269–279.
Taylor, C. (2013). *The discourses of climate change.*
UNFCCC. (1987). *Montreal protocol on substances that deplete the ozone layer.*
UNFCCC. (1997). *Kyoto Protocol.*

UNFCCC. (2015). *Paris Agreement.*

UNFCCC COP28. (2023). United Nations Climate Change Conference or Conference of the Parties of the UNFCCC.

United Nations. (1987). Montreal protocol on substances that deplete the ozone layer.

Van der Heijden, K. (1996). *Scenarios. The art of strategic conversation.* Wiley.

Wack, P. (1985, Sept–Oct). Scenarios: Uncharted waters ahead. *Harvard Business Review*, 73–89.

Weick, K. (1995). *Sensemaking in organizations.* Sage Publications.

Werners, S. E., Wise, R. M., Butler, J. R. A., Totin, E., Vincent, K. (2021, Feb). Adaptation pathways: A review of approaches and a learning framework. *Environmental Science and Policy, 116*, 266–275.

Whittington, R. (2006). Completing the practice turn in strategy research. *Organization Studies, 27*, 613.

Part II
Situational Characteristics of Adaptation

Orientation on Adaptation in the Bangladesh Setting

Chapter 3
The Setting and Need of a Delta Plan

Abstract Before preparing a long-term strategic plan, it is imperative to understand the earlier approaches, plans, and actions, as shown in this chapter under tracking of strategies and the way the Government of Bangladesh (GoB) took action before. Tracking the earlier strategies and plans to learn what the mindset and narratives were and what has been done on planning and execution of projects as well as on maintenance are important parts of adaptive strategy making and transformation. It is felt by the government that the need to deal in time with climate change in coherence with other issues is pressing. Hence, the need for adaptation and the initiative to define scope and objectives of a long-range scenario-based strategic planning project. This is in fact an initiative to formulate a strategic plan for the Bangladesh delta almost covering the whole country. The ways the Bangladesh Delta Plan 2100 is institutionally embedded is elaborated, and the strategic planning process is described in the initial episode. Most of the by GoB selected thematic areas under the Bangladesh Delta Plan (BDP) 2100 are related to water. So, it is important to gather knowledge on the development of the delta and water resources systems in the country. Bangladesh, as riverine country, faces major inter-related delta challenges in water safety, food security and in socio-economic development. It also experiences too much water during the monsoon and too little water in the dry season. The available water resources comprise internally generated surface water through rainfall and runoff and transboundary inflows. The groundwater is also a source of water, but overexploited. Besides this, the main water resources management issues are considered as flood, drought, storm surge, salinity intrusion, waterlogging, river bank erosion, groundwater depletion and sea-level rise. Over time, the issues are exacerbated due to human interventions which contribute to the deterioration of the water quality. Bangladesh has an impressive history in the BDP 2100 important areas.

Keywords Tracking strategies · Adaptation · Transformation · Scenario · Transboundary · Water resources · Flood · Flood control · Drainage · Irrigation · Storm surge · Cyclone · Salinity · Salinity intrusion · Waterlogging · Erosion · Governance · Policy · Plan · Water resources · Water resources management · Groundwater · Groundwater depletion

J. M. de Heer and G. A. Choudhury, *Strategies for Adaptation to Climate Change in a Transformative Approach*, Springer Climate,
https://doi.org/10.1007/978-3-031-88884-7_3

3.1 Tracking of Strategies

3.1.1 Developments in the Past

Water resources management in Bengal (now Bangladesh) has a history that dates back to ancient times. Early rulers introduced an 'overflow irrigation' system to enhance agriculture, with farmers participating in its operation and maintenance. The local landlords, or Zaminders, prioritized water management for both agriculture and navigation, collecting levies from riverine vessels to fund the maintenance of canals and drainage systems. This persisted until the onset of the colonial period (1757–1947), when water management underwent significant changes. The Zaminders were relieved of their water management duties, and the British rulers, being traders and sailors unfamiliar with agricultural irrigation in the delta, focused on revenue collection. Income was frequently prioritized over service to the people, leading to a decline in traditional water management systems and infrastructure. In the post-colonial period (from 1947 onward), water resources development became more institutionalized under government ministries, with large-scale structural interventions (see, e.g., Fig. 3.1) requiring skilled technical expertise.

Fig. 3.1 Embankment Brahmaputra/Jamuna used by local people. *Photo* J. M. de Heer

3.1.2 Key Developments in Water Governance and Related Sectors

3.1.2.1 Initial Measures for Protection

Water resources development in Bangladesh[1] began before independence, in 1952, with three key projects: the Karnaphuli Multipurpose Project (for power generation and flood control in the Chattogram region), the Ganges-Kobadak Project (irrigating Kushtia and Jessore districts), and Cross Dams No. 1 (for land reclamation in Noakhali District). The main focus during this period, extending into the 1970s, was on large-scale flood protection to safeguard agricultural lands and communities.

Following severe floods in 1954, 1955, and 1956, the United Nations' Krug Mission was commissioned to study flood control and water resource development, leading to the establishment of the East Pakistan Water and Power Development Authority (EPWAPDA) in 1959. Following Bangladesh's independence in 1971, the organization was divided into two entities: Bangladesh Water Development Board (BWDB) and Bangladesh Power Development Board (BPDB). BWDB inherited the responsibilities of the previous water wing of EPWAPDA. In 1975, the Bangladesh Agricultural Development Corporation was also established, further focusing on water and agriculture issues. Over time, additional organizations, such as LGED, DPHE, and WASA, were formed to address specific water-related challenges, but BWDB remains the key institution in water resources development and management.

The first National Water Master Plan, prepared in 1964 by the US-based International Engineering Company (IECO), aimed to rapidly develop the country's resources to raise living standards, promote equality, and ensure equitable income distribution. Major flood control, drainage, and irrigation (FCDI) projects were launched to boost agricultural output, but these efforts largely ignored the needs of fisheries and navigation sectors. Issues such as siltation, waterlogging, reduced biodiversity, and declining river navigation routes were noted as unintended consequences of these large-scale interventions. Criticism of large FCDI projects grew due to their long realization times and limited benefits for poorer populations. In response, a 1969–70 study by the International Bank for Reconstruction and Development (IBRD) suggested smaller-scale irrigation projects, but large-scale efforts continued, especially after severe flooding in 1974. This led to the implementation of Early Implementation Projects (EIP), supported by countries like the Netherlands, and EIP Type Projects by other international agencies. The main focus was on flood control and drainage as well as in a limited way on irrigation. The approach was sectoral with limited inter-sectoral communication. The small-scale projects proved vulnerable during floods in 1987 and 1988. The initiatives on small irrigation works and privatization of groundwater use resulted, however, in a considerable increase of irrigated areas.

[1] See also: Rashid & Rahman (2010), on these and earlier water resources developments.

3.1.2.2 Towards National Policy and Plans

In 1983, the Master Planning Organization (MPO) was established, which prepared the first National Water Plan in 1986 followed by several updates. It assessed water availability and future demands, leading to the development of numerical models and strategies. However, the MPO fell short of creating a comprehensive water plan, prompting further studies. In the aftermath of severe 1987 and 1988 floods, the comprehensive FAP studies, and Flood Action Plan (FAP) were between 1989 and 1995 initiated with international support. One of the outcomes of the FAP was formulation of Bangladesh Flood Management Strategy in 1995. FAP sought to address flood management issues but faced criticism for its lack of community involvement and environmental considerations (see Fig. 3.2). The government responded to these challenges by adopting a National Water Policy (NWPo) in 1999, which set guidelines for managing surface and groundwater resources, ensuring equitable access, and promoting sustainable water delivery systems. It emphasized institutional changes and decentralization as well as stakeholder participation and greater involvement of women in water management (Ministry of Water Resources, 1999). The policy also aimed at integrating public and private investments and stressed the importance of environmental sustainability. The NWPo provides herewith the guidelines for agriculture, fisheries, industry, navigation, environment, basin-wide planning, water rights and allocations, public and private investment, water supply and sanitation. The policy underscores the broad principles of water resources development and its rational utilization. It emphasizes both public and private actions and highlights the importance of conjunctive use of ground and surface water, e.g., for irrigation; see Fig. 3.3.

The NWPo stated national goals which formed the basis for the development of the National Water Management Plan (NWMP) Ministry of Water Resources, 2001). The NWMP provided a framework for the preparation of interventions and for setting priorities on investments and project proposals. In this period, the Coastal Zone Policy was introduced in 2005 to manage the unique challenges of Bangladesh's coastal regions. Later on, in 2012, a Master Plan for the Haor Area was formulated to address flash flooding and foster sustainable development of the area.

In 2013, the Bangladesh Water Act was enacted, establishing a framework for the integrated management of all forms of water (surface, groundwater, and rainwater), prioritizing drinking water and domestic use. The Participatory Water Management Act followed in 2014, formalizing stakeholder involvement in water management.

The Perspective Plan of Bangladesh 2010–2021 aims to protect Bangladesh from effects of natural and human-induced hazards, global warming and climate change. Bangladesh is in the Perspective Plan aiming to become a middle-income country shortly and a developed country by 2041. The plan aims to protect vulnerable people from natural calamities, to reduce poverty and inequality, to take action against various forms of pollution, to make the country an ecologically attractive place and to promote environment-friendly tourism.[2] The National Sustainable Development

[2] See for other backgrounds, e.g., Khan et al. (2008).

Fig. 3.2 Harvesting just before flooding. *Photo* courtesy of Amirul Hossain

Fig. 3.3 Irrigation channel, providing freshwater for many functions. *Photo* J. M. de Heer

Strategy (NSDS) for the period 2010–2021 has been prepared to meet the considerable environmental challenges related to the development ambitions and to the challenges arising from global climate change.

Important milestones for water resources planning and development are shown in Fig. 3.4. This planning and development started with fielding of the Krug Mission, establishment of EPWAPDA, preparation of the Master Plan in 1964, establishment of BWDB in 1972, preparation of National Water Plans, establishment of FPCO in 1990, commissioning FAP Studies, (1990–1995), establishment of WARPO in 1992, approval of National Water Policy (1999), formulation of NWMP (2004), enactment of the National Water Act (2013), and formulation of Bangladesh Delta Plan 2100 (2018).

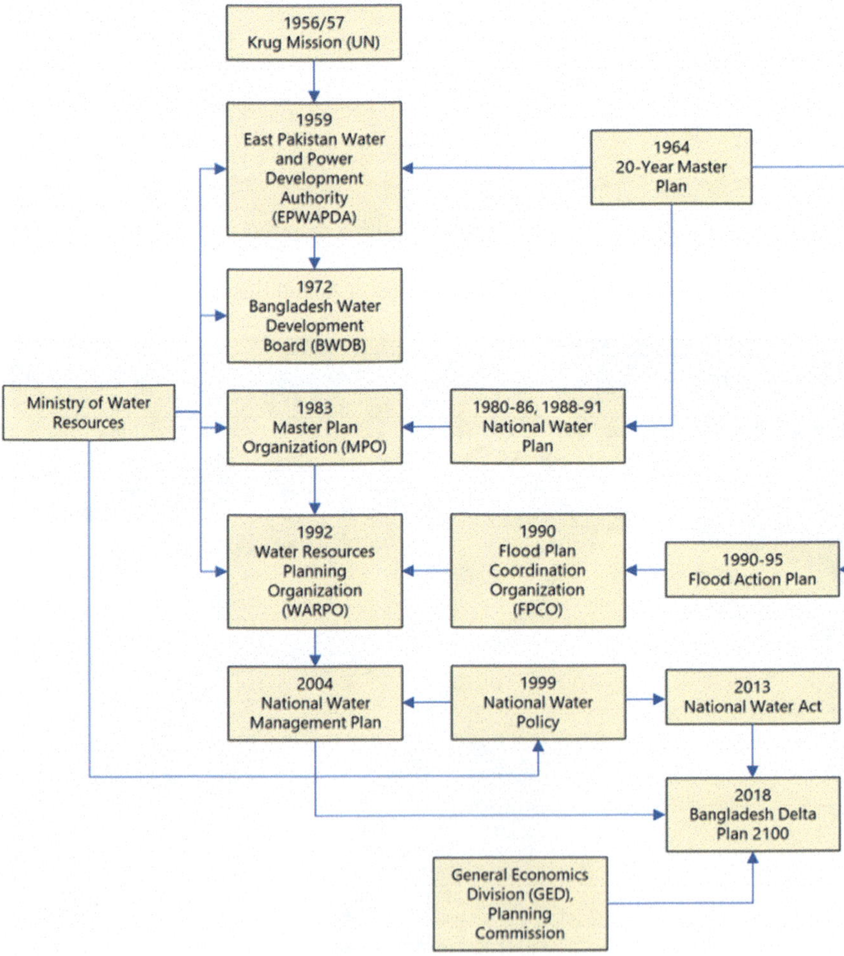

Fig. 3.4 Milestones for water resources planning and development

3.1.3 Observations

Water resources planning in the country initially focused on meeting local needs and addressing specific areas. After the three consecutive floods during 1954–1956, large-scale flood control projects were undertaken. In the following decade, political and socio-economic changes led to a need for updating the planning processes that had been adopted during the earlier Master Plan period. This resulted in the preparation of the National Water Plan and the development of new analytical tools. Further changes in planning occurred after the severe floods of 1987 and 1988, which prompted the Flood Action Plan (FAP) studies. However, these studies were criticized for lacking a multi-sectoral approach and for minimal community participation. The FAP led to the development of a strategy paper for water and flood management in 1995, followed by the National Water Policy in 1999 and the National Water Management Plan (NWMP) in 2001. Although the NWMP was an integrated, multi-sectoral plan developed through extensive consultations, progress was limited due to budget constraints, administrative gaps, and a shortage of skilled professionals.

The good things that came out of various phases of water resources planning in the country are that (1) the planning approach has undergone many stages and refinement, responding to the socio-economic needs of the people; (2) a lot of data and studies became available, which is an asset till today; (3) many project ideas and proposals or programs were developed. Despite challenges in systematic implementation, around 800 projects have been realized over the last 65 years. These projects have provided flood management for more than 6.3 million hectares of land, irrigation for 1.6 million hectares, drainage canal facilities, town protection measures, and embankments. The Bangladesh Water Development Board (BWDB) has improved water and food security through these initiatives, contributing to flood-free agricultural land, enhanced safety for people and infrastructure, increased crop yields, and better fisheries. Notable projects include the Ganges-Kobadak and Teesta Barrage Projects, with the Teesta Barrage standing out for being planned, designed, and executed with local expertise and national funding.

Flood control, drainage, and irrigation systems also offer multiple services beyond flood protection, such as serving as roads, providing space for plantation, and enabling fish culture. However, some projects have had negative impacts, including reduced fish habitats, restricted navigation, and waterlogging. Bank protection, particularly along major rivers like the Brahmaputra, the Ganges, and the Padma, remains a significant challenge, with maintenance and repair costs often exceeding the initial investment. On smaller rivers, protective structures have generally been more successful.

Participatory Water Management, although not initially prioritized, became important after several water development projects failed to meet local expectations (Ministry of Water Resources, 2000). While the concept has been applied in donor-funded projects like the Early Implementation Project and Delta Development Project, achieving smooth cooperation between government agencies and local communities remains a challenge. Recent projects like the Water Management Improvement Project and the Blue Gold initiative have further emphasized participatory approaches.

Over the past 65 years, substantial investments have been made to protect citizens from natural disasters like floods, droughts, and storm surges, creating an environment that supports high-yield crops, fisheries, and industrial development in floodplains. When water resource development began in the 1950s, the population was around 35 million, with widespread food shortages. Today, Bangladesh feeds over 180 million people, and the water sector has played a crucial role in this progress.

However, as climate change brings increased flooding, droughts, and sea-level rise, further investments in water infrastructure are required. Urbanization and industrialization in flood-prone areas also demand enhanced safety measures to protect economic investments. Unlike in the 1950s and 1960s, when floods mainly damaged crops and homes, today's floods have far-reaching economic impacts, halting operations and causing major infrastructure losses. The development of a reliable Flood Forecasting and Warning System (FFWC) by the BWDB since 1972 has become crucial for managing flood hazards, with timely dissemination of forecasts through media channels.[3] Many vital infrastructures developed by BWDB and others require regular maintenance to remain functional. Earthen embankments are susceptible to erosion from wind, rain, human activity, and animal grazing. Drainage channels often silt up during monsoons, while sluice gates corrode over time. Unfortunately, BWDB receives only about 10% of its required maintenance budget, leading to deteriorating infrastructure and rising repair costs. As a result, most of the maintenance works suffer due to lack of adequate budget. Consequently, the cost of repair keeps rising and the condition of the structures deteriorates.

The institutional development and the capacity building of the two principal water resources planning and implementing organizations such as BWDB and WARPO could be enhanced. It is obvious that BWDB has been weakened considerably due to a lack of institutional development over the years for various reasons as budget and manpower constraints and limited modernization. Capabilities eroded over the years. Reasons for decline are also that many experienced and qualified engineers have retired, and the new intake was limited due to working conditions and competition with other agencies or companies. In order to move forward, the country has to ensure a reasonably high level of safety from natural disasters. Moreover, the conservation of water resources is also of prime importance to meet the growing demands of industrialization and urbanization.

3.2 Considering a Long-Term Strategic Plan

Against the backdrop of the before described water plans and developed infrastructure, and agricultural growth, Bangladesh has achieved a remarkably consistent average economic growth rate of 6% over the last two decades before COVID-19.

[3] See achievements of BWDB https://bwdb.portal.gov.bd/

While the textile and garment industry, and remittances, continue to contribute to the GDP growth, industrialization and innovation in many areas are increasingly gaining importance in the economy as providers of employment, quality products and consumables. Since COVID-19, the country struggles with less economic growth but it still may realize its ambitions to become upper middle-income country in the near future. The growth rates have been achieved in spite of extreme adverse climate variability, with frequent storm and tidal surges, flooding, and droughts. However, many challenges lie ahead, the most important being population growth and pressure on land use, environmental protection, governance, climate change, globalization, and macro-economic development. Given the ambition to attain continuity of societal and economic processes and a better national income status, it is considered important to address water issues and the expected impacts of climate change with a long-term integrated approach to future land and water management in relation to water safety, food security, and socio-economic development. In addition, the sector- and department-wide composite governance could be strengthened, more integrated and together with institutional development be focused on addressing Bangladesh's future potential and challenges. Bangladesh has already developed a lot in terms of economy, food security, poverty reduction and natural resources strategies and plans. The challenge lies in safeguarding the achievements with consistent adequate long-term strategies as well as efficient coordination and effective implementation of required interventions. GoB found that, fitting in the described planning tradition, a holistic plan with a long-term time frame that is commonly supported would provide ideas and measures for improvement and innovation. Such an integrated and coordinated planning approach, including facilities for better spatial planning and disaster management is seen as desirable for managing the natural resources and to fulfill the demands of sectors such as agriculture, fishery, forestry, livestock, industry, water supply and sanitation, environment, and navigation.

Institutional development and good governance are important factors for achieving safety and security of life, livelihood, and economy of the Bangladesh delta. The recent and future anthropogenic changes in the hydrological cycle due to, e.g., climate change, construction of dams and barrages in the upstream countries (Rahman et al., 2019) in combination with increasing water demand are expected to make future water governance and management even more challenging. Developing an adaptive holistic approach in planning and funding would help to implement the programs as designed under different sectoral plans in a more integrated way. The development culture could be transformed more profoundly from a sectoral project approach to a coherent program approach with a strong emphasis on integrated or at least well-coordinated implementation.

With a view to these factors, the Government of Bangladesh (GoB) has discussed this with the Government of the Netherlands (GoN) which already prepared a long-term delta plan and assistance has been arranged accordingly on the formulation of a Bangladesh Delta Plan 2100 (BDP 2100). The Government of the Netherlands provided this support through its Embassy (EKN) in Dhaka. The formulation of a BDP 2100 is seen as of significance for the future growth of Bangladesh and as fundamental to addressing its vulnerability for water and climate change issues. The intentions are further explored in the preparatory work in 2011–2012 of the Delta Plan Preparatory Team (Choudhury et al., 2012) and agreement on the findings was formalized in the Memorandum of Understanding (MOU) between the two countries, signed in May 2012. This was followed by preparation of a project document and tendering process for selection of a private sector party who could deliver the required technical assistance. The proposal of a consortium called BanDuDeltAS[4] was selected and its interdisciplinary team of national and international experts assisted GoB in its ambition to develop BDP 2100. Comprehensive knowledge of and experience with challenges in the Bangladesh delta and deltas around the world (Mekong, Jakarta, Manila, New Orleans, The Netherlands) has been utilized in the formulation of the Plan. The Delta Plan activities are being undertaken under the mentioned MoU (2012), an Administrative Arrangement (2014) for continued collaboration between Bangladesh and the Netherlands and a GoB approved contract between the EKN and Consortium BanDuDeltAS, signed on March 12th, 2014. The decision was made by GoB to host the project at the General Economic Division (GED) of the Planning Commission, Ministry of Planning, as the GED is mandated for holistic long-term planning at the national level.

3.3 Fitting in National Ambitions and Development Goals

From the start with the preparatory work and tendering preparations as well as selection of assistance, it was clear that Bangladesh wanted its own type of delta plan. The development priorities of Bangladesh[5] include among others: ensuring broad-based growth and reducing poverty; ensuring effective governance and sound institutions; addressing regional cooperation; providing security for development and welfare; building a sound infrastructure and managing urban challenges; mitigating the impacts of climate change; and promoting innovation in a knowledge-based society. Because GoB expected that the formulation of BDP 2100 would contribute to the aforementioned development priorities, the MDGs and SDGs of the United Nations, relevant national policies, strategies, and plans have been reviewed by the assistance team.[6] Bangladesh has integrated the MDGs within the economic and social targets of its 6th Five-Year Plan (2011–2015). GED has been monitoring and reporting the MDG status through the Millennium Development Goals: Bangladesh Progress Report. Bangladesh has been moving towards achieving most of the goals, while some require enhanced efforts.[7] These goals are relevant to BDP 2100, especially the goals on governance, disaster risk reduction, climate change, food security and environment which have been further explored in the Baseline Studies.

---

[4] BanDuDeltAS is the international BDP 2100 research and formulation consortium, an abbreviation meaning: Bangladesh Dutch Delta Advisory Services.

[5] See, e.g., Bangladesh Perspective Plan 2021–2041.

[6] During BDP 2100 formulation the MDGs were used and later on the SDGs.

[7] See also: Shamsul Alam (2015).

Besides inspiration from other delta plans and from the MDGs/SDGs, the setup of BDP 2100 was also inspired by and linked with the Perspective Plan of Bangladesh 2010–2021 (Ministry of Planning, 2012). This was formulated in 2012 and envisages an outlook and guidelines for realization of the Vision 2021. The Perspective Plan of Bangladesh 2010–2021 was prepared with the goal to implement strategies to protect Bangladesh from impacts of natural and human-induced hazards, global warming and climate change. The plan indicated actions to protect the vulnerable people from natural calamities, to take actions for the prevention from industry and transport-related air pollution and to dispose of wastes in a scientific manner. The measures were suggested to make the country an ecologically more attractive place and to promote environment-friendly tourism. The perspective plan provides the road map for accelerated growth and lays down broad approaches for eradication of poverty, inequality, and human deprivation. Specific strategies and the task of implementation have been articulated through the 6th Five-Year Plan 2011–2015 (Ministry of Planning, 2011) and 7th Five-Year Plan 2016–2020 (Ministry of Planning, 2015). Development priorities are distilled from statements formulated in the mentioned Plans and MDGs/SDGs to take Bangladesh to where it wants to be in the year 2021, given its human potential and natural resource capabilities. Those development priorities cover more or less the same vision elements as for the MDGs/SDGs: ensuring broad-based growth and reducing poverty; ensuring effective governance and sound institutions but creating a caring society; addressing globalization and regional cooperation; providing energy security for development and welfare; building a sound infrastructure and managing the urban challenge; mitigating the impacts of climate change; and promoting innovation in a knowledge-based society. These thematic approaches shaped and formed the foundation on which specific strategies are developed. The GoB follows a Five-Year Planning (FYP) scheme which envisages the sectoral development goals of the country. National Five-Year Plans of Bangladesh envisage the perspectives of the sector medium-term goals. Article 15 of the Constitution of the People's Republic of Bangladesh articulates that a planned economic growth is required for the development of the country for which the National Five-Year Plans are prepared and practiced since independence. So, between FY 1973 and FY 2002 Bangladesh implemented 5 successive Five-Year Plans and one interim Two-Year Plan (1979–1980). The 6th Five-Year Plan was prepared for 2010–2014. GED is basically responsible for preparation of the National Five-Year Plans and in preparing these, GED follows a participatory process having different inputs from stakeholder ministries and related organizations. The 6th Five-Year Plan was formulated having different distinctive characteristics as follows: (a) first strategic and indicative plan of the country; (b) shows an indication of the resources to achieve planning goals through annual development programs; (c) encourages goal achievement through public–private partnership and participation of private sector; (d) articulated a results framework to realize the economic growth; (e) gender, women and environmental sustainability are given high importance; (f) the plan is considered as living document and will be annually monitored by the GED.

The status of achieving the targets of this 6th Five-Year Plan and lessons learned from that has been considered in the BDP 2100 Baseline Study on Governance and

Institutional Development. BDP 2100 delivers input to be used in the Five-Year Plan and already has done this for the 7th and 8th FYP and a guideline for the 9th FYP. The perspective plan and Five-Year Plan of Bangladesh emphasize the diversified sectoral uses of water resources and promote among others an approach guided by the concept of Integrated Water Resources Management to define water resources functions and uses. Giving utmost importance to water safety and security and to agricultural water use for food, the plans also recognize other uses of water for socio-economic development, such as for industry, forestry, fishery, navigation, drinking, and domestic use including for drinking. In fact, principles, policy, and action guidelines for integrated water resources management and people's participation go back to the National Water Policy, 1999.

References

Administrative Arrangement. (2014). Agreement between the Governments of Bangladesh and the Netherlands on formulation of BDP 2100.

Bangladesh Delta Plan 2100. (2018). *Delta plan, ministry of planning*. Government of Bangladesh.

Choudhury, G. A., Terwisscha van Scheltinga, C., Van den Bergh, D., Chowdhury, F., De Heer, J. M., Hossain, M., & Karim, Z. (2012). *Preparations for the Bangladesh Delta Plan*. Alterra Report 2300. ISSN 1566-7197.

Ministry of Planning, General Economics Division, Bangladesh Planning Commission. (2011). *Sixth Five Year Plan FY2011-FY2015 Accelerating Growth and Reducing Poverty*.

Ministry of Planning, General Economics Division, Bangladesh Planning Commission. (2012). *Perspective Plan of Bangladesh 2010–2021*.

Ministry of Planning, General Economics Division, Bangladesh Planning Commission (2015). *Seventh Five Year Plan FY2016–FY2020 Accelerating Growth, Empowering Citizens*.

Khan, S. I., Islam, A. A., & Haque, M. I. (2008). *Political Culture, political parties and the democratic transition in Bangladesh*. UPL.

Ministry of Water Resources Bangladesh. (1999). *National Water Policy*.

Ministry of Water Resources Bangladesh. (2000). *Guidelines for participatory water management*.

Ministry of Water Resources Bangladesh. (2001). *National Water Management Plan*.

Ministry of Water Resources Bangladesh. (2005). *Coastal Zone Policy*.

MOU. (2012). Memorandum of Understanding between the governments of Bangladesh and the Netherlands.

National Water Policy. (1999). *Water policy developed by ministry of water resources*. Government of Bangladesh.

National Water Act. (2013). Water Act 2013 of Government of Bangladesh.

NWMP. (2001). *National water management plan developed by ministry of water resources*. Government of Bangladesh.

Rahman, K.S., Islam, Z., Navera, U.K., Ludwig, F. (2019). A critical review of the Ganges water sharing arrangement. *Water Policy, 21*(2), 259–276.

Rashid, S., & Rahman, R. (Eds.). (2010). *Water resources development in Bangladesh Historical Documents*. UPL.

Shamsul Alam. (2015). The Economy in Transition: Challenging Issues.

Chapter 4
Framing of a Long-Term Adaptive Strategy

Abstract Bangladesh's coastal regions, which lie at low elevations, face significant vulnerability to frequent monsoon floods, exacerbated by tides, and to strong tropical cyclones. Additionally, salinization impacts drinking water quality and soil salinity restricts food production in these coastal areas. Flooding further contaminates drinking water, causing outbursts of infectious diseases. This situation is expected to get worse due to a higher temperature, increasing sea level and changing weather pattern as impacts of climate change. Besides, land subsidence, which is a natural process in this delta area, and the growing population complicate the situation. Population growth may drive suburbanization and economic activities in the coastal areas possibly bringing environmental degradation as well. The gradually changing climatic conditions could over time render large areas of Bangladesh highly vulnerable to flooding and by 2050, potentially displacing millions of people in coastal areas. For other locations, like big cities, it is expected that by 2050 or 2100, huge numbers of residents, may face serious challenges to essential services like electricity, safe drinking water, transportation and a healthy living environment. Bangladesh has made substantial progress in economic growth, poverty reduction, and natural resource management. The challenge now is to preserve these achievements through a long-term, multi-sectoral approach, and effective, timely implementation of necessary measures. This requires a dedicated delta plan supported by key political parties, with coordinated efforts in water resources and disaster management, food production, spatial planning, and supporting socio-economic development. In this way, adaptation to climate change in an integrated approach is still an enormous task in this country but essential for managing natural resources and meeting the demands of water safety and security, food security, industry navigation, and last but not least the environment and ecology.

Keywords Bangladesh delta plan · Adaptive strategy · Temporal domain · Spatial domain · Vision · Scope · Objectives · Goals · Mission · Thematic area · Baseline study · Framework · Holistic · Adaptive delta management · Interactive strategy process · Governance · Content · Stakeholders · Framework · Project organization · Consortium

J. M. de Heer and G. A. Choudhury, *Strategies for Adaptation to Climate Change in a Transformative Approach*, Springer Climate,
https://doi.org/10.1007/978-3-031-88884-7_4

4.1 Shaping the Contours of the Delta Plan

The initiative to undertake a strategic planning endeavor has been described in Sect. 1.2. The Technical Assistance team started the project by a signing ceremony for the Delta Plan contract for the period from March 12, 2014, till March 12, 2018. A suitable office was hired and the team of the project organization was mobilized soon. The project consisted of 12 Dutch professionals and 32 Bangladesh professionals who would be active mostly on intermittent basis, except management, and support staff. Time was spent to establish relationships with officials and representatives of relevant institutions. Many team sessions with team members required to start the project took place in the start-up period with creative discussions, strategic conversations, and prototyping sessions to understand and structure the process ahead and to make a first outline of the Delta Plan according to the in the bidding process approved proposal and additional preferences. The Inception report covers the main items as foreseen at that time and explores the practical working situation as well as contours of the Delta Plan. Thereafter, the Baseline studies were started on the project document indicated 19 themes. During the early preparations, government officials stressed the importance to tackle the challenges mentioned above in an integrated and coherent manner and take concrete short- and long-term steps towards developing a safe, resilient and prosperous delta, the backbone of which is formed by a robust water system. In addition, in view of the major uncertainties the future will bring, aims have been underlined to develop the capacity to adapt flexibly and productively to future climate and socio-economic changes as well as to international developments in the river basins.

4.1.1 Temporal Domain

Considering the above, the Bangladesh Delta Plan 2100 is seen by GoB as a 'long-term adaptive, holistic, techno-economic water-centric strategic plan' to help achieve safety, security, and sustainable development. A widely accepted and approved delta vision, which guides the formulation of the plan with short-, medium-, and long-term strategies while being aware of uncertainties over time, is therefore of utmost importance. The long-term strategies are meant as strategic direction for achieving the delta vision, keeping major developments concerning climate, transboundary issues and cooperation, growth of population and economy, modernization and their uncertainties in mind. The medium-term strategies are targeted towards achieving different immediate goals and targets in compliance with the country's vision for 2021 and later for 2041. This means contributing to the country's growth targets, enabling the country to become an upper middle-income country and a further developed country by 2041. Avoiding or reducing setbacks from climate change impact, monsoon floods, or natural disasters is an important challenge here. The short-term strategies guide a set of 'no-regret' measures to help the country achieve comprehensive benefits within the five-year planning horizons of the 7th (2015–2020), 8th (2020–2025), and 9th Five-Year Plan (2025–2030).

4.1.2 Spatial Domain

Initially, the delta was defined as coastal zone and low-lying areas, the rivers and flood plains, urban areas, haor areas, and drought-prone areas. Given the specific conditions and problems of hilly areas, also the Chattogram Hill Tracts were included in the delta approach. Later on, the sea area of the Bay of Bengal under Bangladeshi control needed to be considered as well in terms of aspects and potentials of the 'Blue Economy'. The BDP 2100 covered by doing so practically the whole of Bangladesh. Another requirement is that besides the broad issues of national importance, e.g. concerning floods or freshwater supply, special attention has to be paid to regional and local characteristics and problems, which is the basis for the latter identified Hotspot areas.

4.1.3 BDP 2100 Thematic Areas

GoB intended from the start to understand the delta processes in a holistic way and to prepare an integrated adaptive delta plan accordingly. A total number of 19 different thematic areas have initially been identified, for which individual studies have been conducted. These so-called Baseline Studies investigate the characteristic issues, challenges and potential for development within the Bangladesh delta. The expected strategies and 'no-regret' measures and projects cover different issues linked with concerned thematic areas, while the emphasis is put on the water resources which is of great importance in relation to the mentioned setbacks. The thematic areas for Bangladesh Delta Plan are listed in Table 4.1.

Table 4.1 Initial and added thematic areas for the baseline studies. *Source* BanDuDeltAS, 2014

Initial thematic areas	
1. River systems management including morphological dynamics of Bangladesh delta and transboundary cooperation 2. Climate change and issues 3. Disaster management 4. Water resources 5. Coastal polder issues 6. Land resources management 7. Agriculture and food security 8. Fisheries and livestock 9. Ecological settings 10. Forests and biodiversity	11. Environmental pollution 12. Urbanization and settlements 13. Socio-economic and demographic conditions 14. Public health, water supply, and sanitation 15. Sustainable transportation and infrastructure 16. Growth of population and management 17. Information and knowledge management 18. Institutional framework arrangements 19. Regional cooperation
Added thematic areas	
20. Groundwater 21. Meghna Estuary 22. Socio-economic characteristics of the Chattogram hill tracts	23. Advancing the blue economy 24. Renewable energy 25. Earthquakes 26. Sixty years of water resources development in Bangladesh

4.2 Scope and Objectives of the BDP 2100 Formulation Process

To define scope, objectives, process as well as work plan, and schedule, the TA-Team started with Inception. The objective of the Inception stage and Report is to achieve consensus on the actual situation, approach, and work plan among the core parties GoB/GED, EKN, and BanDuDeltAS Consortium. Important is to discuss the whole strategy process as intended and also differences between contracted activities based on the project document and the actual situation: changes may have occurred, preferences shifted, or specific circumstances could require attention. Because of the broad scope and long-term perspective, diagnostics of problems and knowledge gaps follow in the Baseline Studies while the strategies with possible interventions are elaborated during the Strategy Formulation. The Inception Report as such also has a communicative function and explains distinguished episodes, practices to be applied and methods and tools to be used in the BDP 2100 formulation process. Besides, it provides information about guiding committees and stakeholders and their participation.

With consideration for climate change, BDP 2100 was supposed to cover a long duration up to 2100 because of the long-term climate change models and expected impact over time in terms of temperature, sea-level rise, and weather patterns. The long duration offers lead time while improving safe living and sound economic development in the Bangladesh delta. Considered as a multi-sectoral umbrella and investment plan interacting with other, often sectoral plans, the Bangladesh Delta Plan 2100 absorbs a lot of input as well as offers a long-term vision, strategy and coherent set of measures. Because of the multi-sectoral approach of vision, goals and strategies, active participation of stakeholders was intended to get an increasing degree of involvement and trust between parties and ownership on analysis of issues, formulating robust strategies and proposing appropriate measures and investments.

During the Inception phase, the requirement came to include the Chattogram Hill Tracts (CHT) in Baseline Studies as well as strategy formulation. The CHT shows specific problems of hilly areas such as soil erosion, landslides, and forest degradation. Point of attention here is the cultural setting and specific governance position of the CHT which are different from other areas. Transboundary issues form also a specific point of attention given the issues of transboundary rivers in a downstream country and of international borders also in the Bay of Bengal. The CHT and transboundary issues have been explored in Baseline Studies and are also considered in the institutional framework, with respect to international treaties and water diplomacy.

Over the years, Bangladesh carried out many studies and prepared interesting plans and programs concerning water resources management, agriculture and disaster management as described in Chap. 3. Many of the methods and plans that have previously been developed can be applied for the Baseline Studies and during the BDP 2100 formulation process. By enabling better integration of climate change adaptation and development-related issues and challenges, the Delta Plan serves as a long-term reference frame, which is meant to be frequently updated and revised when needed for both short-term and future planning activities. BDP 2100 provides thereby contributions to the Five-Year Planning cycle and other national plans.

The overall objective of BDP 2100 is defined by GoB as: 'to realize sustainable and a commonly agreed upon strategy, with all relevant stakeholders, for an optimum level of water safety and food security as well as sustained economic growth of Bangladesh and a framework for its implementation'. This indicates that GoB wanted to make a Delta Plan and undertake investments for adapting to climate change as well as supporting the development of the country.[1] This 'adaptation by development' approach is shown in the elaboration of the detailed objectives of the Delta Plan setup in terms of supporting enabling conditions for the Plan, its strategies, measures and its realization; taking care of an open knowledge base, preparing institutional reform measures; formulating a long-term vision and goals in interaction with stakeholders; promoting private sector participation as well as stimulating local involvement and improvements.

[1] See 'Project Document' (2013). This is a document prepared by EKN together with GED-GoB; it served as basic document for the tender procedure.

4.3 Expected Role of BDP 2100 in National Strategies and Plans

4.3.1 Bangladesh Planning and Programming Framework

BDP 2100 is based on the long-term development vision of the Government and is as such related with the 6th Five-Year Plan FY2010-FY2015, the Perspective Plan of Bangladesh 2010–2021 and other existing sectoral plans, policies, and strategies of the government. The objective is to recommend strategies which contribute to disaster risk reduction, climate change resilience and adaptation, water safety, food security, environment safety, and economic development of the country.

Plan preparation and formulation of BDP 2100 as such umbrella plan, with its specific institutional and financial Delta Framework, are taking place within the framework of the national programming cycle; see Fig. 4.1.

Next to the described role in the country programming cycle, the role of BDP 2100 should be considered as reference in the long-term adaptive holistic planning process that leads to planning and programming taking both development issues and climate change challenges into account. In this role, the development of new insights in adaptive strategies and measures is important. Policy coordination and a more integrated allocation of funds for implementable projects are also of crucial importance to make the required progress in time. The Bangladesh National Water Policy (NWPo) provides a useful basis for the Institutional Framework, as it outlines an integrated approach towards sustainable water resources management. The National Water Management Plan (NWMP), approved in 2004, is meant as an operationalization of the NWPo. It is a comprehensive water resources' plan in which 13 ministries and

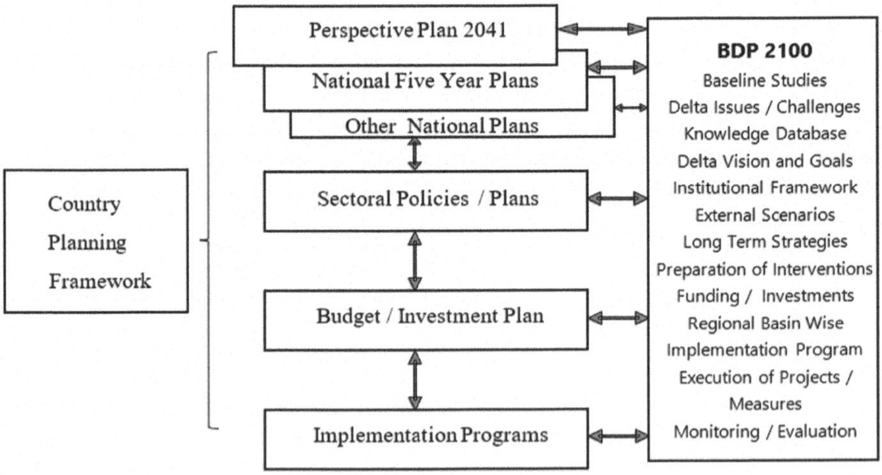

Fig. 4.1 Country programming/planning framework. *Source* BanDuDeltAS, 2015

more than 30 agencies are involved. Within the institutional context at the time, the realization of the NWMP suffered from implementation issues, which are interesting to analyze and learn from.

Given its emphasis on cooperation and integration of important issues, the Master Plan for Agricultural Development of the Southern Region (Ministry of Agriculture, 2013) provides valuable input. This is also the case for the 6th Five-Year Plan that demonstrates the whole planning and implementation strategy used by the Ministry of Planning. Reevaluated are pertinent initiatives and programs connected to the Bangladesh Delta Plan. The intention is to use valuable knowledge and fit in suitable measures and interventions as good as possible. While existing planning documents show many ideas and proposals, an adaptive and comprehensive strategic plan addressing climate change over a related time span was seen as urgently required. Given the uncertainty of climate change and its impacts in various sectors flexibility will thus have to be an important feature of the Delta Plan in order to adapt effectively and efficiently. Besides, innovation and technological development can offer solutions over time which were not possible before. Based on a long-term vision, scenarios show various narratives of possible futures for which strategies and institutional transformation can be considered. The Delta Plan was therefore intended to offer useful contributions already for the 7th Five-Year Plan in 2015. The proposal for the institutional framework (including a draft Delta Act to establish the Framework) was expected to comprise broadly supported interventions to improve governance with respect to implementing and frequently updating and revising the Delta Plan.

4.4 Design Aspects of the BDP 2100 Strategy Process

4.4.1 Points of Inspiration

In addition to the policy and planning context and administrative and institutional setting, the formulation of BDP 2100 and its implementation are in terms of methods and literature based on insights of strategic management under conditions of uncertainty as explained in Chap. 2. A number of lessons learnt have inspired the design and practices of the strategy process such as lessons from the Delta Program in the Netherlands, and their developed ideas about 'Adaptive Delta Management' (ADM).[2] The experience suggests the following.

---

[2] See e.g.: Timmermans et al. (2015); Dewulf and Termeer (2015); Government of the Netherlands, Delta Commissioner (2018), Zevenbergen et al. (2018); De Heer and Aartsen (2019); De Heer (2023).

4.4.1.1 Considering Main Long-Term External Trends and Drivers of Change with Their Uncertainties and Impact Characteristics Elaborated in Scenarios

Due to the uncertainties with respect to climate change and socio-economic development over longer term, the linear planning approach, which has been used for a long time, is being enriched with adaptive strategy making and adaptive design and implementation. Rather than business as usual approaches with short-term action or standalone intervention, external influences need to be considered and flexible strategies and programs with consistent measures are needed. Strong institutions and a good knowledge infrastructure are also required to allow policy makers and stakeholders to anticipate and decide on the most appropriate investments. External scenarios play therefore a role in understanding how the future may look like and how the vision and goals can best be achieved under various uncertain conditions and adaptive strategies.

4.4.1.2 Holistically Analyzing Issues, Challenges, and Knowledge Gaps Related to Multi-sectoral Planning

The Bangladesh delta is a highly complicated and dynamic delta. Only a combination of dedicated disciplines with a large number of themes to be studied can successfully explore the mechanisms and highlight the real issues and challenges in a proper way. Holistic understanding and a multi-sectoral approach with involvement of the stakeholders are needed to address the challenges in the right way and to build ownership. Alignment with other processes is a must to generate coherence and synergy.

4.4.1.3 Connecting Short-Term Action and Investments with the Long-Term Vision, Developments, and Higher-Level Policies

Creating a strategic vision of the future is required given the time frame of developments such as climate change, manifestation of impacts, intervening in physical and societal processes. Translating the long-term vision to the present, e.g., by back casting makes it possible to link properly to short-term actions and establish a framework to guide future actions. Such a vision has, e.g., in relation to the development of climate change over a long time period, a longer time horizon than usual in planning activities (hence BDP 2100) to capture the effect of long-term processes. So instead of focusing on short-term 'trial-and-error' actions and projects, the idea is to keep the long-term vision in mind while prioritizing short-term actions towards achieving the vision.

4.4.1.4 Elaborating on Path Dependency: Elaborating Adaptation Pathways

In the way, a delta has been formed, developments can be observed which, once started, cannot easily be changed or adapted to new conditions. When applied to policy development, path dependency is the extent to which a policy action is limited by actions implemented in the past or by actions already planned to follow in the future. Learning from the past and recognizing that one cannot predict the future means that such lock-ins need to be avoided. One way to do this is to explore adaptation pathways, i.e. the flexible sequencing of different policy actions over time to achieve a given set of objectives. Adaptation can be shaped like this. Interesting is that indications ('signposts') can be utilized to signal in time that certain measures are (for whatever reasons) no longer effective to achieving the vision and goals and that a switch to other measures in the pathway may be needed. Point of attention is that it is attractive for decision-makers to know that such a pathway can be followed, but that in the administrative practice pathways are often difficult to handle because of budget constraints, political changes, or conflicts of interest between sectors.

4.4.1.5 Avoiding Over- and Under-Investments

Insight in adaptive programs and investments (in adaptation pathways or not) and in adaptive design of interventions is not only relevant for maintaining flexibility in adaptive plans, but also to limit the risks of over- or under- as well as too early- or too late investments. Underinvestment occurs if it turns out that the solutions are not adequate and may lead to calamitous outcomes. Overinvestment on the other hand happens when measures are overdimensioned or ultimately unnecessary, which can prove costly. Adaptive programming and adaptive design are meant to take suitable action given the short–medium-term circumstances with the possibility to build further on the first intervention if the situation becomes worse.

4.4.1.6 Connecting Public and Private (Investment) Agendas

Another idea is to actively search for windows of opportunity to combine different investment agendas, either within the public domain or between public and private investments. This way, measures may be easier (and cheaper) to implement and yield more added (societal) value. Problem is often that investment agendas are not well known or are even kept secret.

4.4.1.7 Developing an Adequate Institutional Framework for Mainstreaming BDP 2100, Coordination and Monitoring Its Realization

A strategy process for preparing a delta plan will, given its long-term vision and goals, multi-sectoral planning and impact as well as large related investments be closely related to the core governance of the country. For Bangladesh, mainstreaming of BDP 2100 in the Five-Year Planning Cycle and a close relationship with the then upcoming Perspective Plan 2041 are of crucial importance for effective implementation and consistency over time. A clear institutional framework is needed to this and to facilitate decision making, funding as well as multi-sectoral planning and implementation and to provide room for dedicated people's participation. Moreover, BDP 2100 is not another connotation of 'Integrated Water Resources Management' because the Delta Plan goes far and beyond IWRM with respect to time horizon, holistic approach of the delta with the multi-sectoral elaboration, focus on climate change impacts, etc. BDP 2100 focuses, in other words on water safety and security, facilitates food security and economic growth, and contributes to national-level goals promoting prosperous development of the nation.

The above-mentioned items of Adaptive Delta Management do not provide a clearcut methodology for preparing a strategic adaptive plan as BDP 2100. The methodology to prepare BDP 2100 was developed in the early stage of the process together with GoB and EKN. At the start of the strategy process, the ADM-items formed points of inspiration and attention given Dutch and Vietnamese experiences and played as such a role in the design of the strategy process and in some applied practices as scenario development.

4.4.2 Design of the BDP 2100 Strategy Process

The BDP 2100 formulation process may be seen as the vehicle to climate change inclusive strategy making and to the development of a required institutional framework in which government agencies, stakeholders and possibly the private sector play an active role. Figure 4.2 gives an overview of the BDP 2100 approach. In the strategic planning process, content, governance, and stakeholder participation form basic elements in the interactive strategic planning process. The picture indicates the strategy process during the application of practices and preparing drafts and modifications to come to the expected results. The episode of 'preparing the ground' has been expanded with seven themes and Baseline Studies, so it took longer than indicated. Besides, the interactive strategy process took more time with respect to the additional Chattogram Hill Tracts area and Blue Economy while also more time has been spent on Delta Ateliers and consultation workshops with stakeholders all over the country for interaction with the local people and developing ownership. Moreover,

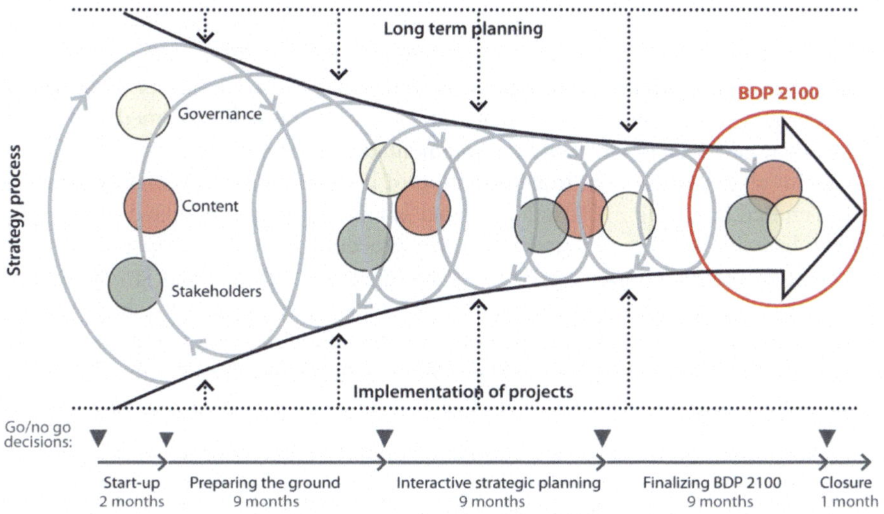

Fig. 4.2 Interactive strategy process. *Source* BanDuDeltAS, 2014

GoB included external review on all Baseline Study and Strategy Reports and extensive consultation at the end of the strategy process. Duration of the whole strategy formulation process up to approval by GoB was from March 2014 till September 2018.

4.4.2.1 Adaptation Content

The planning context and a sectoral description of issues in the dynamic delta of Bangladesh are described in Chap. 5. In the Interactive Strategic Planning process, generating and sharing knowledge of and interaction about the physical dynamics, drivers, problems and challenges are important in relation with a vision and scenarios in order to formulate strategies with promising interventions. The Baseline Studies show the complexity of the delta in a multi-disciplinary way on 26 themes. The studies needed to be based on contemporary and state-of-the-art knowledge and methods; additional fundamental scientific research was not the intention. The results of the Baseline Studies formed together with content-related input from experts and local stakeholders the basis for strategy formulation and measures with proposed interventions.

4.4.2.2 Governance

In Fig. 4.2, governance refers to the process of adaptive decision making and performance of the administrative system, especially focusing on for adaptation required institutional development and transformation, multi-sectoral coordination, implementation approach and investment planning, including PPPs, and funding also by GoB, dedicated funds, development partners, and IFIs. Emphasis on adequate governance is important to ensure that BDP 2100 can be mainstreamed within the government and society and will be implementable and acted upon.

This involves institutional transformation of involved ministries and agencies as well as embedding BDP 2100 in the relevant planning, institutional, and stakeholder context. Developing common understanding, procedures, methods, or approaches is important to this, supported by capacity building and institutional strengthening. From a governance and implementation point of view, the relationship between national, regional as well as local planning requires much attention to gain commitment, but also to align strategies and plans. With respect to governance, the interactive process relates to:

- Seeking socio-political participation and support in a broader network.
- Joint planning of ministries and, e.g., with development partners World Bank, Asian Development Bank and other IFIs.
- Interaction and cooperation with the 'inner circle' of government agencies (guiding committees, focal points, etc.) about issues in their fields of policy and interests.

The Five-Year planning cycle is an important governance instrument and therefore forms an adequate element in the coordination and planning context.

4.4.2.3 Stakeholders

Successful design and application of BDP 2100 depend on broad support and commitment, and active participation of national and international players. Thus, it is necessary to closely integrate public and private partners and to synchronize with the needs of such sectors as water resources, agriculture, navigation, fisheries, forestry, cattle, industry, water supply, sanitation, and environment inside the scope of this project. Early on, IFIs and development partners have been involved in the process, raising awareness and discussing the BDP 2100 approach and options for involvement in the execution of projects.

Formulated during an interactive strategic planning process, BDP 2100 reflects the content of the several themes in which the GoB-agencies and other stakeholders, public and private parties, are active. Learning by all stakeholders about the new adaptive strategy approach is crucial and encouraged by connecting the long-term vision and planning process with short-term suggestions on measures and implementation of projects. Initiating the planning process, mobilizing information, starting engagement, thinking, learning, and action, and obtaining sustainable results all depend critically on developing ownership, commitment, and trust.

The interactions with stakeholders at the local level took place in dedicated analysis and design workshops called 'Delta Ateliers'. Here, large Touch Tables have been utilized with local maps and information to enable interaction between specialists and stakeholders. The emphasis on (combinations of) maps to represent problems and solutions, on spatial planning and conceptual design of interventions, promote valuable contributions and integration of the involved stakeholders. This throws fresh light on the issues and challenges, generates diverse points of view, which lead to new ideas and creative means of adaptation and action. The interactive approach aims to bring innovation, participation and creativity by organizing and facilitating of the Delta Ateliers 'roadshow' at the national scale and for hotspot areas where knowledge has been exchanged about the multi-sectoral issues and challenges and what can be done about it. The Delta Ateliers are organized as one- or two-day workshops in which the issues at stake are analyzed, discussed, and ideas and views to deal with problems are generated and presented while interests of the stakeholders are identified and linked with the discussed issues and are elaborated in supported results. The Delta Ateliers form an interactive strategy making concept which was used during essential moments in the BDP 2100 formulation process.

To realize the BDP 2100 formulation and start the implementation process, seven steps with corresponding practices have been performed as follows.

4.4.2.4 Start-Up with Early Preparations, Scope Definition, and Formal Arrangements

The initial efforts towards a strategic planning process stemmed from the recognition of climate change and the understanding that it could adversely affect Bangladesh, resulting in damage and loss, economic challenges or even climate refugees. Engagements occurred at the prime ministerial level in Bangladesh, resulting in collaboration with the Netherlands to enhance strategic deliberation and action about climate change, as described in Chap. 1. Meanwhile, the Government of Bangladesh incorporated specifications about the delta region and scope, to address not only water safety issues and challenges but also food security, socio-economic development, and environmental aspects. A comprehensive multi-sectoral approach supporting the development of the country was preferred instead of a sectoral approach.

4.4.2.5 Preparing the Ground

The second episode concerned mobilization of the project including high level and practical guiding committees, government organizations and project organization with the consultants support team. Figure 4.3 shows the overall BDP 2100 organization setup and involved organizations and consultant project.

The team of consultants consisted of a multi-disciplinary team of an international consortium with 10 international (of which 8 intermittent) and 26 national consultants, also partly intermittent. The consortium covered the disciplines strategic planning, governance and institutional development, climate change, water resources management, coastal defense and land reclamation, hydrology, morphology, water

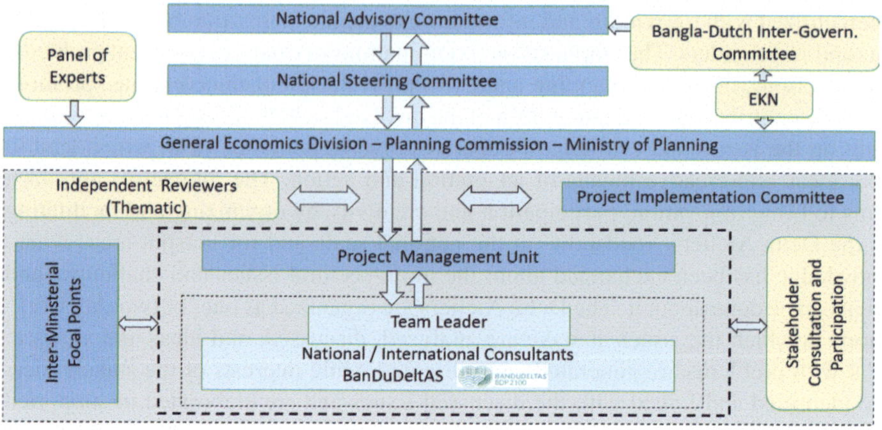

Fig. 4.3 BDP 2100 Formulation project organization and BanDuDeltAS consortium. *Source* BanDuDeltAS, 2014

supply and sanitation, spatial planning and land management, food security, agriculture, fisheries, livestock, environment and ecology, macro- and project economy, financing, project management, and support. Later on, an additional small team of local economists contributed to economic and budgetary aspects at the end of the formulation process, fine-tuning the fit with specific budget and governance aspects. Leadership was performed by the Team Leader (strategic management and institutional specialist) and deputy Team Leader (water resources specialist). Specific points of attention formed close interaction with Bangladeshi officials, cross-cultural management of the project and an optimal fit with the administrative culture and

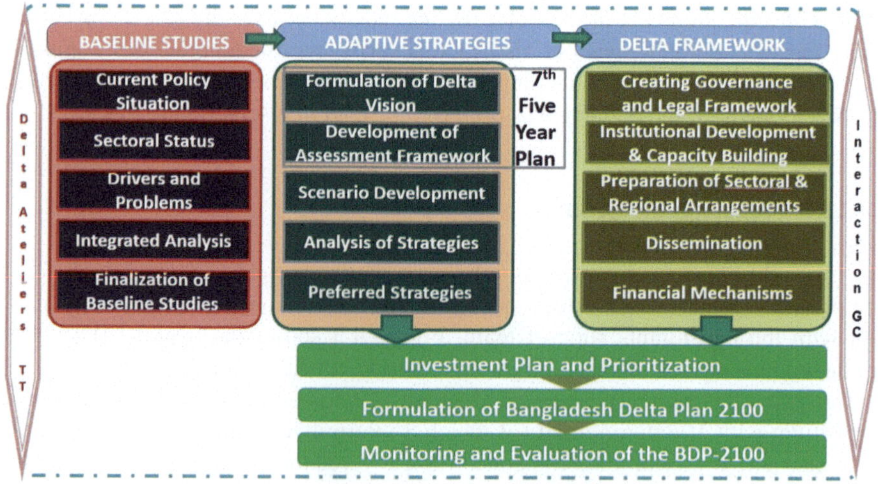

Fig. 4.4 BDP 2100 process overview. *Source* BanDuDeltAS, 2014

processes. Frequent close interaction, next to formal meetings, took place with the General Economics Division of the Planning Commission, Ministry of Planning which hosted the BDP 2100 project because of experience with longer-term planning and also because of the multi-sectoral scope of the project.

The Inception of the project is also about awareness raising and gaining support. Concrete activities comprise:

- Mobilization of the project organization with appropriate work space, mobilizing the team and required vehicles, equipment, and logistics.
- Comparing the actual project situation with that of the earlier prepared project document and contracts while where needed proposing adjustments and compensatory measures.
- Stakeholder identification and meetings: activating the ministerial focal points, the guiding committees as well as in a broader context the knowledge institutions, NGOs and private sector parties.
- Definition of scope, work plan, and timeline of the project, including deliverables and expected results and outcomes of the process.
- Outlining the methodology to carry out the strategic planning process, including proposed practices, tools and working methods. Besides, selection of study items, ways for data collection, maps, documents and other sources for research and planning get full attention to get ready for the next phase.
- Sharing the Inception Report leads to validation and exchange of additional information, also for dealing with the guiding committees, decision making and approval.
- Prototyping of an early outline of the BDP 2100 and its development process. Figure 4.4 presents the main elements of this process.

4.4.2.6 The Tracking of Strategies, Baseline Studies, and Diagnostic Search

The tracking of strategies is an important practice to find out the ways earlier plans were formed, and measures have been taken or projects have been executed in relation to the strategic planning exercise. Fruitful discussions take place about this, offering insight and understanding for traditional approaches, the existing situation and cultural appreciations; see Fig. 4.5. Besides, a lot of useful data, insights, and ideas come forward and form a valuable basis for the diagnostic search and strategy process. Tracking also was done in the field of exploration of existing implementation practices, constraints and possible modifications in implementation procedures.

Another crucial activity is conducting Baseline Studies with joint fact-finding, including content on the themes and institutional matters. The themes are first explored by quick scans. Second, the Baseline Study teams elaborate these in a first run and second run if needed more in depth. Prioritization of problems related with the themes is needed for identification of the most important drivers and trends as well as issues and challenges to be used in the vision, scenarios and strategy formulation. The diagnostic search by Baseline Studies and problem analysis show national issues as floodings and freshwater provision, specific issues defined for six Hotspot areas and a number of Thematic issues. Based on the analysis of issues and challenges, it becomes clear what the topical problems are and how climate change adds new challenges. This makes it possible to indicate ambitions explicitly in the strategic delta vision. Also Mission and six Delta Goals have been formulated in order to achieve the delta vision over time.

4.4.2.7 Interactive Strategy Process and Adaptive Decision-Making Ahead

An interactive planning process followed on the tracking and diagnostic search practices, to integrate sectoral analysis and develop adaptation strategies. Concrete activities comprise the following.

Fig. 4.5 Contributions during Delta Ateliers. *Photo* J. M. de Heer

- Utilizing the Delta Ateliers and Baseline Studies for further problem analysis and visualization to define the impacts of main driving forces and trends as preparation of different delta plan scenarios.
- Developing scenarios concerns a more detailed elaboration with stakeholders of the developments or trends and their driving forces. This is done in relation to issues and challenges, as well as outcomes of the Baseline Studies and overall problem analysis. Especially the economic development, the transboundary developments and climate change show the highest uncertainties with potentially significant impacts. Four scenarios are elaborated as narratives of possible futures where these three developments with their uncertainties and impacts play a dominant role.
- Because of the Five-Year planning cycle and its timeline, input from the BDP 2100 strategy process was already in this stage expected for the content of the 7th Five-Year Plan. This is seen as important because BDP 2100 would be embedded in the core governance of the country, and it would thereby mainstreamed and widely accepted.
- Exploration of possible strategies and interventions with considerable input from stakeholders led to a large set of potential strategies and measures which have been reduced to realistic ones. The potential, realistic, and finally the preferred strategies can be selected on the basis of the developed vision and scenarios as well as with selection criteria and standards of an assessment framework. The Strategies are clustered in national, Hotspot, and thematic strategies.

4.4.2.8 Developing the Institutional Framework

Managing adaptation to climate change as intended in the multi-sectoral long-term BDP 2100 including adaptive decision making on delta plan investments is regarded as problematic, resulting in the requirement of institutional transformation. Establishing an adequate institutional framework including a draft Delta Act is seen as crucial for facilitating future support and coherence of adaptive decision making. The formulation of a draft Delta Plan, accompanied by extensive participation and consultation processes overseen by official committees, enables the development of an appropriate framework aligned with the adaptive strategy approach. The finalization step involves the refinement on the basis of feedback from consultations and concluding talks over the proposed institutional transformation. The draft Delta Act would get follow-up in a legal process after formal approval of BDP 2100.

4.4.2.9 Developing Investment Planning and Funding

For developing the investment plan, the strategies have been elaborated into a large number of project concept notes with intervention proposals and institutional measures. In collaboration with a World Bank team, an investment plan has been formed. Basin-wise assessment of (pre)feasibility is viewed as needed for selected strategies and measures, along with a re-assessment of the stakeholder needs and ambitions to ensure the right track has been followed. This has been done under the SIBDP project in terms of adaptive programming.

4.4.2.10 Monitoring and Reflective Evaluation Followed by Updating and Adjustments

Monitoring and reflective evaluation are foreseen for considering the strategies and measures against new information on climate change and other major developments. Transformation is also here proposed since progress of BDP 2100 implementation needs to be result based monitored in a multi-sectoral way with judgment of efficacy in relation to climate change impacts. Moreover, a good linkage with external developments is regarded as very important in relation to the progress made and to adaptive decision making on possible adjustments of measures and interventions.

References

BanDuDeltAS. (2014). Inception Report BDP 2100 Formulation Project.

BanDuDeltAS. (2015). Baseline Study on Governance.

De Heer, J. M., & Aartsen, M. (2019, Mar). Bangladesh prepares itself for a climate resilient future. *Water Governance.*

De Heer, J. M. (2023, Mar). Update to: Bangladesh prepares itself for a climate resilient future. *Water Governance.*

Dewulf, A., & Termeer, C. (2015). Governing the future? The potential of adaptive delta management to contribute to governance capabilities for dealing with the wicked problem of climate change adaptation. *Journal of Water and Climate Change, 6*(4), 759–771.

Embassy of the Kingdom of the Netherlands. (2013). *Project Document. The Formulation of the Bangladesh Delta Plan 2100.*

Government of the Netherlands, Delta Commissioner. (2018). https://english.deltaprogramma.nl/delta-programme/what-is-the-delta-programme/adaptive-delta-management

Ministry of Agriculture. (2013). Master Plan for Agricultural Development of the Southern Region.

Ministry of Planning, Five Year Plans. https://plancomm.gov.bd/

Timmermans, J., Haasnoot, M., Kwakkel, J., & Rutten, M. (2015). Adaptive delta management: Roots and branches.

Zevenbergen, C., Van Alphen, J., Shah Alam Kahn, M., Terwisscha van Scheltinga, C., & Veerbeek, W. (2018). Adaptive delta management a comparison between the Netherlands and Bangladesh delta program. *International Journal of River Basin Management.*

Chapter 5
The Bangladesh Delta and Its Challenges

Abstract This chapter emphasizes the importance of knowledge and understanding of the physical system and content in the strategic planning process, particularly in the context of Bangladesh's unique deltaic environment formed by the confluence of three major rivers before they flow into the Bay of Bengal. It addresses the characteristics and persistent challenges and issues faced by the country, including transboundary concerns and the impacts of climate change. Altogether, the chapter provides the total picture of a time-bound diagnosis for which the strategic planning process needs to find and become (part of) an adaptive therapy. In Bangladesh, the main issues for water resources management are considered as flood, drought, drainage congestion, riverbank erosion, salinity intrusion, groundwater (both contamination and depletion), land subsidence, frequent cyclones, and storm surges, all with their consequences. Over time, some of these issues are exacerbated due to anthropogenic interventions which contributed to the deterioration of the quality of water. On top of these, transboundary problems and climate change impacts including sea level rise add to the complexity of the problems. Unfortunately, the magnitude of these problems is expected to increase in the near and far future. As such, human interventions should be restrained or limited so that the actions do not augment the problems more than what nature is already causing. Besides the physical systems, socio-economic issues and challenges of the country have been highlighted in the perspective of climate change. Important in the shorter term are also the transboundary issues the country is facing. As downstream located delta area, interventions like the construction of dams of the upstream countries India, China, Nepal, and Bhutan are highly relevant but very hard to influence. The felt need for preparing a comprehensive BDP 2100 has been elaborated against the background of the developments taking place with increasing uncertainties for people and businesses in the delta area.

Keywords Delta · Bangladesh delta · BDP 2100 · Water resources management · Surface water · Ground water · Wetlands · Water bodies · Coast · Coastal zone · Estuary · Flood · Drought · Drainage · Drainage congestion · Sedimentation · Erosion · Riverbank erosion · Salinity · Salinity intrusion · Land subsidence · Cyclone · Storm surges · Coastal zone · Transboundary river · Downstream · River

J. M. de Heer and G. A. Choudhury, *Strategies for Adaptation to Climate Change in a Transformative Approach*, Springer Climate,
https://doi.org/10.1007/978-3-031-88884-7_5

systems · Hilly river · Haor · Meandering · Braiding · Protection · Accretion · Land reclamation · Irrigation · Navigation · Inland water transport

5.1 The Bangladesh Delta

The Bangladesh Delta is defined as the delta that includes all administrative districts that face various natural hazards owing to the deltaic formation of Bangladesh and the related interface with the vast river networks, the Bay of Bengal and climate change impacts. The delta comprises the combined areas of the southern coastal belt and the areas surrounding the flows of the three rivers (the Ganges, the Brahmaputra, and the Meghna). The Bangladesh delta, located in a unique geological setting, is also identified as the largest dynamic delta in the world. The Himalayas, the world's highest mountain range, lies in proximity to the north. The Ganges and the Brahmaputra emerge from the Himalayas, and the Meghna from the Lushai Hills. Together, they carry more than one billion tons of sediment per year to the Bay of Bengal building the delta over time, which is still prograding, growing seaward. Further offshore, the sediments are carried down to form the Bengal fan, a deep-sea fan or under water delta. The Bengal delta basin and the Bengal deep sea fan together form the most extensive depositional system in the world. Bangladesh has, therefore, been regarded as a living museum of sedimentary geology.

5.1.1 Surface Geology

Bangladesh is situated in the Bengal Basin, bordered by the Shillong Plateau to the north, the Precambrian Indian Shield to the west, the Tripura-Mizoram Mobile Belt to the east, and the Bay of Bengal to the south. Its geological evolution is closely linked to the uplift of the Himalayan mountains and the formation of a mega-delta by major river systems like the Ganges, the Brahmaputra, and the Meghna. This delta-building process continues today, with sediments being deposited in the Bay of Bengal.

The eastern part of Bangladesh has experienced uplift, forming hilly terrain as part of the Indo-Burman range, due to the collision of the Indian and Asian tectonic plates. Beneath the northern Bengal Basin, pre-Cambrian rocks are tilted along a hinge line formed during the Eocene epoch, dividing the basin into two zones. The northwest zone has relatively thin sedimentary fill (minimum 350 m), while the southeast zone features a significantly thicker sedimentary fill, reaching up to 18 km at Barishal.[1]

[1] Alam et al. (2003). Alam et al. (2018). See also Brammer (1996).

5.1.2 *Topography*

The topography of Bangladesh, primarily fluvial-deltaic, is characterized by low and flat land, with the exception of hilly regions in the southeast and northeast. The country is traversed by a vast network of rivers. Due to the flat landscape, the major rivers—Ganges, Brahmaputra, and Meghna (GBM)—exhibit very gentle gradients of just four to eight centimeters per kilometer. The topography is categorized into three major physiographic units (see Fig. 5.1):

- Hills, occupying 12% of the land area, are located in the southeast, and, to a minor extent, in the northeast. The elevation of ridges range from 300 m to almost 1000 m above sea level.
- Terraces, occupying 8% of the land, are known as Barind Tract in northwest and Madhupur Tract in northcentral regions. These are older alluvial deposits from previous river systems related to higher sea levels during the Pleistocene and situated a few meters above adjacent floodplains.
- Floodplains, covering 80% of Bangladesh, the floodplains have been shaped by sediment deposition from the GBM-river systems. The elevation gradually decreases from about 90 m in the northwest to less than three meters along the coast. Formed during the Holocene epoch, the floodplains exhibit diverse subunits, including the Piedmont plain, active river floodplains, estuarine plain, tidal plain, and coastal plain.[2]

5.2 River Systems

Bangladesh is a riverine country. There are more than 815 rivers identified using Sentinel-2 remote sensing images of 10 m resolution.[3] The number of rivers hydrological regions-wise are: northwest region: 251; northcentral region: 76; northeast region: 175; southeast region: 47; southwest region: 235; and eastern hilly region: 31. Of these, 57 are transboundary rivers: 54 from India and 3 from Myanmar.

These rivers may be broadly classified into four systems (see Fig. 5.2): the Ganges-Padma system, the Brahmaputra-Jamuna system, the Meghna system, and the Eastern Hilly Rivers. These river systems cover a combined catchment area of 1.72 million square kilometers, the second-largest riverine drainage basin globally. This extensive area spans across India (64%), China (18%), Nepal (9%), Bangladesh (7%), and Bhutan (3%), representing around 1% of the world's land surface but home to approximately 10% of the global population.

---

[2] Ibid (2018).

[3] See: CEGIS; www.cegisbd.com.

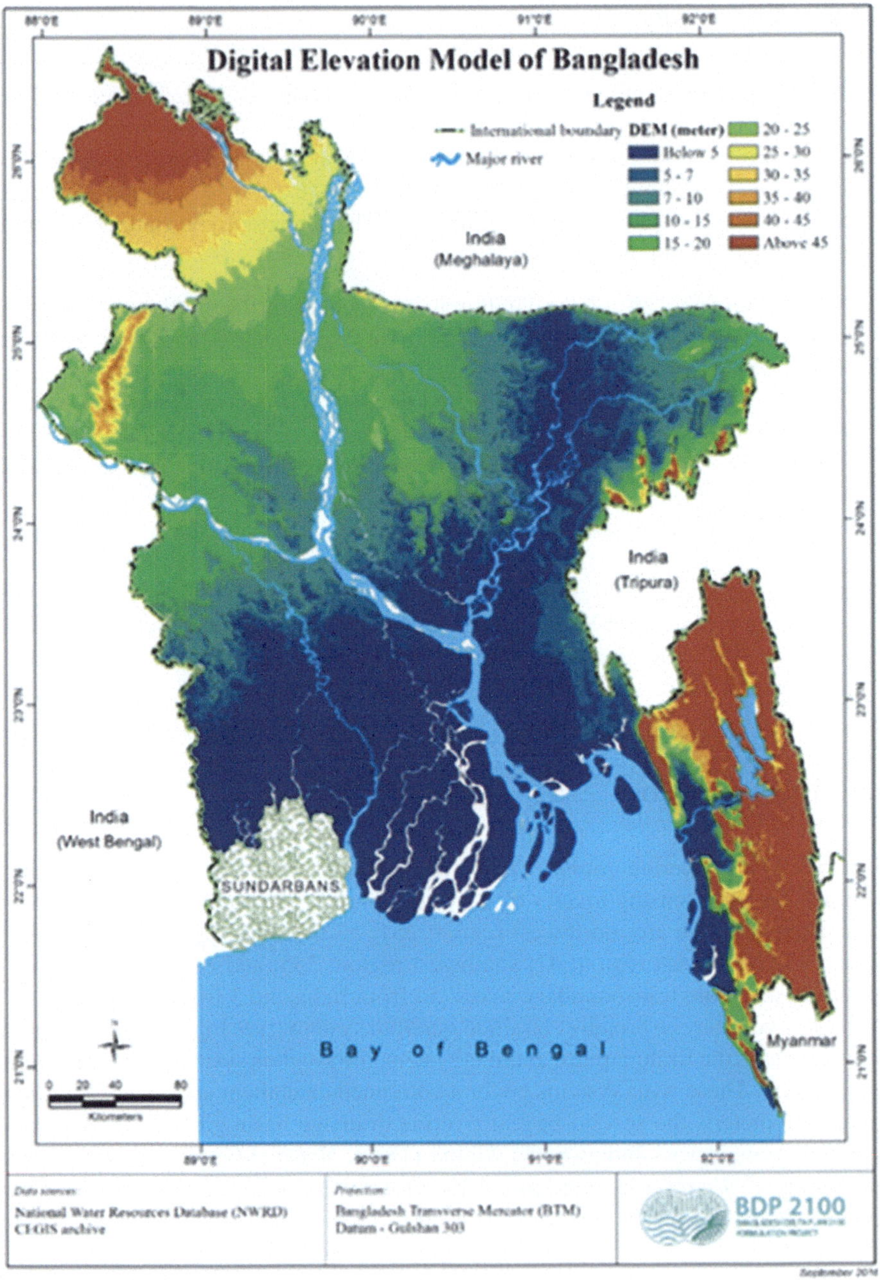

Fig. 5.1 Digital elevation model of Bangladesh. *Source* NWRD

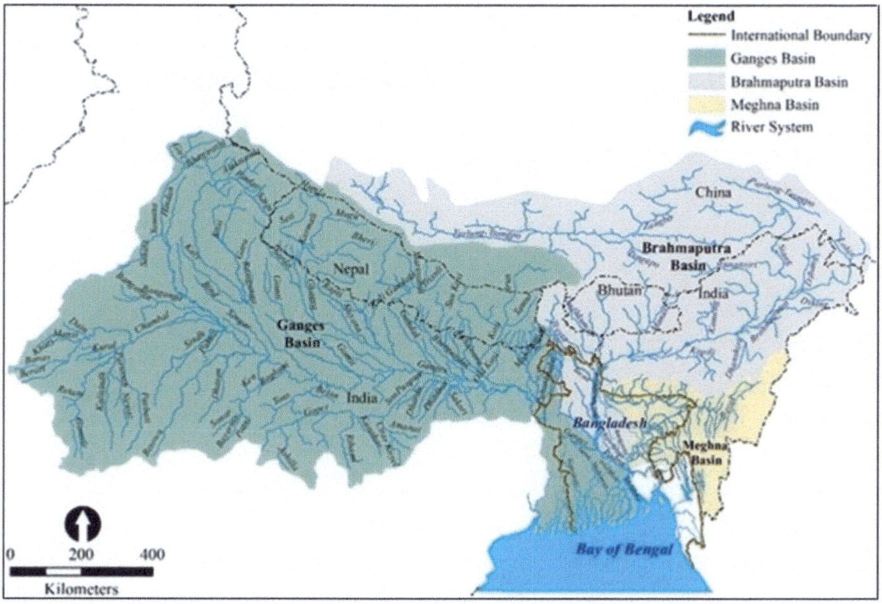

Fig. 5.2 Ganges, Brahmaputra, and Meghna basins (GBM-basins). *Source* SIBDP Consortium

Bangladesh, located in the lower deltaic region, accounts for only 7% of this catchment area, yet it houses about 20% of the basin's population. This dense population is supported by rich ecosystems and agricultural irrigation systems, making these basins vital for sustaining some of the world's highest population densities and extensive biodiversity.[4]

5.2.1 The Brahmaputra/Jamuna River

The Brahmaputra River, known as Jamuna within Bangladesh, originates in the Kailash range of the Himalayas in Tibet, China, at an altitude of 5150 m, where it is locally known as the Yaluzangbu. Flowing eastward for approximately 1700 km parallel to the Himalayas, the river takes a sharp turn southward and southwestward, entering India through Arunachal Pradesh in the northeastern corner.

The key tributaries include the Dibang, Lohit, Subansiri, Dhansiri, and Manas rivers before the Brahmaputra crosses into Bangladesh. After it enters Bangladesh, the river is joined by many tributaries like Dudhkumar, the Dharla, the Teesta, the Ghagot, and the Hurasagar (see Fig. 5.3).

[4] See also Ahmad et al. (2001) on GBM-region and its sustainable development.

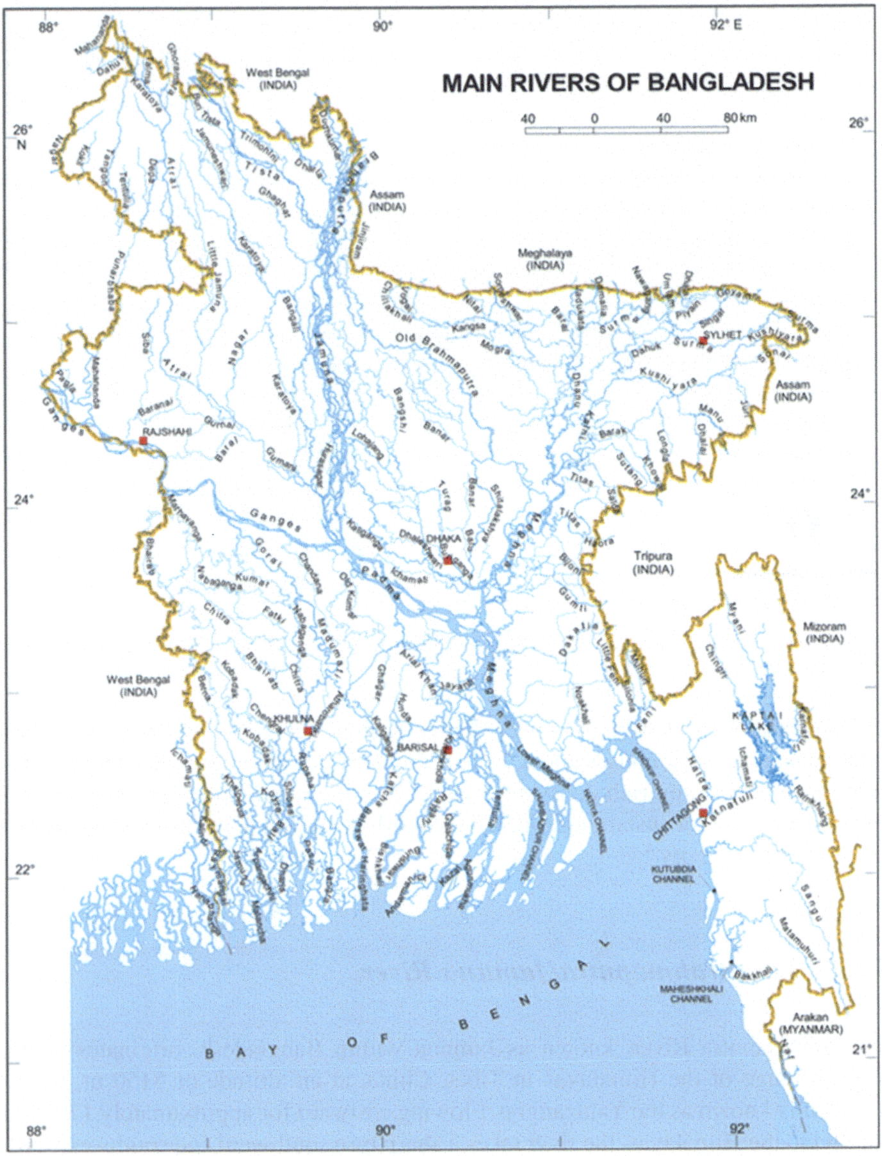

Fig. 5.3 Main rivers of Bangladesh. *Source* BWDB

The important distributaries are the Old Brahmaputra and the Dhaleshwari. The Old Brahmaputra flows southeasterly to join with the Upper Meghna near Bhairab Bazar. The course of the Old Brahmaputra was the original channel of the Brahmaputra before its avulsion in the late eighteenth century. The Brahmaputra (Jamuna) flows southward and joins the Ganges, and then the combined flows travel in a south-easterly direction as the Padma and ultimately joins with the Upper Meghna, draining into the Bay of Bengal as the Lower Meghna.

The Brahmaputra spans approximately 2900 km from its source to its confluence with the Ganges, with around 230 km flowing through Bangladesh. Covering a catchment area of 570,000 km^2, only about 7% of this lies within Bangladesh. The Brahmaputra, is a large braided, multi-channel river with an average width exceeding 12 km during the flood season. Its banks, composed of loose and erodible materials, make it highly susceptible to severe bank erosion. However, recent satellite images indicate that the river's tendency to widen has stabilized since 2010.

According to BWDB, the maximum flood flows recorded in the Brahmaputra at Bahadurabad was 102,535 cumec in September 1998, while the lowest recorded flow was 2036 cumec in March 2013. Average flood flows are estimated at around 68,000 cumec, and average low flows at 3430 cumec. The river typically experiences two peak water levels each year: the first between May and July, and the second, generally higher, during August to September. The historical highest water level of 20.84 m PWD was recorded in August 2017, with the lowest water level of 11.92 m PWD in April 2015 at Bahadurabad. Low flows generally occur toward the end of February or early March. The significant difference between high and low discharges poses a major challenge for water management in the country (see Fig. 5.4).

The Brahmaputra, or Jamuna, is among the world's largest braided rivers (see Fig. 5.5) and remains in ongoing development, making it highly unstable.

Figure 5.6 illustrates the Jamuna's bank line migration from 1830 to 2010 across various phases. Between 1914 and 1953, the river continued its significant westward migration, expanding in width and transforming from a meandering to a braided river. By 1973, the average width slightly decreased, although the rapid westward shift persisted. From 1973 to 1992, the average width of the river expanded significantly, even as the rate of westward migration slowed. Over the past four decades, the river has continued to widen at an exceptionally high rate, though westward center-line migration has effectively stopped. Instead of the center line, right bank migration now primarily drives right bank erosion and overall widening.

The avulsion of the Brahmaputra towards the Jamuna on the alluvial floodplain has led to extensive riverbank erosion and accretion along both banks (see Fig. 5.7). Since 2005, CEGIS has pioneered erosion prediction tools to identify areas vulnerable to erosion along the Jamuna. Over recent decades, erosion has remained the primary morphological process affecting the river. The average erosion rate increased to 4820 ha per year between 1980 and 1989 but decreased to 1500 ha per year from 2000 to 2010. This decline in erosion can be attributed to both natural processes and human interventions, notably the construction of major riverbank protection structures beginning in the mid-1990s. According to CEGIS estimates, between 1973 and 2020, erosion along the Jamuna amounted to 94,000 ha, while accretion totaled 15,500 ha.

The river training activities along the Brahmaputra/Jamuna began in the 1970s in the form of construction of groynes, spurs, and bank revetment. BWDB implemented several bank protective measures to protect Sirajganj town during the 1970s and 80 s.

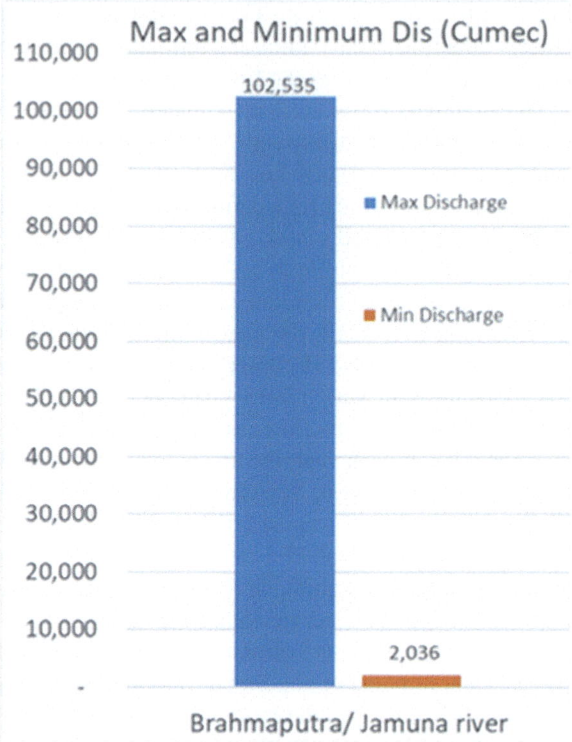

Fig. 5.4 Discharge characteristics of the Brahmaputra/Jamuna. *Source* BWDB, Hydrology

In the aftermath of the devastating flood of 1988, the Brahmaputra Right Embankment Study (FAP 1) was commissioned. It is about implementation of river training works at Sirajganj, Sariaandi, Mothurapara, Betil, Kazipur and Fulchari. FAP21 selected those critical locations for test structures for taking protection measures against erosion along the right bank of the Jamuna. In the early 1990s, a China-Bangladesh Joint Expert Team also addressed the possibility of training the Jamuna. During the 1990s, several significant bank protection structures were built, including the Kamarjan Impermeable Groynes, guide bunds for the Bangabandhu Multi-Purpose Bridge, and revetments for the protection of Sirajganj town. Other notable constructions included the hard points at Sariakandi and Mathurapara, the Kalitola Groyne, and revetments at Bahadurabad and Bhuapur. The hard points at Sariakandi and Mathurapara were constructed, and the Kalitola Groyne was reinforced as part of the Right Bank Protection Project. These structures were completed between 1995 and 1998. Furthermore, BWDB constructed several low-cost structures, such as spurs, along the right bank of the Jamuna (see Fig. 5.8). Initially, bank protection works along the Brahmaputra/Jamuna used concrete blocks. In the early 2000s, BWDB began to construct bank revetment with earth-filled geo-bags, providing an alternative approach to riverbank stabilization.

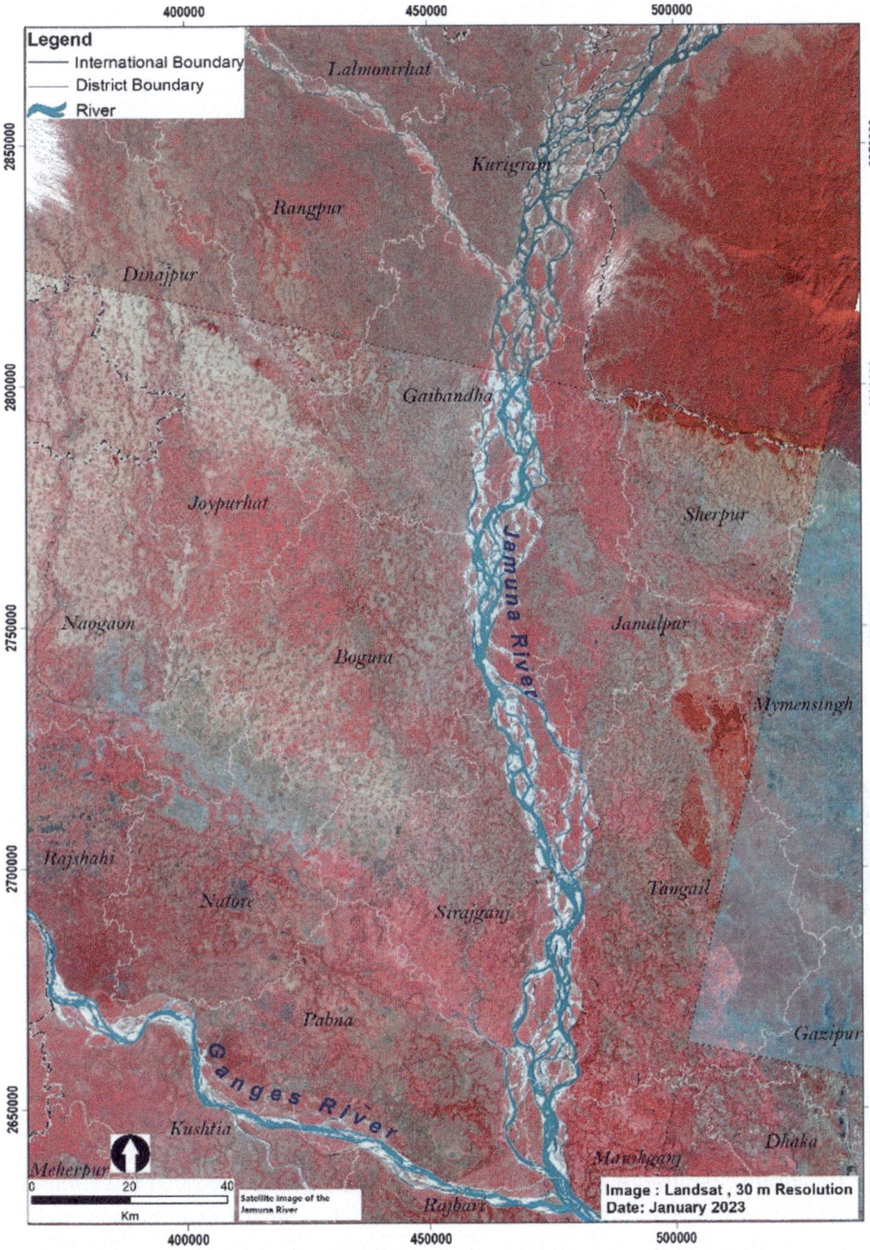

Fig. 5.5 Satellite image of Brahmaputra/Jamuna river (January 2023), showing multi-channels and chars land within the river. *Source* SIBDP Consortium

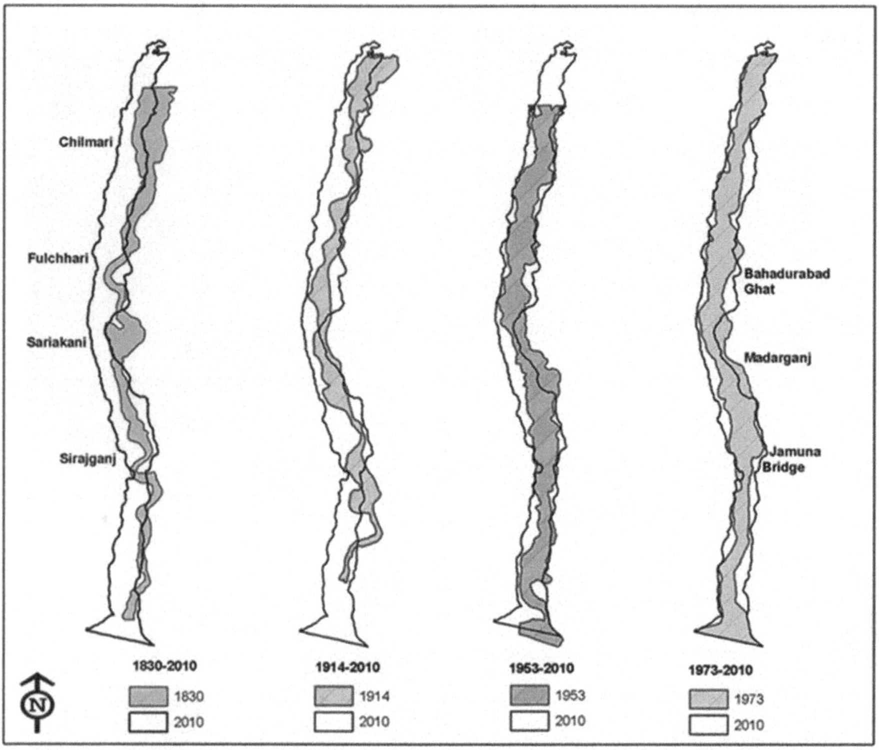

Fig. 5.6 Bankline migration of the Jamuna River from 1830 to 2010, showing the widening of the river in phases. *Source* SIBDP Consortium

The various bank protective structures along both banks of the Brahmaputra/Jamuna are constructed to safeguard the Brahmaputra Right Embankment, which extends 217.5 km from Kurigram to Pabna. The specific locations of these structures are organized by BWDB divisions, as outlined in Table 5.1.

5.2.2 The Ganges River

The Bhagirathi, the traditional source of the Ganges, originates from the Gangotri Glacier in the Himalayas at an elevation of 7010 m. The Ganges spans approximately 2200 km, with only the lower 230 km flowing through Bangladesh. The river flows southeastward, traversing the Indian states of Uttarakhand, Uttar Pradesh, Bihar, and West Bengal before bending around the Rajmahal Hills to enter Bangladesh at the western edge of Nawabganj district (see Fig. 5.2). The Ganges has a catchment area of about one million square kilometers, home to over 400 million people. Three major tributaries of the Ganges, the Karnali, the Gandaki, and the Kosi, originate in China

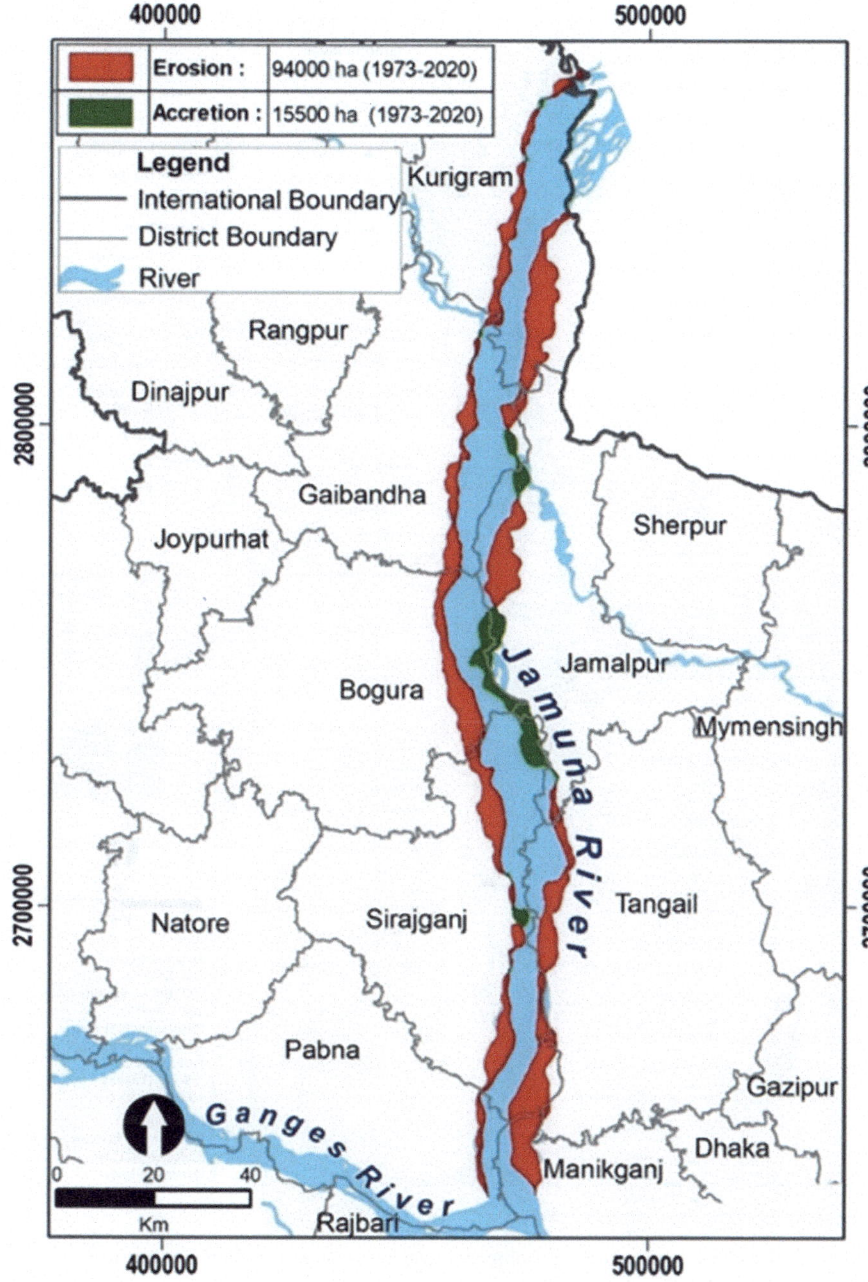

Fig. 5.7 Erosion–accretion along the Jamuna. *Source* BanDuDeltAS

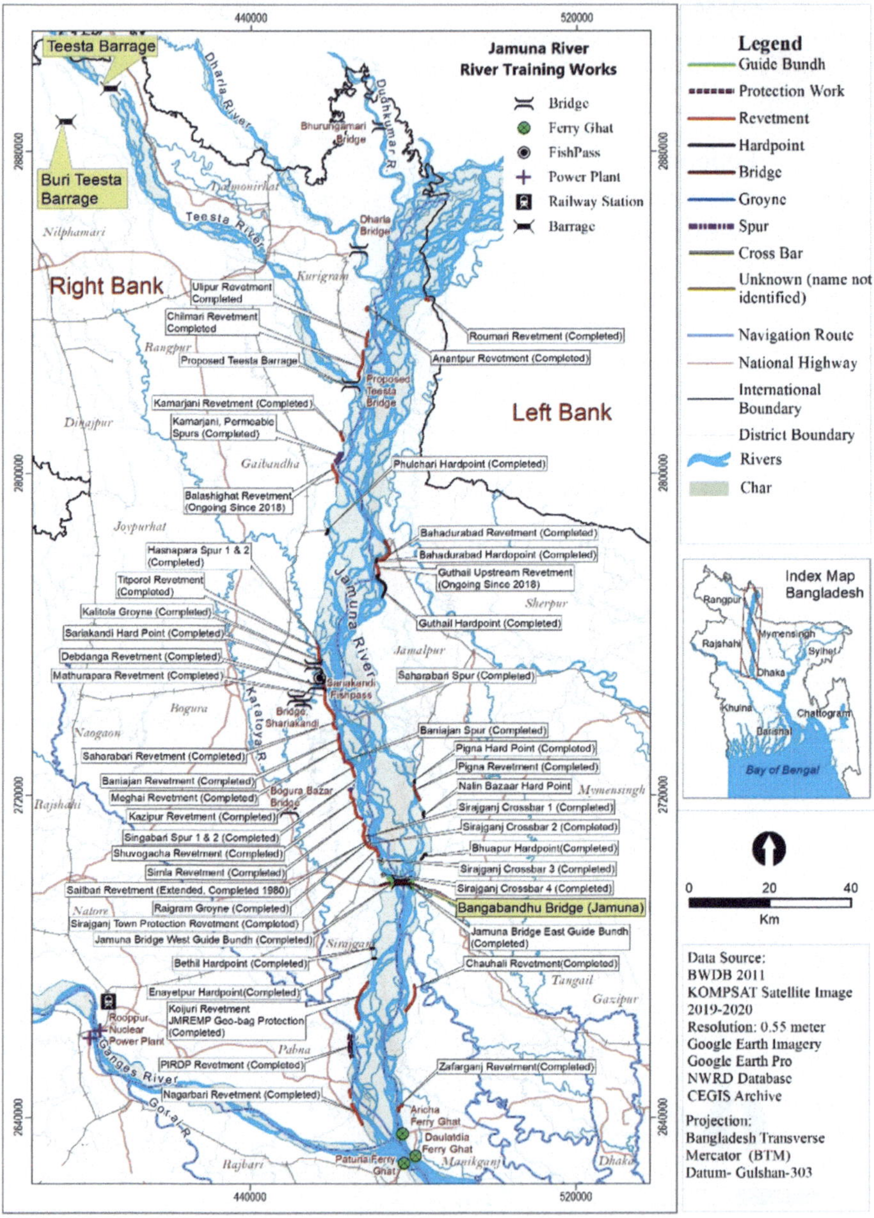

Fig. 5.8 Bank protective measures on Brahmaputra/Jamuna. *Source* SIBDP Consortium

Table 5.1 Bank protection/revetment works along the Brahmaputra/Jamuna until June 2021. *Source* SIBDP Consortium

Name of the BWDB division	Riverbank protection work (km)	
	Right bank	Left bank
Kurrigram O&M division	16.44	12.765
Gaibandha O&M division	34.50	
Bogra O&M division	19.00	
Serajganj O&M division	43.00	7.0 (at Chowhali)
Jamalpur O&M division		34.57
Tangail O&M division	22.38	
Bangabandhu Jamuna Bridge[a]	5.865	5.00
Total	141.185	59.335

[a]Bangabandhu Jamuna Bridge is under the Bridge Authority

and flow through Nepal before merging with the Ganges. These left-bank tributaries account for approximately 71% of the river's natural dry season flows and 41% of its total annual flows. Notable right-bank tributaries of the Ganges are the Ramganga, the Yamuna, the Ton, and the Sone. About 23 km downstream of the Farakka Barrage in West Bengal, the Ganges forms a common boundary between Bangladesh and India for about 110 km. At the Farakka Barrage, the dry season flows (from January to May) are diverted through the Bhagirathi-Hoogly River system towards Kolkata port.[5]

After entering Bangladesh, the Ganges flows southeastward and merges with the Jamuna at Aricha. The combined flow is referred to as the Padma River, which continues southeastward to join the Upper Meghna near Chandpur. The Mahananda is the only significant tributary of the Ganges along its left bank within Bangladesh. In contrast, the right bank features numerous distributaries and spill channels, with the Gorai River being the most prominent. The Gorai is crucial lifeline for the livelihoods and environmental health of southwest Bangladesh (see Figs. 5.3 and 5.9).

In recent decades, the use of Ganges water has increased significantly within the Indian side of the basin. In the 1960s, the Indian government began developing large-scale irrigation projects, including dams and barrages. Currently, there are about 784 dams, 66 barrages, and 92 weirs in the Ganges Basin, with an average water potential of 525.02 billion meters^3 (BCM). The Ganges basin features at least 144 major irrigation projects and 334 medium-sized projects. Consequently, the availability of Ganges flows in Bangladesh has been declining over the years, particularly during the dry seasons.

In 1975, India commissioned the Farakka Barrage with the stated purpose of diverting 40,000 cusecs of water through the Bhagirathi–Hoogly systems during the dry period (January–May). This diversion has led to significant hydromorphological changes in the Ganges system within Bangladesh, notably resulting in reduced dry

[5] Alam et al. (2018).

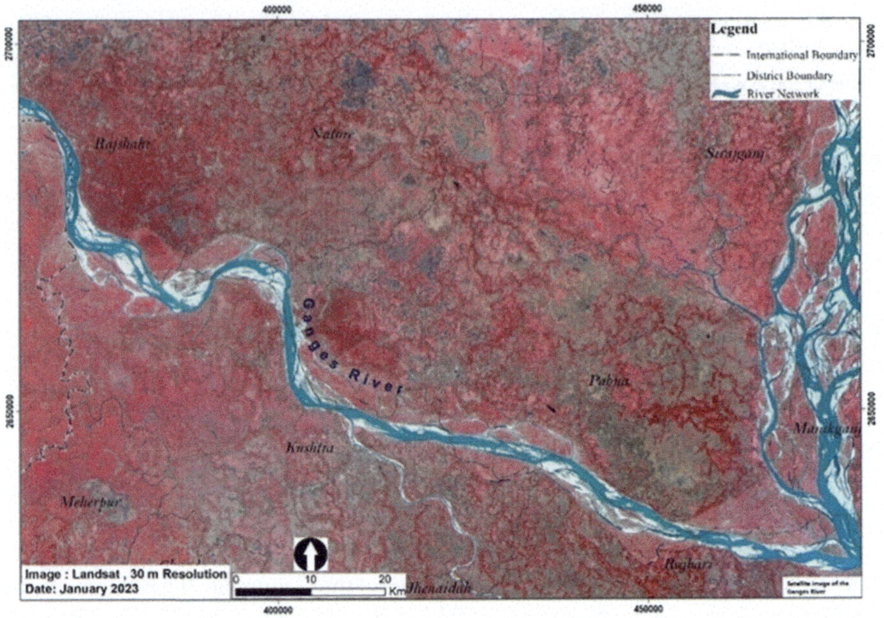

Fig. 5.9 Satellite image showing different bends along the Ganges. *Source* SIBDP Consortium

season flows and increased siltation across the entire river system in the southwestern region, including the Gorai Basin. Furthermore, the flood regime has been altered. In 1996, Bangladesh signed the Ganges Water Sharing Treaty with India, which is set for a duration of thirty years and regulates the dry season flow from January to May; this treaty is scheduled to expire in 2026.[6] Since the commissioning of the Farakka Barrage, the annual dry season flows and discharges have decreased primarily due to the diversion of water at the Farakka barrage.

The Ganges begins rising at the end of June and attains a peak at the end of August through September. The historical recorded highest discharge was 76,033 cumec in September 1987, and the historical lowest discharge at Hardinge Bridge was 261 cumec in March 1993 (Fig. 5.10). The lowest discharge was recorded in 1993 before the signing of the Ganges Treaty, the non-treaty period. The estimated annual mean discharge of the Ganges is 49,835 cumec. The huge difference between high and low discharges is a major challenge for water resources management during wet as well as dry season. During the dry season, the availability of water for irrigation is reduced greatly, hindering navigation significantly.

The Ganges is a very dynamic river in Bangladesh eroding and accreting its banks. The river's sediment load is about half a billion tons per year.

This is also a large meandering river with alternating expansions and contractions. At the very wide expansions, the pigeon-belly-shaped bends demonstrate a

[6] Alam et al. (2018).

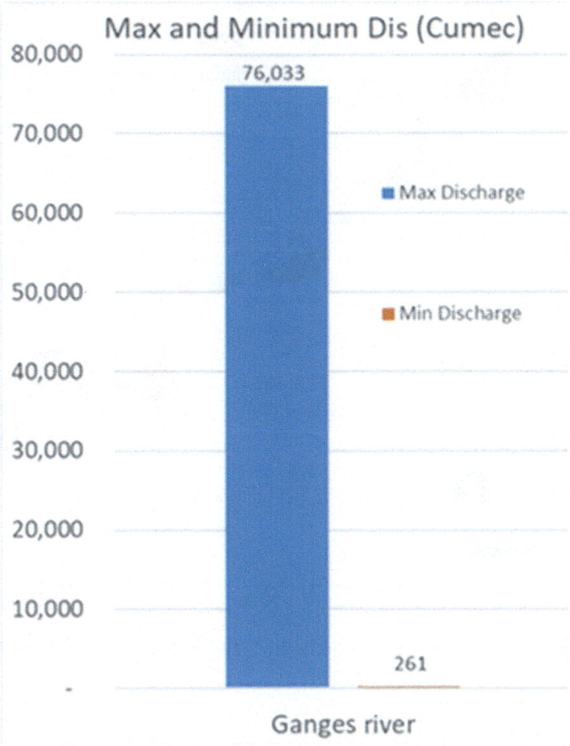

Fig. 5.10 Maximum and minimum discharge of Ganges river. *Source* SIBDP Consortium

braided nature with multiple constituent shallow and deep channels separated by sandy islands. Erosion is a very severe socio-economic issue. It is one of the major causes of poverty in the area. People are displaced, lands and infrastructure are lost annually causing social disruption. Figure 5.11 shows erosion and accretion of banks of the Ganges during the period of 1973 to 2020. During this period, about 40,000 ha of land was eroded and about 34,000 ha accreted.

The riverbank protection and stabilization works are the responsibility of BWDB. A summary of the riverbank protection works along both banks of the Ganges undertaken by the different BWDB Divisional offices is given in Table 5.2.

The left bank of the Ganges has been protected for a length of 77.17 km up and the right bank protected for a length of 16.25 km up to 2022. See also Fig. 5.12.

Moreover, at Pakshi the Hardinge Railway Bridge and at Lalon Shah a Road Bridge are located over the Ganges nearby the Ruppur Nuclear Power Plant which is located on the left bank.[7] These structures are protected by bank revetment structures. The Pabna Irrigation Project is located on the left bank. The Ganges-Kobadak Project is located on the right bank and irrigated with the Ganges waters (Fig. 5.13).

[7] Alam et al. (2018).

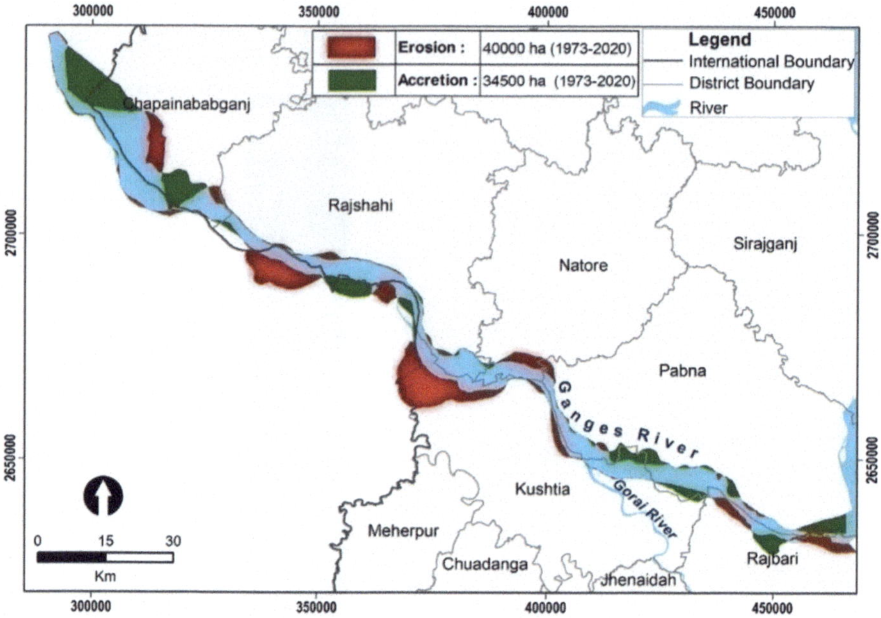

Fig. 5.11 Erosion/accretion in the Ganges during 1973–2020. *Source* CEGIS

Table 5.2 Summary of bank protection/revetment works along the Ganges until June 2021. *Source* SIBDP Consortium

Name of the BWDB division	Riverbank protection works in Ganges (km)	
	Right bank	Left bank
Chapa Nawabganj O&M division		27.37
Rajshahi O&M division		46.30
Natore O&M division		3.50
Kushtia O&M division	6.75	
Rajbari O&M division	9.50	
Total	16.25	77.17

5.2.3 The Padma River

The Padma River is formed by the combined flows of the Brahmaputra/Jamuna and the Ganges. The current course of the Padma is relatively recent; between 1830 and 1860, the combined flows of the Jamuna and Ganges breached the old alluvium known as Chandina Alluvium (GSB, 1990), leading to its merger with the Upper Meghna near Chandpur (see Fig. 5.14). The Padma is highly dynamic, exhibiting braided and straight patterns that shift between intermediate meandering states both spatially and temporally. In recent decades, the Padma has undergone significant changes in its planform and shifted its course southeastward gradually.

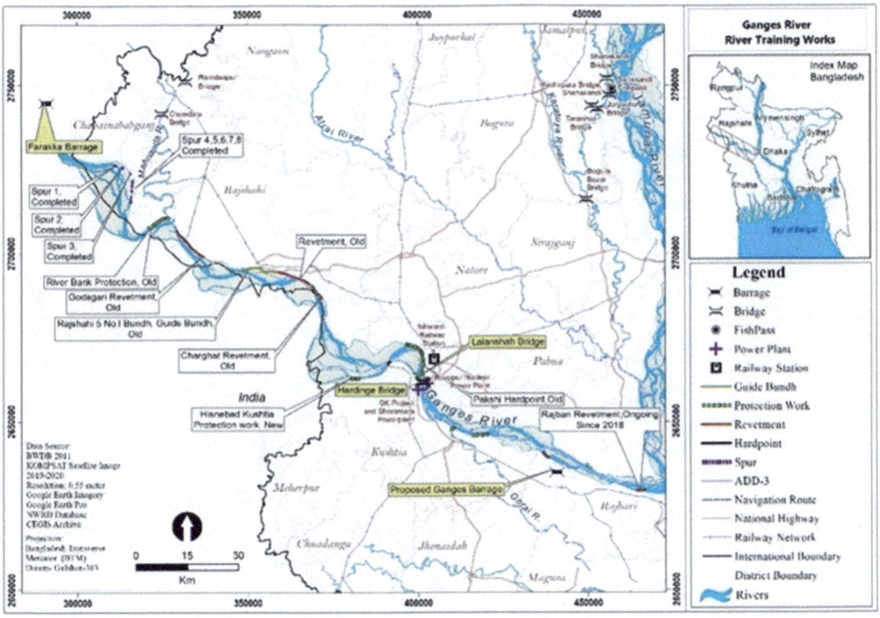

Fig. 5.12 Structures/river training works on the Ganges, SIBDP Consortium

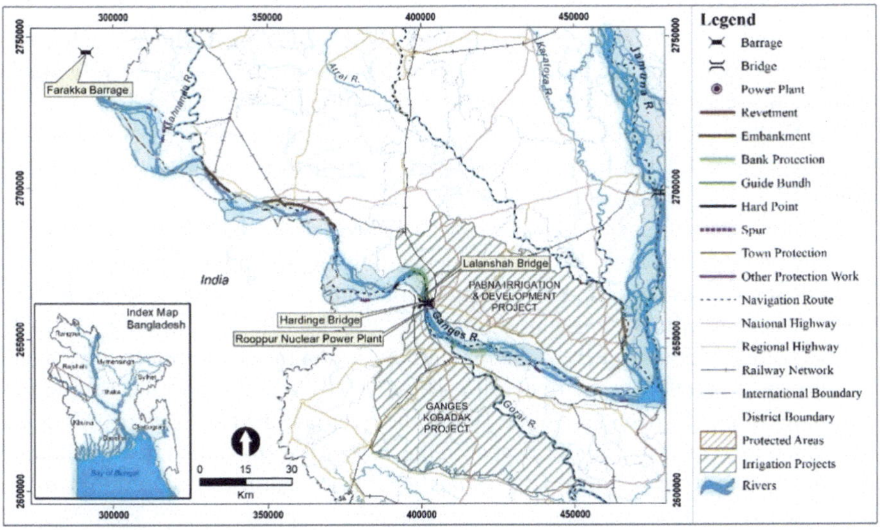

Fig. 5.13 Ganges–Kopotaksha project and Pabna Irrigation project on the Ganges. *Source* SIBDP Consortium

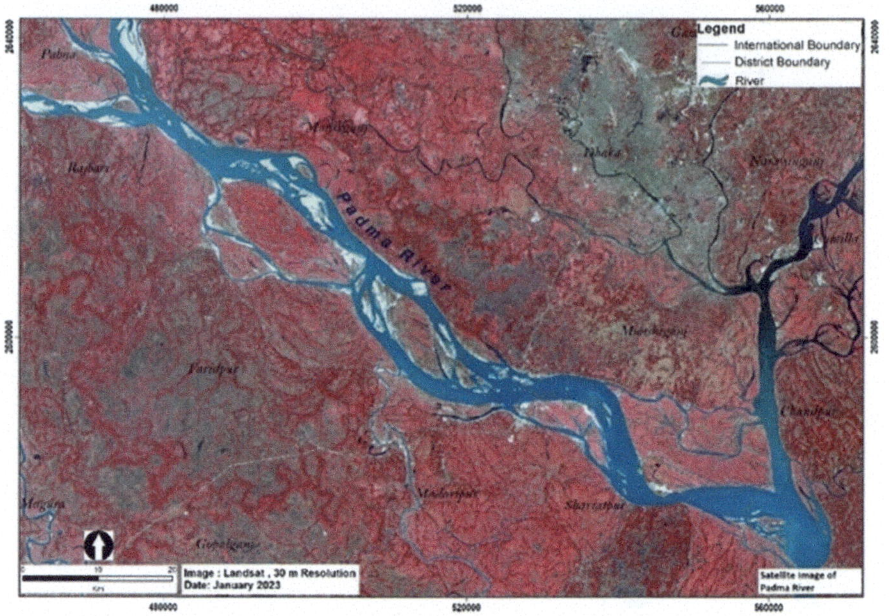

Fig. 5.14 Satellite image of the Padma river. *Source* SIBDP Consortium

Due to the cohesive bank material found at the Paturia and Mawa reaches (on the left bank) and Naria (on the right bank), the river maintains a relatively stable course in this reach within a defined corridor, transitioning its planform from braided to straight through meandering and back again. The catchment area of the Padma is approximately 1.57 million square kilometers, which is equal to the combined catchments of the Jamuna and the Ganges.

At present length, averaged width of the river is about 10.5 km. The banks of the river can be divided into two classes depending on erodibility. The bank erosion characteristic of less erodible banks is different from that of highly erodible banks.

The annual maximum erosion rate may exceed 1000 m/year in case of highly erodible bank materials, whereas the maximum erosion rate will limit to few meters in case of less erodible bank materials. The rate of erosion along the Padma has varied over time. For instance, in the 1980s, the river widened at a rate of 160 m per year, which increased to 230 m per year in the 1990s. Recently, the rate of widening has reduced to 140 m per year. CEGIS analyzed the erosion and accretion data for the period 1973–2020, revealing a total erosion area of approximately 39,500 ha and an accretion area of 9500 ha (Fig. 5.15). The average erosion rate of erosion for the Padma is estimated to be 840 ha per year.

Figure 5.16 illustrates the highest and lowest recorded flows of the Padma. The highest recorded flows of the Padma at Baruria is 141,939 cumec and recorded lowest flows is 2859 cumec. The mean flow is 30,000 cumec.

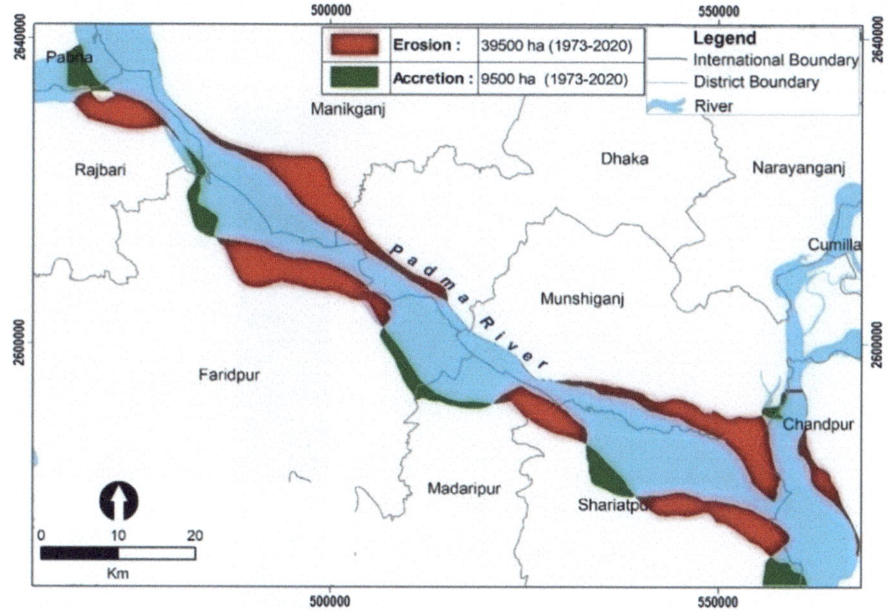

Fig. 5.15 Erosion and accretion along the Padma. *Source* SIBDP Consortium

Bangladesh has constructed the 6.15 km long road-cum-railway Padma Multipurpose Bridge over the Padma at Mawa to connect the southwestern region with Dhaka. The river training is a major component of the bridge. The length of guide bundh on the left bank is 1.6 km and that of the right bank is 12.4 km. BWDB has constructed bank protective works at Harirampur, Dohar, Narisha Meghulia and Mawa on the left bank and at Naria and Zinjira at the right bank of the Padma (see Fig. 5.17).

5.2.4 *The Meghna River*

The Meghna river origins from the Manipur hills in India as the Barak River. The Barak flows through Shilchar and Karimganj districts of Assam, India before entering north-east corner of Bangladesh at Amalshid and bifurcates into two rivers known as the Surma and the Kushiyara. These two rivers flow through the low-lying Haor areas of the northeast region and converge again near Bhairab Bazar and flows southward as the Meghna and ultimately falls into the Bay of Bengal. The Meghna is divided into two sections, the Upper Meghna and the Lower Meghna, due to distinct hydromorphological characteristics. The Upper Meghna refers to the stretch of the river upstream, up to the confluence where it meets the Padma River. From this confluence onward, it becomes the Lower Meghna, which has different flows and sediment

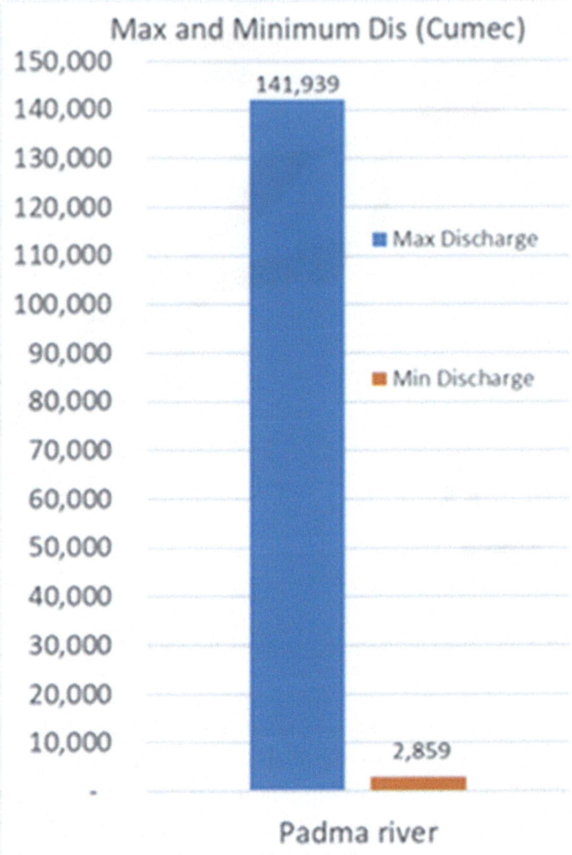

Fig. 5.16 Maximum and minimum discharge of the Padma at Baruria. *Source* SIBDP Consortium

dynamics as it approaches the Bay of Bengal. This division reflects variations in river behavior, including flow velocity, sediment transport, and tidal influences, which impact water resource management and environmental planning in the region.

5.2.4.1 The Upper Meghna River

The Upper Meghna serves as the primary drainage outlet for the northeastern, north-central, and southeastern regions of Bangladesh. This river system is further reinforced by significant tributaries. The Old Brahmaputra, a distributary of the Brahmaputra River, joins the Upper Meghna downstream of Bhairab Bazar, and together they continue southward to meet the Padma near Chandpur. Additionally, the Dhaleshwari River, a right-bank tributary, flows into the Upper Meghna, enhancing its discharge capacity.

The Upper Meghna stretches approximately 900 km, with around 400 km within Bangladesh, covering a catchment area of 77,000 km^2, about half of which lies in

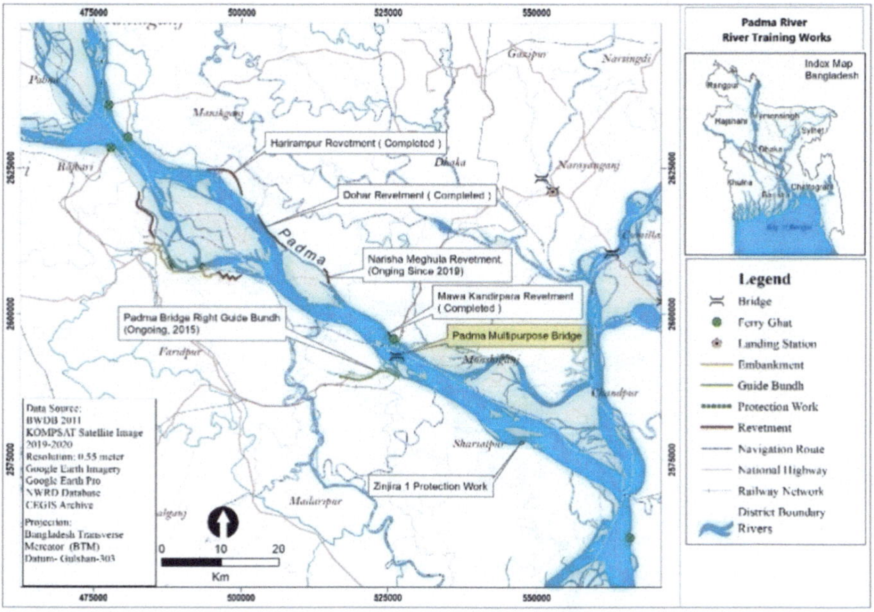

Fig. 5.17 River training works on the Padma. *Source* SIBDP Consortium

Bangladesh. This catchment drains the Meghalaya Hills, an area of intense rainfall averaging around 12,000 mm annually. Consequently, the river's estimated annual mean discharge is substantial at 6572 meters^3 per second (cumec). Historical records show that the Upper Meghna has experienced significant fluctuations in water levels. At Bhairab Bazar, the historical peak water level reached 7.65 m above the Public Works Datum (PWD) in August 1974, while the lowest recorded level was 0.82 m above PWD in February 1984. This variability illustrates the river's sensitivity to seasonal rainfall patterns, impacting flood management and water resource planning.

The Upper Meghna, particularly the section downstream of Bhairab Bazar, was morphologically active in the past, but once the Brahmaputra avulsed, it became a less active river. The river's flood slope has nearly flattened to 2 cm/km. The Upper Meghna can be better classified as an anastomosing (multi-threaded low energy) river, and its planform is relatively stable. Erosion and accretion rates in this river are relatively low compared to other major rivers. Notably, accretion is the primary morphological process, with a net gain in land over recent decades. Between 1980 and 2020, the river experienced approximately 2865 ha of erosion, while accretion reached 6975 ha, indicating a net accretionary trend over this period (see Fig. 5.18).

The Upper Meghna River is experiencing a gradual decline, and this trend is likely to continue. Climate change and rising sea levels could exacerbate this process, particularly since sediment supply from upstream is insufficient to counterbalance any rise in sea level by raising the riverbed level. As a result, the Upper Meghna may face future challenges related to reduced sediment deposition, potentially impacting

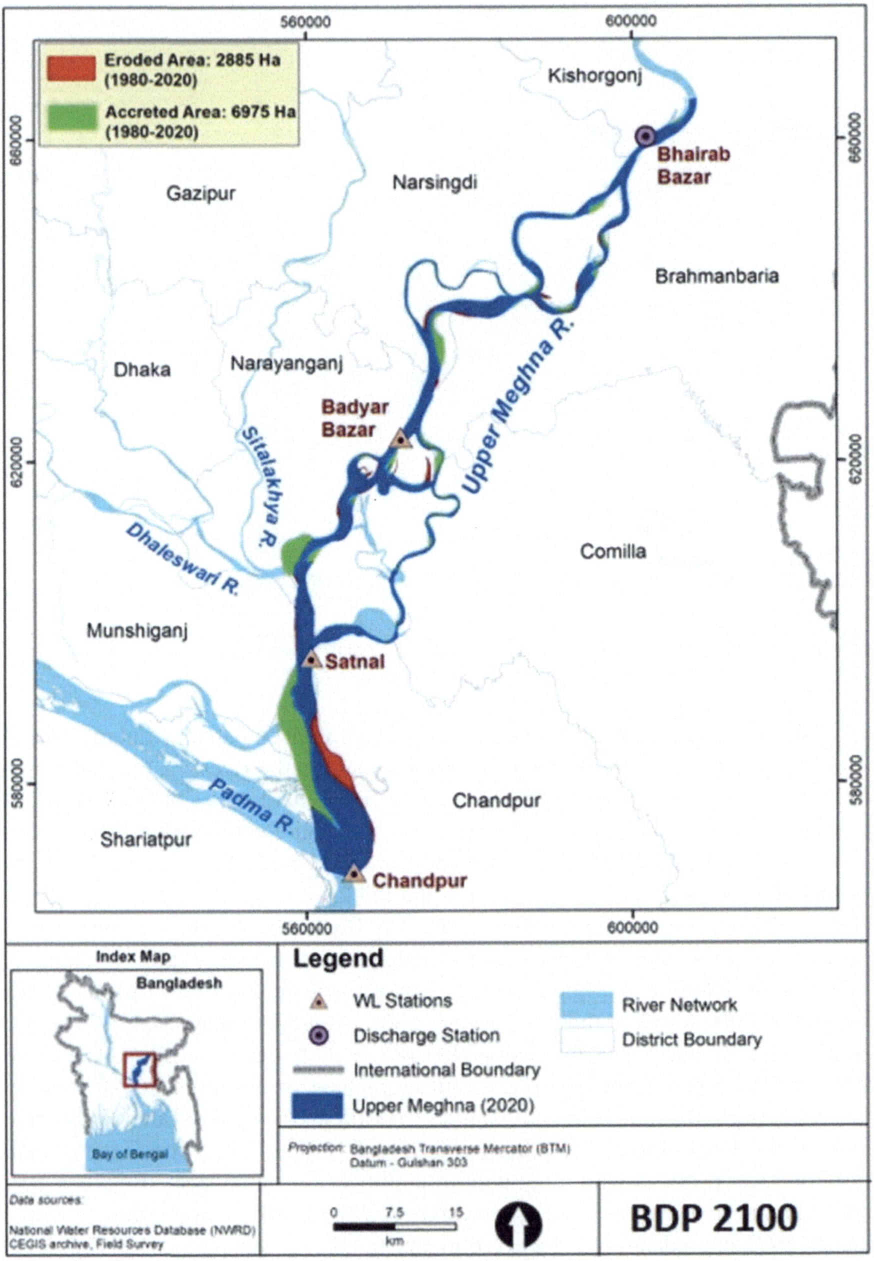

Fig. 5.18 Map showing erosion and accretion site in the upper Meghna. *Source* SIBDP Consortium

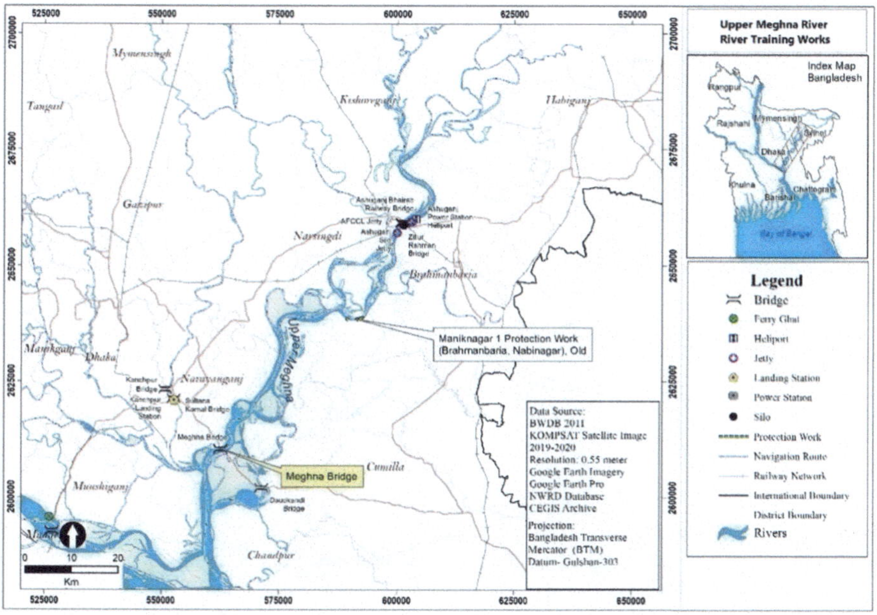

Fig. 5.19 Structures/river training works in the upper Meghna. *Source* SIBDP Consortium

its morphological stability. Despite these challenges, the Upper Meghna remains a critical navigational route due to its stability and year-round navigability. This characteristic opens significant development opportunities, as both banks of the river are ideally positioned to support industrial growth. Leveraging the river as a transportation artery could bolster industrialization in the surrounding areas, fostering economic expansion along this crucial waterway.

The Upper Meghna River at Bhairab Bazar is spanned by two parallel railway bridges and a road bridge. Additionally, there is a road bridge at Meghna Ghat on the Dhaka-Chattogram highway. To safeguard these bridges from erosion, protective measures have been implemented around the bridge structures. There is also a protective structure located at Maniknagar to mitigate erosion (Fig. 5.19).

5.2.4.2 The Lower Meghna River

The Upper Meghna merges with the Padma near Chandpur and then flows southward into the Bay of Bengal as the Lower Meghna. At their confluence, the river widens significantly, reaching approximately 11 km across, extending from Chandpur to Bhola Island. The Lower Meghna then traverses the vast Meghna estuary, one of the largest estuarine systems worldwide. It carries an immense sediment load of over one billion tons annually, effectively transporting nearly the entire flow of the GBM-basins to the coastal waters.

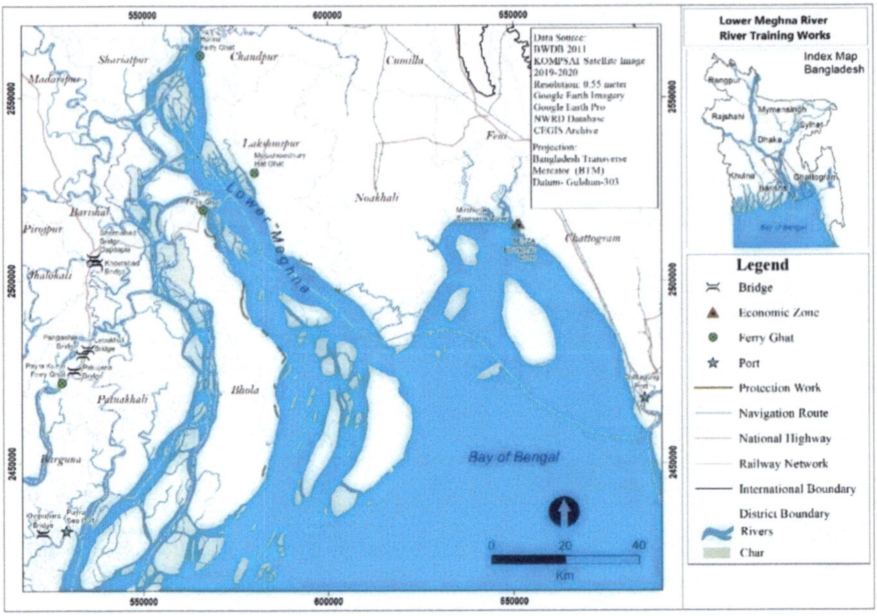

Fig. 5.20 Lower Meghna river. *Source* SIBDP Consortium

The Lower Meghna is a highly dynamic river system, ranked third globally in terms of water volume. This river is characterized by significant bank erosion and frequent development of chars (islands), which are primary morphological features. When chars form, they obstruct the flow, leading to intensified bank erosion as the river adjusts to these new obstacles. Consequently, these riverine chars play a dominant role in shaping the river's morphology, leading to continuous shifts in its course and sediment patterns (Fig. 5.20).

The Lower Meghna is fully influenced by tidal movements. At Chandpur, the tidal range reaches a maximum of 1.5 m, with an average spring tide range of 1.25 m during the dry season, increasing further downstream. The highest recorded water level at Chandpur was 5.23 m PWD in August 1955, while the lowest was 0.56 m PWD in February 1986. As a crucial navigational route, the Lower Meghna is classified by the BIWTA as a Class-I route, supporting regular transit of large ships and cargo vessels. The river's considerable width and depth facilitate the movement of these large vessels, making it an essential artery for inland water transport.

5.2.4.3 The Meghna Estuary

The Meghna estuary is the downstream extension of the Lower Meghna, where it gathers the flow and sediment from the GBM-river systems. The eastern boundary started from Chattogram coast and followed the coastline from the left bank of the

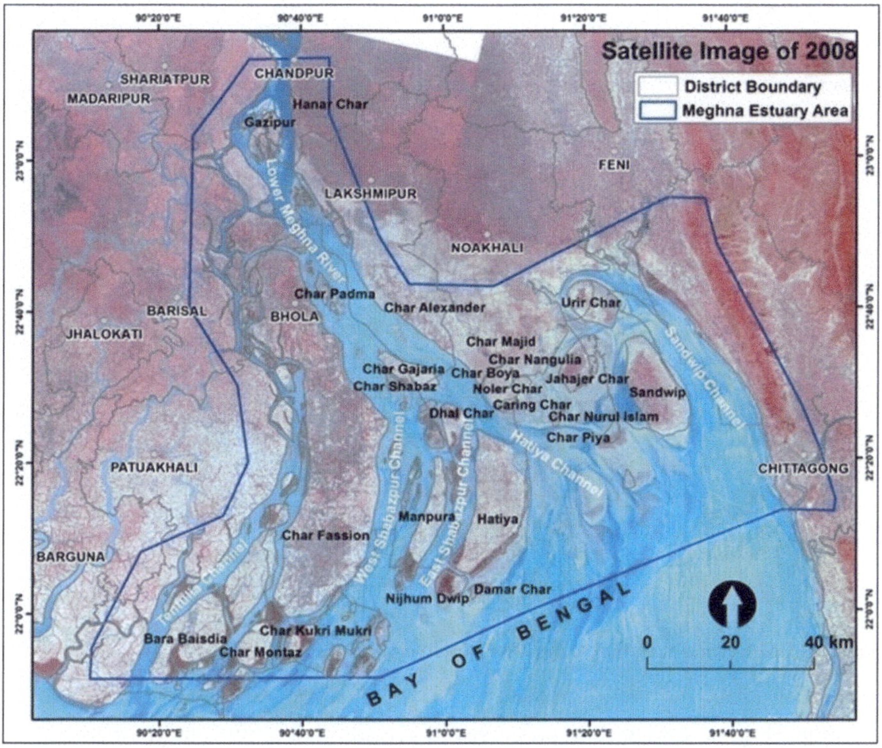

Fig. 5.21 Meghna estuary bounded by blue line. SIBDP Consortium

Karnafuli River at its outfall in the Bay of Bengal. The western boundary followed the right bank of the Lower Meghna, the Tentulia River and the coastline to the sea (Fig. 5.21).

The Meghna estuary features three main distributary channels: Tentulia, Shahbazpur, and Hatiya. The flow distribution within the estuary is primarily influenced by the interplay between tidal forces and freshwater inflow from the Lower Meghna, which receives water from the entire GBM-river system. The Shahbazpur and Tentulia channels are primarily characterized by freshwater flow, while the Hatiya (or Sandwip) channel is dominated by tidal influences. This dynamic balance shapes the hydrology and ecology of the estuary, contributing to its rich biodiversity and importance for local communities.

The Meghna estuary is an active estuary of the GBM-basins, where the delta is prograding vertically and horizontally towards the sea. Prograding means the delta grows seaward, into the Bay of Bengal over time, filling more space with deposition. The degree of progradation varies due to (1) differences in incoming volumes of sediment versus volumes of sediment lost in sediment transportation to other (deep sea) areas; (2) subsidence; (3) sea level rise and (4) erosion (e.g., Saito, 2005).

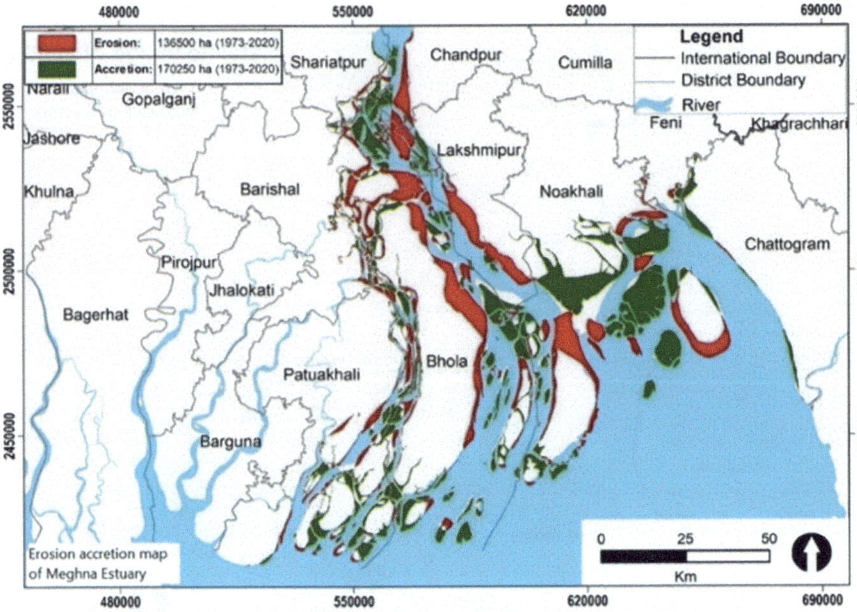

Fig. 5.22 Erosion and accretion in the Meghna estuary. *Source* CEGIS

In the Meghna estuary, the processes of erosion and accretion are ongoing and dynamic. From 1973 to 2020, the net erosion in this area amounted to 136,500 ha, while the net accretion measured 170,250 ha. The accretion is the primary process occurring in the Meghna estuary. The most of the accretion occurs south of Noakhali, while significant erosion has occurred in the districts of Lakshmipur, Chandpur, and Bhola (see Fig. 5.22).

The rate of erosion in the region varies depending on different time scales and is significantly influenced by environmental and natural stresses. Historical records show that the sizes, shapes, and locations of chars (islands) in the Meghna Estuary are on constant change. Between 1776 and 1943, the average erosion rate was only 4.5 km^2 per year, which increased to 42 km^2 per year between 1943 and 1973. In the last four decades, the rate of net accretion has decreased to 17 km^2 per year. It is also observed that the delta is prograding. Over the centuries, the river course has shifted from west to east, and more recently, the rivers have changed direction towards the southwest, indicating a westward shift of the active delta. Sarker et al. (2021) have identified the drivers behind different net accretion rates on various time scales and expressed concerns about the future, particularly regarding reduced or no sediment coming from upstream. Under this scenario, the estuary system may enter a retreat phase. Understanding the ongoing changes in the Meghna estuary is crucial for effective management and planning under climate change conditions.

5.2.5 *The Hilly River Systems in Eastern Hills*

The Eastern Hilly region of Bangladesh, is characterized by many hilly discrete rivers draining into the Bay of Bengal. The important rivers are the Muhuri-Feni, the Karnafuli, the Sangu, the Matamuhuri, and the Bakkhali (Fig. 5.23).

In the Muhuri-Feni system, the Muhuri originates in the South Tripura hilly areas of India and enters Bangladesh at Belonia. The Feni also originates from the South Tripura hills and passes by the Indian town of Sabroom and the Bangladesh town of Ramgarh, and for a length, it forms the common border between India and Bangladesh. Both rivers join and flow to the Bay of Bengal as the Feni. The length of the Muhuri within Bangladesh is 62 km, and that of Feni is 153 km. The highest discharge of the combined flows of the Muhuri and the Feni is 10,800 cumec. The Muhuri Irrigation Project gets irrigation water from the reservoirs created by operating a 40-vent Feni Regulator at the outfall of these rivers during dry season.

The Karnaphuli River, originates from the Lushai hills in the Indian state of Mizoram, flows through Rangamati Hill district and Chattogram district before falling in the Bay of Bengal. The length of the river is 131 km. The principal tributaries are Rainkhiang, Kasalong, Thega, Ichamati, and Halda and the distributaries are Saylok and Boalkhali. The man-made Kaptai Lake, lone earth-filled dam of Bangladesh, is situated on this river at Kaptai to generate hydroelectricity and to reduce flooding downstream. The Chattogram port is situated at the mouth of this river.

The Sangu River originates from the north Arakan hills of Myanmar. The river enters Bangladesh in the Bandarban Hill district and outfalls into the Bay of Bengal. The length of this river is 294 km. The river forms a border between Myanmar and Bangladesh for a distance. The Matamuhuri River also originates from Arakan hills of Myanmar and has a length of 146 km. The river has bifurcated into two channels, ultimately draining to the Maheshkhali channel. The bifurcated channels towards the Maheshkhali have a tidal effect. Tributaries of the Matamuhuri are the Bamu Khal, Lama Chara, and Popa Chara. There are some rubber dams constructed on this river, which may have affected the river morphology. A Matarbari deep sea port is under construction at Maheshkhali Island at the outfall of the Matamuhuri.

The Bakkhali River originates from the hilly areas of Bandarban district. Flowing through Ramu and Cox's Bazaar upazilas, the Bakkhali finally outfalls into the Maheshkhali Channel. The Bakkhali is about 70 km long. A rubber dam has been constructed on the river for irrigation purposes.[8]

[8] Alam et al. (2018).

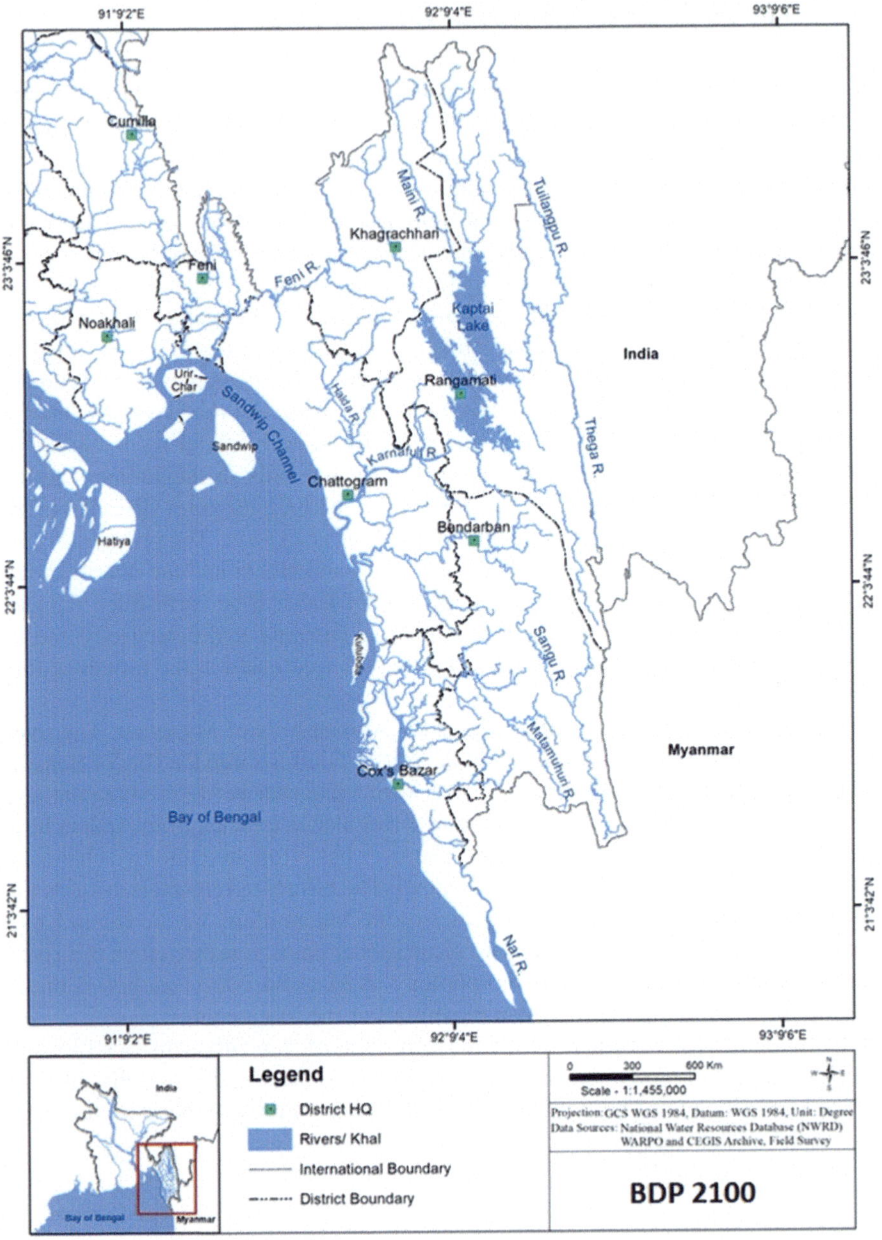

Fig. 5.23 The Muhuri-Feni, the Karnafuli, the Sangu, the Matamuhuri, and the Bakkhali Rivers. *Source* SIBDP Consortium

5.2.6 River Characteristics Related to Navigation

Bangladesh, with its extensive network of rivers and waterways, boasts a significant system of navigation routes. Water transportation has long been a natural, environmentally friendly, and cost-effective mode of transport. Inland waterways play a crucial role not only in maintaining transport links between remote areas of the country but also in facilitating the movement of export–import cargo.

Over the decades, the navigability of many rivers has deteriorated, particularly during the dry season, compounded by inadequate maintenance of inland waterways. The navigability becomes especially critical during the dry season on various river routes and ferry crossings. It is estimated that approximately 12,000 km of waterways are navigable throughout the country. During the monsoon season, about 4000 km are navigable, while this figure drops to 2000–2500 km during the low water period. BIWTA has identified 65 main river navigation routes that are essential for passenger and freight transport. The classified navigational routes are depicted in the map in Fig. 5.24.

5.3 Water Resources

Water resources are key to the country's sustainable development. The National Sustainable Development Strategy (NSDS) published by the Government in 2013 identifies five priority development sectors, they are agriculture, industry, energy, water transport, and human resource development. Water resources are particularly important for the agriculture sector, which includes the crop sector, livestock, fisheries, and forestry sub sectors. An estimated 90% of the country's water resources are consumed by agriculture. Agriculture provides employment to approximately 70% of the population and contributes some 19.5% of the GDP. The industry sector consumes 0.08 km^3 water per year, it is expected that industrial water to rise approximately to 0.35 km^3 by 2050. At present the energy sector's water consumption is not much. Next important sector is transport sector. The water transport subsector supports the livelihood of 6.4 million people. Healthy human resources development is also affected by the availability and quality of water resources and a safe sanitation system in particular concerning poverty reduction, gender, and health improvement. Arsenic contamination and depletion of the groundwater table in Dhaka and northwest of the country are in this respect of great concern.

The water resources available in the country consist of both internally generated surface water resources, through rainfall and runoff, transboundary inflows, and groundwater. During the monsoon season, the country has too much water and during the dry season too little. Moreover, the country is prone to water-related disasters like flood, drought, cyclones, storm surges, salinity, etc. These make water resources management highly complicated.[9]

[9] See also: Hodgson (2014).

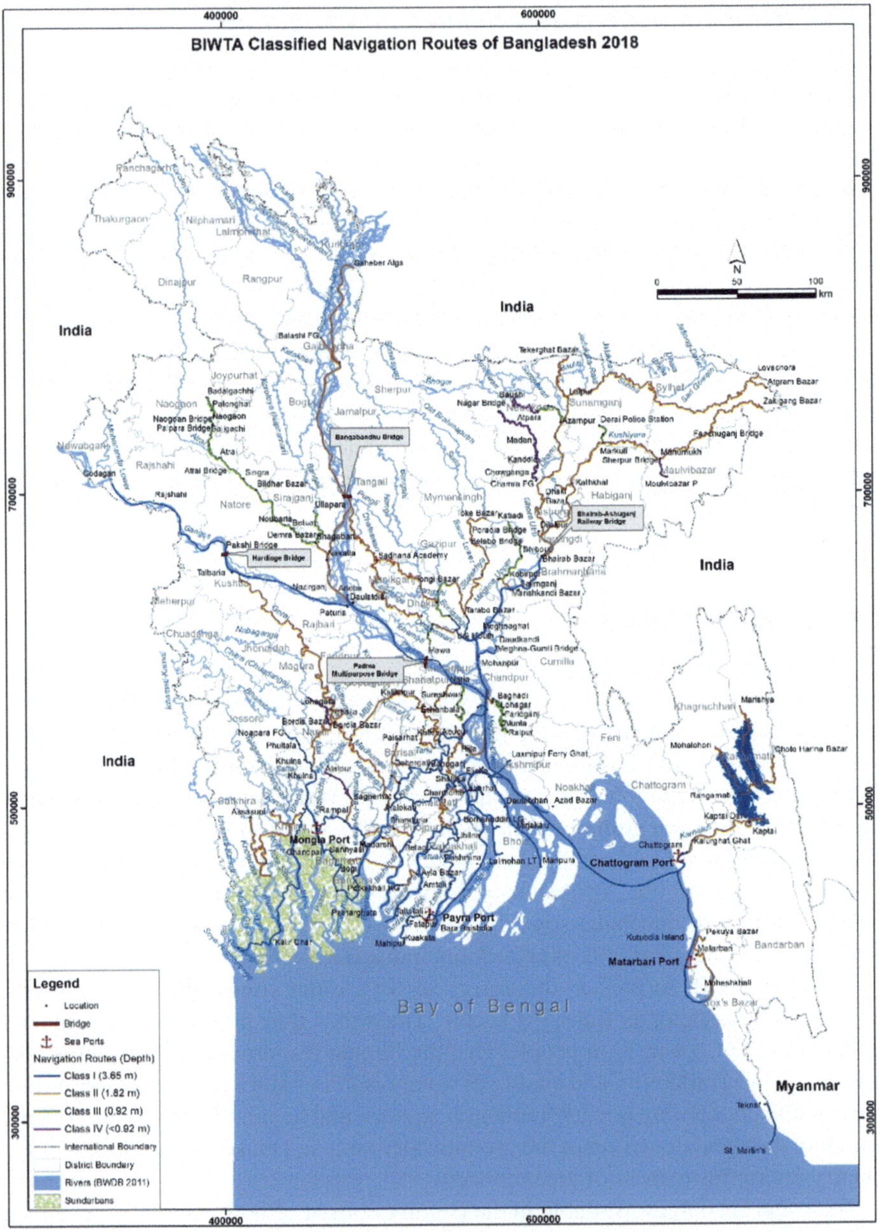

Fig. 5.24 BIWTA-classified navigation routes. *Source* SIBDP Consortium

5.3.1 Surface Water Resources

Bangladesh is as shown before, a riverine country. The country is traversed by a vast network of three major rivers, the Ganges, the Brahmaputra/Jamuna, and the Meghna, their tributaries and distributaries, hilly rivers, and streams, winding seasonal creeks, canals, wetlands, and local water bodies. The surface water resources in Bangladesh comprise of rainfall and transboundary river flows.

The average annual rainfall is 2666 mm, ranging between 1100 mm in the extreme west of the country to some 5680 mm in the north-east. 80% of the rainfall is concentrated in the five monsoon months, with an average volume of rainfall (excluding the Eastern Hills region) is estimated at 284 km^3 (Kirby et al., 2014) based on BMD data. 11% of rainfall falls in the dry season from November to April.

Annual cross-border river flows, entering the river systems and groundwater are estimated to be 1260 km^3, of which the three major rivers contribute some 981 km^3, 85% of which enters the country between June and October. Of the latter amount, some 54% is contributed by the Brahmaputra, 31% by the Ganges, and nearly 14% by the tributaries of the (Upper) Meghna. 1% is contributed by other minor rivers of the Eastern Hills. During the wet season, the rivers receive an additional flow of about 113 km^3 from combined regional (in-country) runoff. Dry season flows are an estimated 186 km^3 or 15% of the total annual river flows (Kirby et al., 2014). Figures 5.2 and 5.3 show the major rivers and estuaries of Bangladesh. Key features of these rivers are provided in Table 5.3.

Out of the total transboundary flows, the contribution of India is 64%, China 18%, Nepal 8%, Bhutan 3%, and Bangladesh 7%.

Table 5.3 Features of the GBM-rivers. *Source* BDP 2100, Baseline Study, 2018

River	Ganges	Brahmaputra/Jamuna	Meghna
Length (km)	2510	2900	912
Catchment (km^2)	1,087,300	552,000	82,000
Average annual rainfall (mm)	1200	1900	4900
Average annual discharge (m^3/s)	11,000	20,000	4600
Maximum discharge (m^3/s)	78,091 (Sept 1998)	102,535 (Sept 1998)	20,000
Sediment transport (mln ton/year)	550	590	13

5.3.1.1 Status of Surface Water Resources Development

In the context of Bangladesh, the water sector is a prime sector given the abundance and importance of water in this delta are; it witnessed a huge development over the last eight decades. Bangladesh Water Development Board (BWDB) is the prime organization for water resources development and management. BWDB is mandated to implement and manage projects above 1000 ha and Local Government Engineering Organization (LGED) implements and manages projects less than 1000 ha. BWDB implemented about nine hundred seventy projects during the last seventy years, out of these 533 projects are flood control, drainage, and irrigation projects. These projects provided flood management facilities to 6.6 million ha land. These facilities provide flood-free land for agriculture, an enhanced degree of safety to human lives, livestock, settlement, industry and infrastructures, multiplicity of cropping intensity, higher crop yields, culture fisheries and controlled hazards due to floods and surges. Out of these projects, 140 projects are providing surface water irrigation facilities to 1.65 million ha land in the country and land reclamation of 1086 km^2. Some of the major flood control, drainage and irrigation projects are: (i) Karnafuli Multi-purpose Project, (ii) G. K Project (197,186 ha), (iii) Dhaka–Narayanganj–Demra Project (6070 ha), (iv) Brahmaputra Right Bank Embankment Project (240,180 ha), (v) Coastal Embankment Project (1,380,000 ha), (vi) Chandpur Irrigation Project (54,036 ha), (vii) Meghna-Dhonagoda Project (19,021 ha), (viii) Muhuri Irrigation Project (40,081), (ix) Pabna Irrigation Project (196,680 ha), (x) Gumti Project (37,440 ha), (xi) Khowai River Project (25,790 ha), (xii) Manu Irrigation Project (22,580 ha), and (xiii)Teesta Barrage Project (132,000 ha).

Notably, there are four barrages over the Teesta, the Buri Teesta, and the Tangon in the northwest and the Manu in the northeast. These barrages are instrumental to distribute irrigation water through gravity flows. Additionally, few small barrages are situated in the northwest region. In the dry season, the Muhuri-Feni water is stored to create a temporary reservoir by operating a 40-vent regulator at the river outfall. Their performances demonstrate that the construction of barrages are feasible over the Ganges, the Brahmaputra, the Meghna, the Dudhkumar, the Dharla, the Sonai-Bardal, the Jury, the Khowai, the Gumti as well as other smaller rivers.

LGED has provided flood, control, drainage, and irrigation facilities in 1118 schemes (less than 1000 ha), which benefited an area of 665,416 ha. LGED has constructed 52 rubber dams throughout the country for utilization of surface water for irrigation. BADC has constructed 12 rubber dams on small rivers.

5.3.1.2 Status of Future Planning

BDP 2100 has proposed the construction of the Ganges (Padma) Barrage over the Ganges at Pangsha, upstream of the confluence of the Ganges with the Brahmaputra/Jamuna. The main objective is to increase irrigation facilities, restore ecological balance, and increase livelihood opportunities for the people living in the Ganges Dependent Area. With this project, the irrigation coverage will increase from resent

7400–141,000 km^2 in the southern portion of the northwest as well as the southwest and the southcentral regions. There is another proposal to build a barrage over the Brahmaputra at Bahadurabad for providing irrigation facilities to the northwest, northeast, and northcentral regions.

The implementation of a small dam on the Mohamaya Chara, a hilly stream in Chattogram demonstrates the feasibility of developing such type of small reservoirs on the perennial hilly streams in suitable locations of Chattogram, Cox's Bazar, Bandarban, Khagrachari, Rangamati, Sylhet, Sunamganj, Moulvibazar, and Habiganj districts. Similarly, in the coastal region, there is a possibility of developing freshwater reservoirs for irrigation purposes like the ones developed in the Muhuri, Matamuhuri, Bakakhali and Idgaon rivers in past years.

5.3.2 *Wetlands and Water Bodies*

A network of water bodies like haors, baors, beels, lakes, khals, and ponds exist in the country, comprising a rich source of surface water. Some of them remain under water throughout the year, while others submerge in the wet season.

Haors are bowl or saucer-shaped depressions, located in the north-eastern region. The Haor area constitutes a mosaic of wetland habitats, including rivers, streams, and canals, beels, and large areas of seasonally flooded cultivated plains. This zone contains about 373 haors, varying in size from a few hectares to several thousand hectares. Within the haors, there are smaller permanent water bodies, which remain submerged, to a lesser, extent, even during the dry period. These are also called beels, which occupy the lowest part of the depressions. The haors are a vital resource for fisheries, irrigation water, eco-system functioning, and seasonal navigation. Beels are usually depressions or topographic lows, produced by geographical processes. These are generally smaller in size and distributed throughout the country. Occasionally, beels are the remains of a river that has changed its course. Some beels dry up in the dry season and expand during the monsoon season into large and shallow water bodies. Beels are generally smaller than haors, but there are also large beels like Chalan Beel in the north-western region, through which the Atrai passes. It has shrunk over the years and now occupies an area of 26 km^2, comprising some 93 smaller water bodies, in the dry season.[10] A third category is the baor or oxbow lake, mostly found in the moribund delta as in the south-central and south-western regions. There are more than 12 baors ranging from half a square kilometer to 13 km^2. The notable baors are Sagarkhali, Jaleshwar, Khedapara, Rampur, Pathanpara, Kathgara, Jogini Bhagini, Ichamati, Joydia, Marjat, Harina, and Arial.

Kaptai Lake is the largest man-made lake in the country, covering an area of around 777 km^2, created after the construction of the Kaptai Dam for hydropower generation. The average annual flow in the reservoir is approximately 15,646 million m^3. The flood absorption capacity is 1024 million m^3. Another prominent lake is Bogakine

---

[10] BDP Hotspot Delta Atelier, Rajshahi, BanDuDeltAS Consortium, June 2015.

Lake, which is a natural lake situated in the Bandarban District. Other lakes include Rinkhyongkine (400 km^2) in Bandarban and Mohamaya Lake and Foy's Lake in Chattogram District.

The natural drainage channels are referred to as khals. They are a common feature in all areas and serve several purposes, including seasonal navigation, fisheries, water retention and drainage.

An estimated 1.3 million ponds exist in the country, covering some 151,000 ha or 11% of the permanent inland water area (FAO, 2011). In general, the size of the ponds varies between 0.02 and 20 ha with an average of 0.30 ha. In Bangladesh, the highest number of ponds exists in the Barisal district (12.11%), followed by Comilla (9.36%), Sylhet (9.10%), Chattogram (8.02%) and Noakhali (7.75%) (BBS, 2002, in FAO, 2011). Before the introduction of tube wells in the sixties and seventies, ponds were an important source of drinking water in the rural areas. Up to 80% of women make use of ponds for their needs, including water for cooking and bathing.

The wetlands and water bodies are vital for the ecological functioning of the delta and provide a key source of livelihood for many rural poor in the country. Over 70% of the rural households in the flood plain catch fish either for income or for food (Thompson et al., 2003) and about 60% of animal protein consumption comes from fish (BBS, 1999).[11]

5.3.3 Groundwater Resources

The groundwater resources play a crucial role in Bangladesh as a primary source for irrigation, safe drinking water, and industrial use. Around 80% of irrigation, especially for Boro rice cultivation, depends on groundwater, a key factor in achieving food self-sufficiency over the last two decades (FAO, 2011). Major concerns surrounding groundwater include arsenic contamination in shallow aquifers, saltwater intrusion, and emerging industrial pollution in specific zones. Intensive groundwater extraction has caused a gradual decline in water tables, particularly in the north-west and north-central regions and in Dhaka, where 80% of drinking water relies on groundwater. Dhaka has seen levels drop by as much as 75 m due to high abstraction rates outpacing natural recharge, emphasizing the need for sustainable management practices in these regions.

The groundwater usage began in the 1960s, initially as an alternative to contaminated surface water sources containing pathogenic micro-organisms, which had traditionally supplied drinking water through ponds, dug wells, and rainwater. The 1970s saw a widespread installation of hand-operated tube wells, providing pathogen-free groundwater for drinking purposes. Although groundwater irrigation started modestly in the 1960s in the northwestern regions, it gained popularity in the 1980s, ultimately establishing groundwater as a primary source of irrigation water. Today, groundwater remains integral to the country's irrigation system and water supply.

---

[11] Alam et al. (2018).

The groundwater aquifer system has been studied extensively in Bangladesh. BWDB-UNDP (1982) study classified a three-layer regional model. In 2013 BWDB identified a complex, multi-layered system down to 350 m in the coastal zone. Using isotopic analysis, Aggarwal et al. (2000) classified groundwater at various depths into four groups and defined three primary aquifer tiers. Refining earlier studies, DPHE (2001), delineated a three-tier aquifer structure:

- Upper Holocene Aquifer (Shallow Aquifer): Extends to 50–100 m in depth, located under a thick clay and silt layer and composed mainly of fine sand with clay lenses. This aquifer, known as the '1st Aquifer', contains water approximately 100 years old (Aggarwal et al., 2000; Zahid et al., 2009).
- Primary Water-Bearing Zone (Mid-Holocene Aquifer): Found at depths between 250 and 350 m, typically bounded by silty clay. It consists of semi-confined or interconnected unconfined aquifers composed primarily of fine sand. Known as the 'Main Aquifer' (BWDB-UNDP, 1982), '2nd Aquifer' (Aggarwal et al., 2000; Zahid et al., 2009), or 'Lower Shallow Aquifer' (DPHE, 2001), the water here dates to around 3000 years old. This zone is a primary source for groundwater extraction.
- Deep Aquifer (Late Pleistocene-Early Holocene Aquifers): Situated at 300–350 m depth, typically beneath a silty clay layer, it comprises fine, dark-gray sand with occasional clay lenses. It aligns with the 'Deep Aquifer' from BWDB-UNDP (1982) and the '3rd Aquifer' (Aggarwal et al., 2000; Zahid, 2015). This aquifer's water is approximately 20,000 years old.

Bangladesh's groundwater system is shaped by a unique combination of hydrogeological factors, including the proximity to both the Himalayas and the sea, seasonal climatic variations, sediment geochemistry, and the presence of major rivers and numerous smaller water courses. The hydrogeological cross-section from north to south illustrates the Quaternary aquifer system and groundwater flow patterns (DPHE-BGS, 2001, Fig. 5.25). This complexity results in a diverse and region-specific groundwater flow, influenced by both natural and climatic factors, making water management a critical challenge for the country.

Bangladesh's geological history has been significantly shaped by cycles of sea regressions and transgressions, coupled with the influence of numerous river systems. The country's predominantly low elevation throughout its geological past has rendered it highly sensitive to sea-level changes, impacting geological processes such as weathering, erosion, and sediment deposition. These shifts have contributed to the complex layering of sediments and have greatly influenced the formation and characteristics of the region's aquifers and soil structure.

The estimated quantity of groundwater resources available for extraction is around 21 billion meters^3, as per MPO (1987). Additionally, an unknown but significant quantity of groundwater flows horizontally from the Himalayas, which is crucial for recharging deep aquifers and preventing depletion. However, the specific contributions of the horizontal flow remain unquantified. The groundwater recharge estimates vary significantly, highlighting the need for further study. The study by Kirby et al. (2014) estimates an annual average recharge of approximately 47 km^3 to shallow,

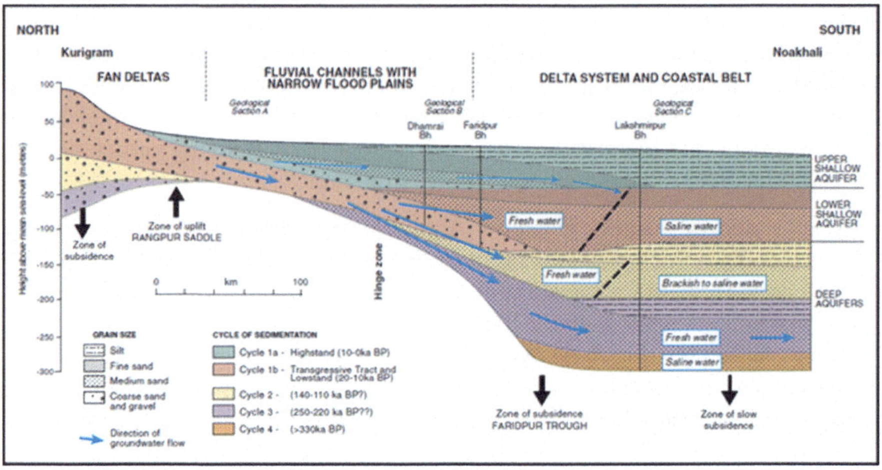

Fig. 5.25 Hydrogeological cross-section from north to south showing quaternary aquifer system and groundwater flow pattern. *Source* DPHE-BGS (2001)

unconfined groundwater (excluding the eastern hills), with about 15 km^3 attributed to percolation from irrigated rice crops, leading to a net recharge of around 32 km^3. Another study by Hodgson et al. (2014) places this recharge at about 28 km^3. These discrepancies emphasize the conclusion by Kirby et al. that groundwater balance components are insufficiently understood and warrant further detailed research.

5.3.3.1 Status of Groundwater Use

Around 87% of the population relies on groundwater for daily needs (BBS, 2010), and the agricultural sector heavily depends on it. The country currently uses about 36,955 deep tube wells, 1,409,639 shallow tube wells, and 204,391 low-lift pumps to extract groundwater and surface water for irrigation (BADC, 2021; BWDB, 2021). BWDB focuses on surface water for irrigation, utilizing structures like barrages, regulators, or vertical lift pumps. BADCs utilize both groundwater and surface water, but largely depend on groundwater for irrigation purposes. The rapid depletion of groundwater has raised environmental and livelihood concerns, leading to this issue being described as a 'silent hazard'.

The groundwater supplies about 79% of water needs across agriculture, livestock, municipal, and industrial sectors, with only 21% from surface water.[12] This high reliance on groundwater underscores the need for rational water use to sustain resources and environmental health, especially with population growth, economic expansion, and climate change pressures. The BDP 2100 emphasizes increasing surface water use over groundwater. In the past two decades, the government has

[12] Alam et al. (2018).

shifted towards surface water supply for urban areas, significantly transitioning Dhaka and Chattogram to surface water sources.

5.3.4 Surface–Groundwater Interactions

The surface and groundwater interactions are essential for effective water resources management. The studies of groundwater levels and river-stage hydrographs indicate that during the monsoon season (May–September), water levels in almost all river channels surpass groundwater levels in adjacent aquifers, allowing for indirect recharge primarily via lateral river-bank infiltration, especially in the early monsoon months (April–June) (Shamsudduha et al., 2011). Shallow aquifers near the Brahmaputra are notably affected by substantial indirect groundwater recharge (Zahid, 2015).

Figure 5.26 highlights the seasonal interactions between the Jamuna River and the surrounding aquifer, showing that the river contributes to aquifer recharge from May–October. Conversely, some discharge occurs from the aquifer back to the river during the dry season. Despite these observations, the specific dynamics of these interactions remain unclear and represent a knowledge gap in understanding regional hydrogeology.[13]

Groundwater recharge data from the pre-irrigation period of 1975–1980 indicate that the highest recharge rates occurred in the Barind and north-west regions, particularly in the districts of Thakurgaon, parts of Kurigram and Gaibandha, Nawabganj, Sirajganj, and Pabna. The mean recharge values range between 250 and 600 mm.

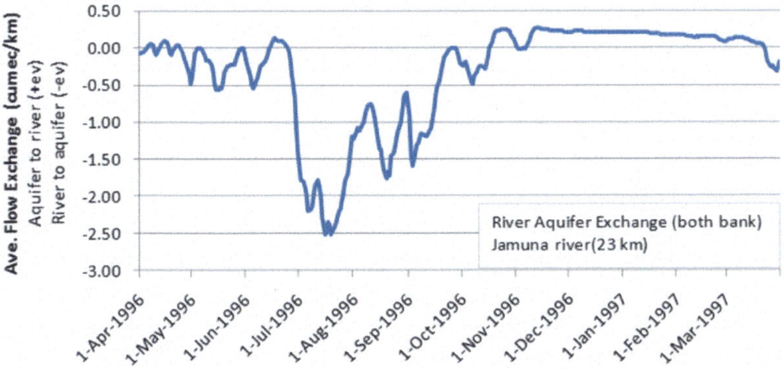

Fig. 5.26 Simulated exchange between the Jamuna and the adjacent aquifer in 1996. *Source* CSIRO, 2014

[13] Ibid.

The least recharge is observed in Naogaon, where it was 125–175 mm. During 2002–2007 and long-term (1985–2007) groundwater recharge increased greatly all over this region ranging between 300 and 600 mm in the all the districts.

In the Dhaka metropolitan area and the north-central region, groundwater recharge rates during 1975–1980 were higher on the western side than on the eastern side. Notably, the districts of Jamalpur, Tangail, and Manikganj experienced recharge rates reaching up to 500 mm, while Mymensingh, Dhaka, Narayanganj, and other nearby districts observed lower recharge rates, ranging between 50 and 150 mm. The recharge patterns in the 2002–2007 and 1985–2007 periods remained similar to those of 1975–1980, though recharge levels increased across all districts. Tangail and Manikganj experienced high recharge values, reaching up to 600 mm.

The Haor and the northeast regions exhibited relatively low groundwater recharge rates from 1975 to 1980, particularly in Sylhet and Kishoreganj districts, where recharge was as low as 10 mm. Other districts in the region, including Sunamganj, Habiganj, and Netrakona, also saw limited recharge, with values ranging between 50 and 125 mm. During the 2002–2007 period, as well as in the long-term data from groundwater-fed irrigation, recharge levels in the Maulvibazar District dropped further to 10 mm, while recharge in the other districts remained relatively unchanged.

In the coastal zone and southern region, the groundwater recharge from 1975 to 1980 was lowest in the southern coastal districts, including Satkhira, Khulna, Bagerhat, and Patuakhali, with recharge values ranging from 10 to 50 mm. By contrast, areas like Faridpur displayed higher recharge levels, between 200 and 350 mm. During the 2002–2007 period, recharge levels increased significantly in Jhenaidah, Magura, and Jashore, reaching up to 600 mm, while remaining low in the southern coastal regions. Similar trends persisted in the long-term analysis for groundwater-fed irrigation from 1985 to 2007.

The southeast region showed relatively higher groundwater recharge compared to other regions, with values generally ranging from 100 to 175 mm. Comilla District itself represented a greater value compared to others with recharge values as high as 175 mm. However, during the 2002–2007 period, as well as in the longer term from 1985 to 2007, the groundwater recharge values decreased significantly in Noakhali District, dropping to just 10 mm, while recharge levels in the other southeastern districts remained stable.

The groundwater recharge data for the period 1975–1980 is mostly absent for the Chattogram Hill Tracts. However, along the coastal districts of Chattogram and Cox's Bazar, recharge was minimal, ranging from 10 to 50 mm. In subsequent periods (2002–2007 and 1985–2007), recharge increased slightly in northern Chattogram, reaching up to 125 mm, while Cox's Bazar saw no significant change.

Among urban areas, Sylhet, Khulna, and Chattogram had the lowest recharge levels, as low as 10 mm from 1975 to 1980, with Rajshahi showing better recharge values between 300 and 600 mm. Notably, only Chattogram experienced an increase in groundwater recharge during the 2002–2007 period and over the long term from 1985 to 2007 compared to other major cities.[14]

[14] Alam et al. (2018).

5.4 The Coastal Zone

The coastal zone is a highly dynamic and diverse region, marked by opportunities and significant risks due to factors like natural disasters, sea-level rise, and climate change. Spanning 710 km along the Bay of Bengal, the coastal zone encompasses 147 Upazilas across 19 districts, covering an area of 47,150 km^2. With a population of around 40 million people (about 35% of the national population), the region has a high population density of 848 persons per km^2 (BBS, 2022). The inland boundary of this zone is defined based on three hydrological factors: tidal influence, salinity intrusion, and the impact of cyclones and storm surges.

The coastal zone is subdivided into four distinct areas, each with unique morphological conditions, challenges, and characteristics; (see Fig. 5.27).

5.4.1 *Ganges Tidal Plain (West)*

The region is characterized as moribund delta formation, having a vast network of rivers of low gradient and very little fresh water flows from the Ganges, mostly through the Gorai. The area is subject to monsoon flooding. This area includes the Sundarbans mangrove forest, a natural coastal buffer zone of 50–60 km inland

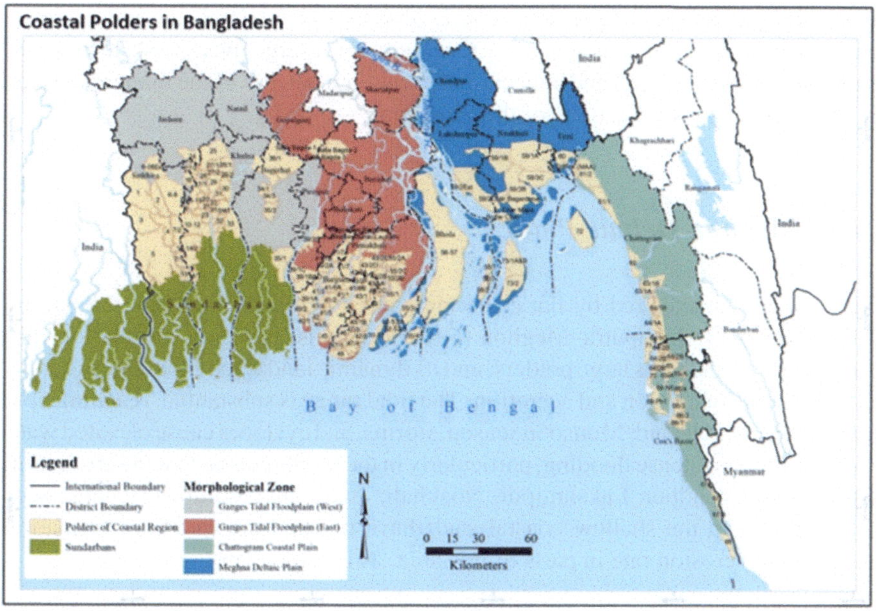

Fig. 5.27 Map of the coastal morphological zone and coastal polders of Bangladesh. *Source* SIBDP Consortium

from the coastline provides the hinterland to the north with a considerable degree of protection from cyclonic storm surges. The semi-diurnal tide has a range up to 3 m at coast. The tidal flows extend far inland, for instance, the tidal range decreases to 1.5 m at Khulna. The salinity has intruded further inland. Several polders have been constructed, some of these are located about 150 km away from the coast. The dominant problem is, therefore, the sedimentation of rivers beds, hindering adequate drainage, and, thereby, causing severe waterlogging in the low-lying areas and affecting the navigation. The area is also attributed with the transverse depressions, called beels, running roughly parallel to the coast. These are difficult to drain for agricultural use but can offer other possibilities, like use as tidal basin for tidal river management for restoring the drainage conditions by increasing the tidal prism.

5.4.2 Ganges Tidal Plain (East)

This region, with its low-lying flat topography just above sea level, represents a younger estuary phase. It is traversed by multiple rivers fed by the Padma River through the Arial Khan River and by the Lower Meghna River. The tidal range along the coast reaches 3.5 m, though it decreases inland. Extensive polder systems extend roughly 60 km inland from the coastline. These sea-facing polders are vulnerable to wave-induced erosion in certain areas, and some polder embankments face erosion due to the migration of river thalwegs. Sedimentation within some rivers is also posing navigation challenges. While river flooding is minimal, the area remains primarily influenced by tidal dynamics. Cyclonic storm surges pose a risk, threatening infrastructure, agriculture, and aquaculture. Scattered patches of the mangrove forest are present along some sections of the coastline.

5.4.3 Meghna Deltaic Plain

This region, characterized by flat estuarine coastal systems shaped by the outflow of the Ganges–Brahmaputra–Meghna (GBM) systems, is highly morphologically active. The area includes large polders, and its dynamic landscape experiences continuous processes of erosion and accretion. The tidal range is substantial, reaching up to 6 m near Sandwip Island. Monsoon season, storms, and cyclones cause elevated water levels, leading to intense flooding, particularly in the Meghna estuary and surrounding districts like Chandpur, Lakshmipur, Noakhali, and Feni. Additionally, large sediment deposits in the shallow coastal shelf have resulted in rapid land formation, exceeding the erosion rate in parts of this area, as illustrated in Fig. 5.28.[15]

---

[15] Furthermore: De Wilde (2022); De Heer (2022); De Heer et al. (2022).

Fig. 5.28 Strong erosion on western side of Hatya Island. *Photo* J. M. de Heer

The land reclamation activities have intensified waterlogging problems in this area, resulting in prolonged inundation. This persistent waterlogging challenges agricultural productivity and infrastructure stability, highlighting the need for sustainable land management and drainage solutions in these vulnerable coastal zones. See also Sect. 5.2.4. on the Lower Meghna and Meghna estuary.

5.4.4 Chattogram Coastal Plain

A relatively narrow area having flat topography extends along the eastern coastline of the Bay of Bengal in Chattogram and Cox's Bazar. The tide ranges from 6 m in the north to 3 m near Teknaf in the south. The area features several small polders along the coastline, directly exposed to Bay of Bengal and vulnerable to cyclone and storm surges. Frequent flash floods occur here, as steep hills to the east cause rapid runoff that drainage channels in the flat coastal plain cannot accommodate, resulting in flood risks. Several rivers in this region also exhibit extensive meandering patterns across the plain.

Rapid industrialization in this coastal area, especially in the northern part of the Chattogram coastline, has created a demand for enhanced flood protection. This has led to the construction of super dykes, designed to provide stronger defenses against tidal flooding, storm surges, and the impacts of sea-level rise, thus supporting industrial growth, while addressing increased vulnerability along the coastline.

Fig. 5.29 New developed land south of Noakhali. *Photo* J. M. de Heer

5.4.5 Coastal Polders

BWDB constructed 139 polders across the coastal zone, starting in the 1960s under the Coastal Embankment Project. These polders aimed to prevent the intrusion of saline water and protect against tidal floods, enabling the development of agriculture in the area. As a result, approximately 1.2 million ha of land became available for agricultural use, as illustrated in Figs. 5.29. The polders also play a crucial role in protecting lives and properties in coastal communities by mitigating the impacts of floods, storm surges, and salinity intrusion.

However, there have been unintended consequences such as river sedimentation and waterlogging that has become increasingly problematic in the last three decades in certain parts of the coastal area.

5.4.6 Inland Water Transport

The coastal areas have a dense network of navigable waterways. The important river ports are Dhaka, Narayanganj, Barisal, Khulna, and Chandpur. The waterways (maximum draft 4 m) connect these ports with Dhaka and Barisal. Three seaports are Chattogram, Mongla and Payra. A deep sea port is being built in Matarbari. A deep sea port is being developed in Matarbari near Cox's Bazar and a sea port at Payra in the Meghna estuary.

5.4.7 Coastal Resources

A majority of the coastal population is highly dependent on coastal resources, especially agriculture, for their livelihood. The major crops are rice, pulses, vegetables, fruits, jute, and sugarcane. The area is rich in freshwater and marine fisheries. Shrimps farming is carried out in the coastal polders around Khulna, Jashore, Chattogram and Cox's Bazar. Salt is produced in Chattogram and Cox's Bazar. Another economic activity is ship-breaking and recycling concentrated in Chattogram. There are some onshore and offshore gas fields. The Sundarbans, covering an area of 6017 km^2 within Bangladesh, is rich in many varieties of mangrove species and abundance and diversity of flora and fauna.

References

Aggarwal, P. et al. (2000). *A report on isotope hydrology of groundwater in Bangladesh: Implications for characterization and mitigation of arsenic in groundwater.*

Ahmad, Q. K., Biswas, A. K., Rangachari, R., & Sainju, M. M. (Eds.). (2001). *Ganges–Brahmaputra–Meghna region. A framework for sustainable development.* UPL.

Alam, S., De Heer, J. M., & Choudhury, G. A. (eds.). (2018). *Bangladesh Delta Plan 2100: Baseline Studies on Water Resources Management. Volume 1. General Economics Division, Planning Commission, Government of Bangladesh.* https://plancomm.gov.bd/

Alam, M., Alam, M. M., Curray, J. R., Chowdhury, M. L. R., & Gani, M. R. (2003). An overview of the sedimentary geology of the Bengal basin in relation to the regional tectonic framework and basin-fill history. *Sedimentary Geology, 155*, 179–208.

BADC. (2021). *BADC Annual Report.*

BanDuDeltAS Consortium. (2015). Delta Atelier Rajshahi Report.

BanDuDeltAS. (2015). Baseline Study on Water Resources.

BBS. (1999, 2002, 2010, 2022). *Statistical Year Book of Bangladesh, Bureau of Bangladesh Statistics, Government of People's Republic of Bangladesh.*

BDP 2100. (2018). Baseline Study on Water Resources.

Brammer, H. (1996). *The geography of the soils of Bangladesh.*

BWDB. (2021). *BWDB Annual Report.*

BWDB-UNDP. (1982). *Groundwater Survey: The hydrological conditions of Bangladesh.* UNDP Technical Report DP/UN/BGD-74-009/1

CSIRO. (2014). *Bangladesh integrated water resources assessment: Final report.*

De Heer, J. M., Oliemans W., & Choudhury G. A. (2022). Meaning of BDP 2100 for the coastal zone: Coastal development in BDP 2100 perspective (Chap. 12). In: de Wilde, K. (ed.) *Moving Coastlines. Emergence and use of land in the Ganges-Brahmaputra-Meghna Estuary.* UPL.

De Heer, J. M. (2022). The Bangladesh Delta Plan 2100 (Chap. 12). In: de Wilde, K. (ed.). *Moving Coastlines. Emergence and use of land in the Ganges–Brahmaputra–Meghna Estuary.* UPL.

De Wilde, K. (ed.). (2022). *Moving Coastlines. Emergence and use of land in the Ganges–Brahmaputra–Meghna Estuary.* UPL.

DPHE. (2001). *Groundwater Studies for Arsenic Contamination in Bangladesh.* Final Report, Rapid Investigation Phase, Department of Public Health Engineering, Government of Bangladesh and Mott MacDonald cs.

DPHE-BGS. (2001). *Arsenic contamination of groundwater in Bangladesh.* Department of Public Health Engineering, Government of Bangladesh and British Geological Survey, rapid investigation phase, Final Report.

FAO. (2011). *Aquastat*.

GSB. (1990). *Geological Survey of Bangladesh*.

Hodgson, G., Ali, R., Turner, J., Ahmed, M., Dawes, W., Masud, M.S., Hossain, M. J., Alam, S., Islam, M. M., & Saha, G. K. (2014). Bangladesh Integrated Water Resources Assessment Supplementary Report. *Water Table Trends and Associated Vertical Water Balance in Bangladesh*. CSIRO.

Kirby, M., Mainuddin, M., Palash, W., Qadir, E., & Shah-Newaz, S. M. (2014). *Bangladesh Integrated Water Resources Assessment supplementary report: Approximate regional water balances*.

MPO. (1987). *Technical Report No 5. The Groundwater Resources and its Availability for Development*. Ministry of Irrigation, Water Development and Flood Control.

Saito, Y. (2005). Uplift of the Himalayas and Tibetan Plateau and large river deltas in Asia: Their characteristics and Holocene evolution. *The Journal of the Geological Society of Japan, 111*(11), 717–724.

Sarker, M.H., Akter, J., Huque, I., Oberhagenmann, K., Kamruzzaman Akand, M. (2021). Formation and Dynamics of Coastal Chars in Bangladesh. In: M. Zaman and M. Alam. (2021). Living on the Edge: Char Dwellers in Bangladesh.

Shamsudduha, M., Taylor, R. G., Ahmed, K. M., & Zahid, A. (2011). The impact of intensive groundwater abstraction on recharge to a shallow regional aquifer system: evidence from Bangladesh. *Hydrogeology Journal, 19*(4), 901–916. https://doi.org/10.1007/s10040-011-0723-4

SIBDP Consortium. (2018, 2023). Regional Implementation Programs.

Thompson, P. M., Sultana, P., & Islam, N. (2003). Lessons from community-based management of floodplain fisheries in Bangladesh. *Journal of Environmental Management, 69*(3), 307–321.

Zahid, A., Qumrul Hassan, M., Imes, J. L., & Clark, D. W. (2009). Hydraulic characterization of aquifer(s) and pump test data analysis of deep aquifer in the affected Meghna river floodplain of Bangladesh. *Environmental Research Journal, 3*(2/3).

Zahid, A. (2015). *Groundwater management aspects in Bangladesh*.

Chapter 6
'Normal' Delta Issues and Challenges

Abstract Bangladesh's main water resource management issues are flood, drought, drainage congestion, riverbank erosion, salinity intrusion, groundwater (both contamination and depletion), land subsidence, cyclone and storm surges, etc. Over time, some of these issues are intensified by anthropogenic interventions, which contribute to the deterioration of water quality and natural systems. On top of these, transboundary issues, add to the complexity of the problems; unfortunately, the magnitude of these problems will only increase over the period in the near future and far beyond due to climate change impacts as described in Chap. 7. As such, human interventions should be restricted or limited so that the actions do not increase the problems more than what nature is already causing. The main issues and challenges of the low-lying Bangladesh Delta are described in this chapter. Bangladesh is a land of rivers originating from the great Himalayan flows through Bangladesh, creating the active alluvial delta country with fertile land. Plenty of rainfall during monsoon and fertile land is the key parameter for developing the country as an agrarian-based livelihood through centuries. Socio-economic features are presented in relation to the delta to highlight other important aspects such as huge population density and growth and the challenge of food security in a relatively small delta area. The found issues and challenges formed the basis to define Hotspots areas, as analytical problem focused on entities. Structuring the issues and challenges led besides the Hotspots to general national as well as thematic issue and challenges. The categories of issues and challenges formed points of departure for the strategy formulation stage.

Keywords Monsoon · Issues · Challenges · River floods · Flash floods · Coastal floods · Urban floods · Hilly region · Groundwater issues · Erosion · Accretion · Food security · Food production · Population · Standard of living · Health · Poverty · Hydrological region · Hotspot · Extended coastal zone · Barind · Drought prone · Major rivers · Haor · Urban · Chattogram hill tracts

J. M. de Heer and G. A. Choudhury, *Strategies for Adaptation to Climate Change in a Transformative Approach*, Springer Climate,
https://doi.org/10.1007/978-3-031-88884-7_6

6.1 Water and Climate-Related Issues and Challenges

6.1.1 Floods

Bangladesh is the most flood-prone country in the world. Three types of floods occur in the country: river floods, flash floods, and coastal floods. It is important to distinguish between the normal rainwater flooding which annually inundates more than 20% of the country. This type of inundations is caused by high-intensity rainfall during the monsoon. These cause relatively little damage or are even beneficial as the silt and biomass that is deposited on the floodplain increases the fertility of the soil. These floods can, however, be aggravated by inadequate drainage, when water levels in the major rivers are high, or by blockage caused by roads and other infrastructure.

6.1.1.1 River Floods

River floods are caused by heavy rains and snow-melting in the upstream catchment of the major rivers: the Ganges, the Brahmaputra, and the Meghna during the wet season from June to October. Normally on an average 20% of the country is inundated annually and, on extreme event, it could be as high as 68%. The flood water rises slowly, spills over the riverbanks, stays for some time and decreases slowly. Catastrophic floods happen when the peak of high flows of the major rivers coincides. The flood inundations map of the 1998 flood is shown in Fig. 6.1.

According to the Flood Forecasting and Warning Centre (FFWC), the floods inundating 30% or more areas occurred on 14 times since 1954, i.e., in 1955, 1974, 1987, 1988, 1998, 2004, 2007, 2015, 2016, 2017, 2019, 2020, 2021, and 2022, while inundation over 25% or more occurred in 22 times as shown in Fig. 6.2.

The extent of flood inundations during the 1987, 1988, 1998, 2007, 2017, and 2020 were 39%, 61%, 68%, 42%, 42%, and 40%, respectively. The flood damages in 1988, 1998, 2004, 2007, 2017, and 2020 were $1.2 billion, $2.8 billion, $6.6 billion, $1 billion, $900 million, and $94 million, respectively. While flood-related fatalities are decreasing, economic losses are on increase over the years (Table 6.1).

An exceptional flood occurred in the north-western region including the Teesta and the Dharla basins in 2021 (Fig. 6.3). The flood in the Teesta in the last week of October 2021 was of very high magnitude; the significance of this extra-high flood was that it was a flood beyond the monsoon or flood season. In previous records, no such flood was reported at the end of October, causing huge losses of properties, livestock, crops, etc. The poor and ultra-poor people living in the char are suffered dramatically. Again, three northern districts of Kurigram, Nilphamary, and Gaibandha experienced flood in 2022.

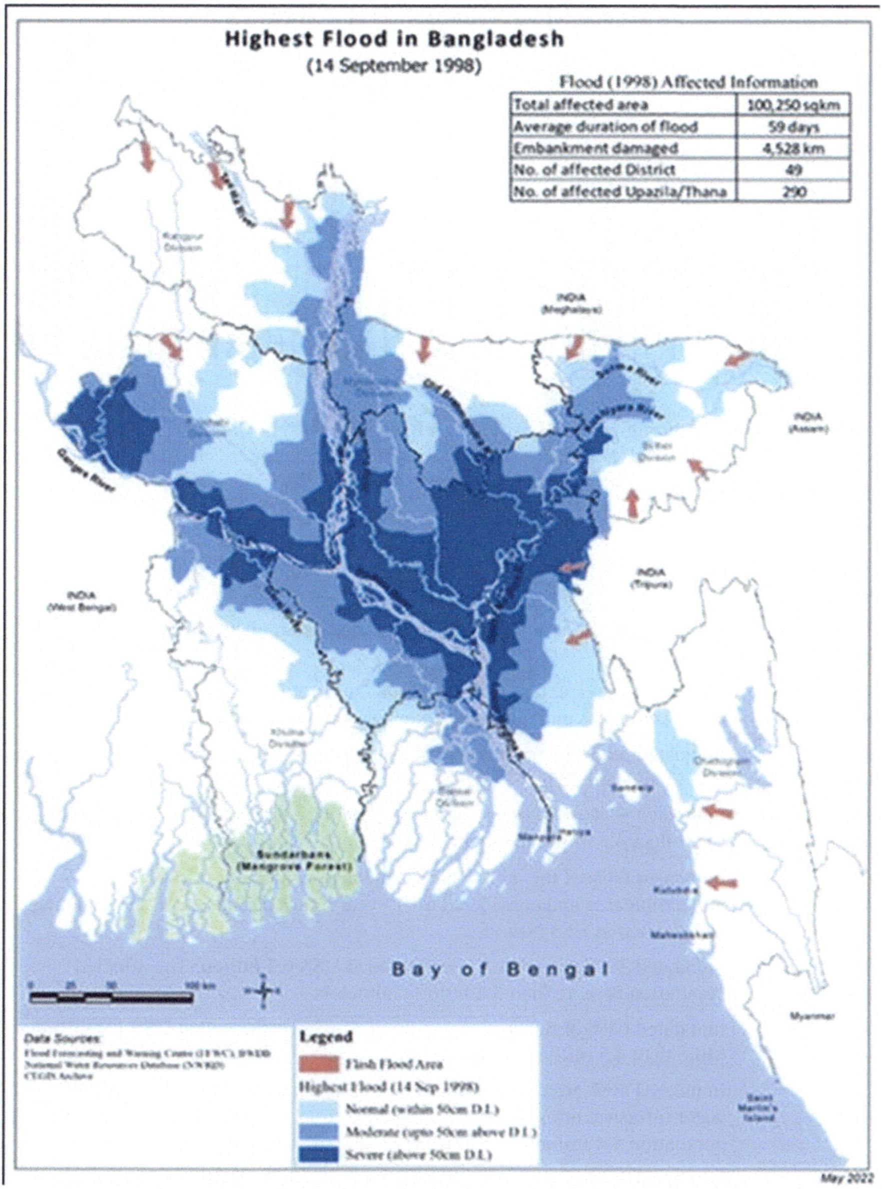

Fig. 6.1 Map showing the areas inundated during the highest ever flood of 1998. *Source* SIBDP Consortium

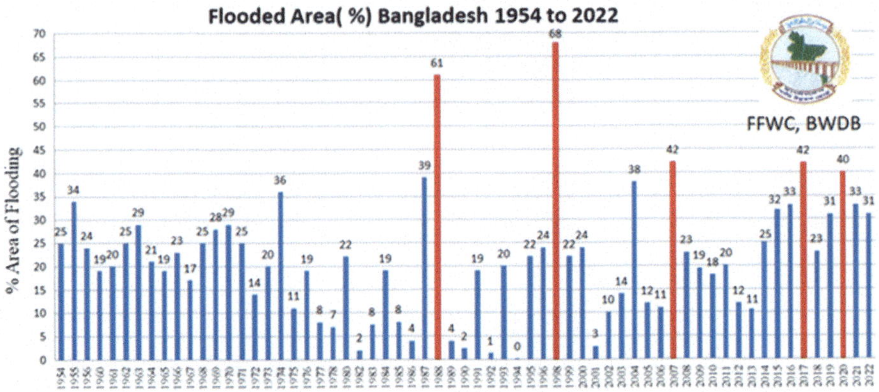

Fig. 6.2 Flooded area during 1954–2020. *Source* FFWC, BWDB

Table 6.1 Notable flood years, impacts and damages

Event	Impact
1954 Flood	Affected 25% area of the country. Reported death of 112 persons
1955 Flood	Affected 35% area of the country. Reported death of 129 persons
1974 Flood	Moderately severe, reported death 1987, affected 38% area of the country, huge Amon loss and after flood famine with over 30,000 death
1987 Flood	Inundated 57,300 km^2 area, direct flood damage estimated at US$ 1.0 billion. Reported death 2055
1988 Flood	Inundated 61% of the area, direct flood damage estimated at US$ 1.2 billion. More than 4.5 million are homeless
1998 Flood	Inundated 68% of the area, which is the highest flood considering the inundation area and considered a 100-year return period. Direct flood damage is estimated at $2.8 billion
2004 Flood	Inundated 38% area, damage estimated at US$ 6.5 billion. The affected population is more than 3.8 million homeless
2007 Flood	Inundated 61% of the area, direct flood damage estimated at US$ 1.2 billion. More than 4.5 million are homeless
2020 Flood	Inundated 40% area, Crop damage 0.15 million ha worth US$ 42 million, water infrastructures damage estimated at 5212.95 million taka. Affected population 5.4 million in 158 Upazila of 30 districts. Reported death 41
2022 Flood	Inundated 31% area, damage estimated at 45,456.13 million taka (including damage to Water infrastructures, Embank erosion, etc. 4725.45 million Tk). Crop damage worth 12,910 million Taka. Livestock losses reached US$ 27.84 million. Affected population 7.2 million. People evacuated to safe places or shelters 390,078

Source FFWC, BWDB

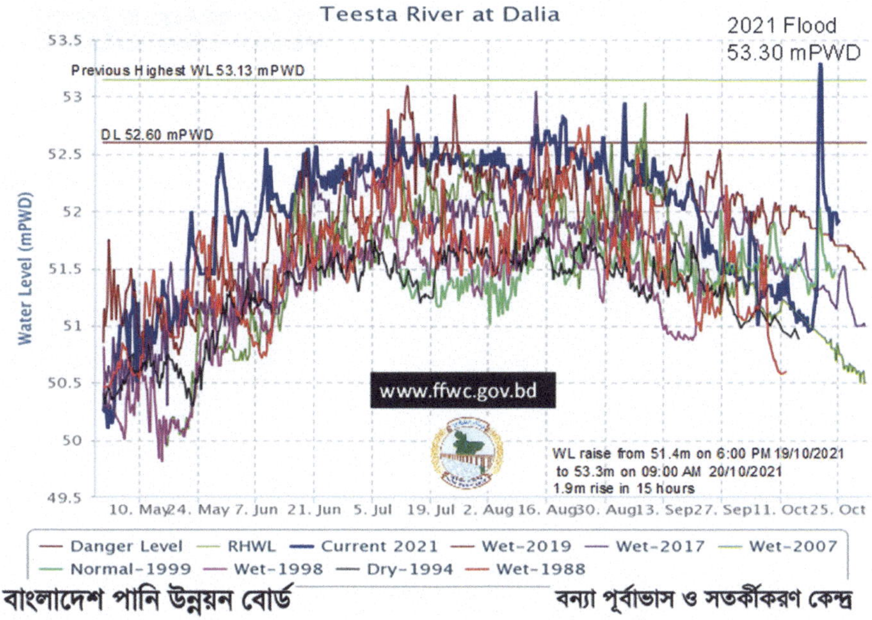

Fig. 6.3 Teesta flood situation in October 2021. *Source* FFWC, BWDB

6.1.1.2 Flash Floods

Flash floods occur mainly due to heavy rainfall in the upstream hilly areas, especially in the north- and south-eastern regions, as well as in the eastern hilly region. Flash floods have a relatively short duration, but generally have high velocities and a rapid increase in water levels, causing widespread destruction. Some case studies of these types of floods are described below.

North-Eastern Flash Floods: The north-eastern region, including haor areas are prone to flash floods regularly, damaging the standing boro crops and infrastructure. These floods occur almost at regular intervals of three to four years. A flash occurred in the region in 2017, it was estimated about \$1.49 billion worth of crops were damaged besides damages to the infrastructure. The north-eastern region experienced one of the worst floods in 2022 inundating about 90% of the areas. The highest water level was recorded on the Surma at Sylhet. This flood had four peaks during April, May, and June (Fig. 6.4). This flood affected about 7.2 million people, and caused damage to 1133 km^2 croplands, 44,254 ponds, 44,254 water ponds, and 49,885 sanitation facilities. The loss of livestock was estimated to be US \$27.84 million. Around 3600 schools suspended activities, and 480,000 people were displaced.

Flash floods in the hilly region: The eastern hill regions are also very prone to flash floods. From 1985 to 2023, about 15 flash flood events occurred in the region. In the first part of August 2023, a prolonged severe flash flood occurred repeatedly inundating the township of Bandarban, and damaging the croplands, roads, bridges,

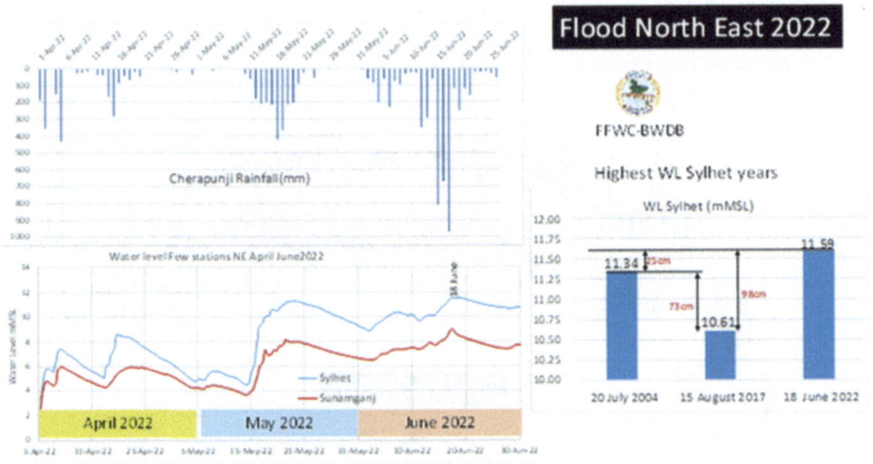

Fig. 6.4 Flood 2022 in the north-east. *Source* FFWC, BWDB

railroads, and houses. About 1.3 million people were affected in four districts: Chattogram, Cox's Bazar, Rangamati, and Bandarban. The peak water level 19.25 m MSL was recorded on August 8, 2023 on the Sangu at Bandarban (Fig. 6.5). The previous highest water level occurred in July 2019.The ever-highest water level 15.44 m MSL was recorded at Lama on the Matamuhuri in August 2023 (Fig. 6.6). The previous highest water level was recorded in 2019.

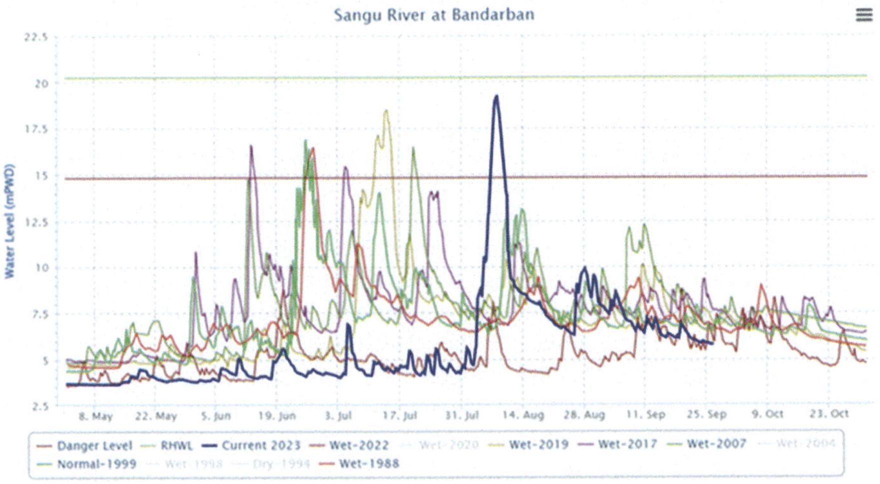

Fig. 6.5 Highest water level at Bandarban was in August 2023. *Source* FFWC, BWDB

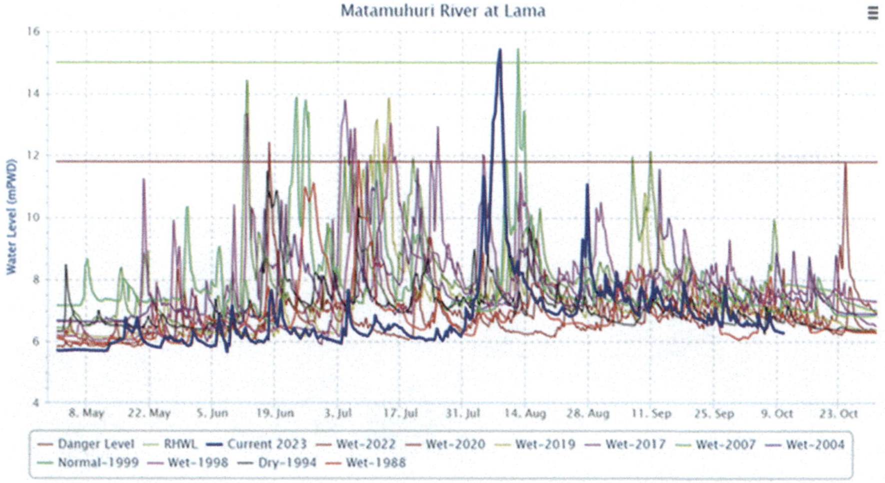

Fig. 6.6 Highest water level at Lama on the Matamuhuri was in August 2023. *Source* FFWC, BWDB

6.1.1.3 Coastal Floods

Cyclone-induced tidal surges occur in the coastal areas consisting of large estuaries and low-lying lands. The surges mainly occur during the pre- and post-monsoon periods. The coastal flood could be catastrophic when high tides and cyclonic storm surges occur at the same time. Tidal/coastal floods last for a short period, and block inland drainage. Only the western part is relatively better protected because of the extensive mangrove forest 'Sundarbans' naturally reducing wave heights. Tidal floods, for example during spring tide, may cause severe loss on unprotected coastal lands as well as sometimes breach earthen embankments, leading to potentially more harm. The northern and eastern part of the coastal zone is also prone to riverine floods, often aggravated by coincident high tides. The Chattogram coastal floodplain is particularly prone to flash floods, coming from the Hill Tracts.

6.1.1.4 Urban Floods

In recent years, urban floods have become an environmental and economic concern for Bangladesh. With growing urbanization, more cities are prone to urban floods as many are in low-lying floodplains, and rainfall extremes are increasing. In 2004, Dhaka saw record-high daily rainfall of 341 mm in 24 h, causing widespread flooding. The event affected over 80% of the city and over 5 million people. Climate change

Fig. 6.7 Two additional pump houses to keep Dhaka dry. *Photo* J.M. de Heer

projections indicate an increasing trend of short-term heavy rainfall in urban catchments, intensifying the risk of urban flooding and waterlogging.[1] Two additional pump houses have been constructed in Dhaka; see Fig. 6.7.

6.1.1.5 Flood Forecasting

The FFWC-BWDB has carried out flood-related inundation and forecasting activities. Annual flooding area has been recorded by FFWC using the model-generated Flood Map of Bangladesh. Since 1954, the big mega flood numerical is presented in Table 6.1. The table indicates the maximum area of inundation that occurred during 1998, that was 68% and next to it was 1988, that was 61%.

Figure 6.8 shows an affected area in percentage, in different return periods of floods.

[1] Ministry of Environment, Forest, and Climate Change (2022).

Fig. 6.8 Affected area percentage in different return period flood years. *Source* FFWC, BWDB

Table 6.2 Major waterlogged areas in the southern region

Districts	Total area of District (ha)	Total area of the upazilla affected area	Water-logged area		
			No. of upazilla	Area (ha)	% affected upazilla
Jessore	260,694	94,911 ha	3	15,700	16.5
Khulna	438,900	51,245 ha	2	19,023	37.1
Satkira	381,729	96,298 ha	2	33,470	34.8

Source FAO 2015

6.1.2 Drainage Congestion and Waterlogging

Waterlogging is a problem over large parts of the coastal zone, especially in the southwest (Satkhira, Jessore, Khulna and Bagerhat) and southeast (Noakhali, Feni) coastal zones (Alam et al., 2018c). The reason for this is complex. Numerous morphologically active tidal rivers characterize the southwest region, the main drainage networks for coastal polders and low-lying beels. This region is characterized by a vast network of morphologically active tidal creeks, the main drainage route for coastal polders and low-lying beels. Thus, the natural drainage pattern of the area is predominantly characterized by the influence of the incoming tide from the sea. Tidal flow brings a huge quantity of silt from the sea into the river systems of the coastal area.[2] The distribution of waterlogged areas of eight selected upazilas under three districts is stated in Table 6.2.

It may be mentioned here that, before polderization in the early 1960s and 1970s, major parts of incoming silt deposited naturally on the low-lying depression areas (beels). On the other hand, there was a significant amount of freshwater flow from the

[2] Alam (2018b).

Ganges, which helped to maintain the perennial tidal rivers of the southwest region. The continued freshwater flow from the Ganges helped flush the incoming sediment with the high tide from the sea; thus, the proper drainage capacity of these tidal creeks was maintained naturally. This natural process seemed to be hampered significantly after polderization and a significant reduction of freshwater flow from the Ganges, especially during the dry period. Also, the reduction of dry season flow downstream of the Ganges and its distributaries due to the withdrawal of water upstream by operation of Farakka Barrage contributed to increased sedimentation in the river beds and tidal creeks significantly in the southwest part of the country. This reduction of fresh water had increased the salinity concentration, proportionately increasing sediment concentration and simultaneously silt deposition in the peripheral rivers during ebb tide.

6.1.3 Drought-Prone Areas

The regions located along the western border of Bangladesh are most vulnerable to meteorological droughts. The area vulnerable to drought is shown in Fig. 6.9. Drought is expressed as the number of days without rainfall. Although droughts are not always continuous in any area of Bangladesh, they occur more frequently in the low rainfall zones.

Occurrences of agricultural drought as a major water deficiency-related issue are most profound in the north-west region, which has the lowest average annual rainfall in the country. The average occurrence of drought is once in 2.5 years, and between 1960 and 1991, nineteen droughts occurred in Bangladesh.

6.1.4 Salinity Intrusion

Coastal zone of Bangladesh is susceptible to salinity intrusion. The coastal zone spreads over 19 coastal districts having an area of 38,971 km^2, out of which 6.35 land is below 1 m, 33.67% land between 1 to 2 m, 22.36% land between 2 to 3 m, and the rest 37.62% land above 3 m PWD. The population of the coastal zone is 43.82 million people and average density is 930 per square kilometer (BBS). It includes the largest mangroves forest in the world, the Sundarbans having an area 6017 square kilometer, sandy beaches, coral reefs, dunes, and wetlands. 139 coastal polders were built to protect an area of 1.2 million hectares of lands from tidal flooding and salinity.

At present 1 ppt salinity front exist in the Barguna and Patuakhali districts and 5 ppt salinity near shore areas of Barguna, Patuakhali, and Bhola districts in the south-central region. In case of sea-level rise, the extent of saline front will move further inland. 1 ppt salinity front will move further 9–13-km inward in the south-west and south-central regions, while 5 ppt salinity front will move 25–37 km inward in the

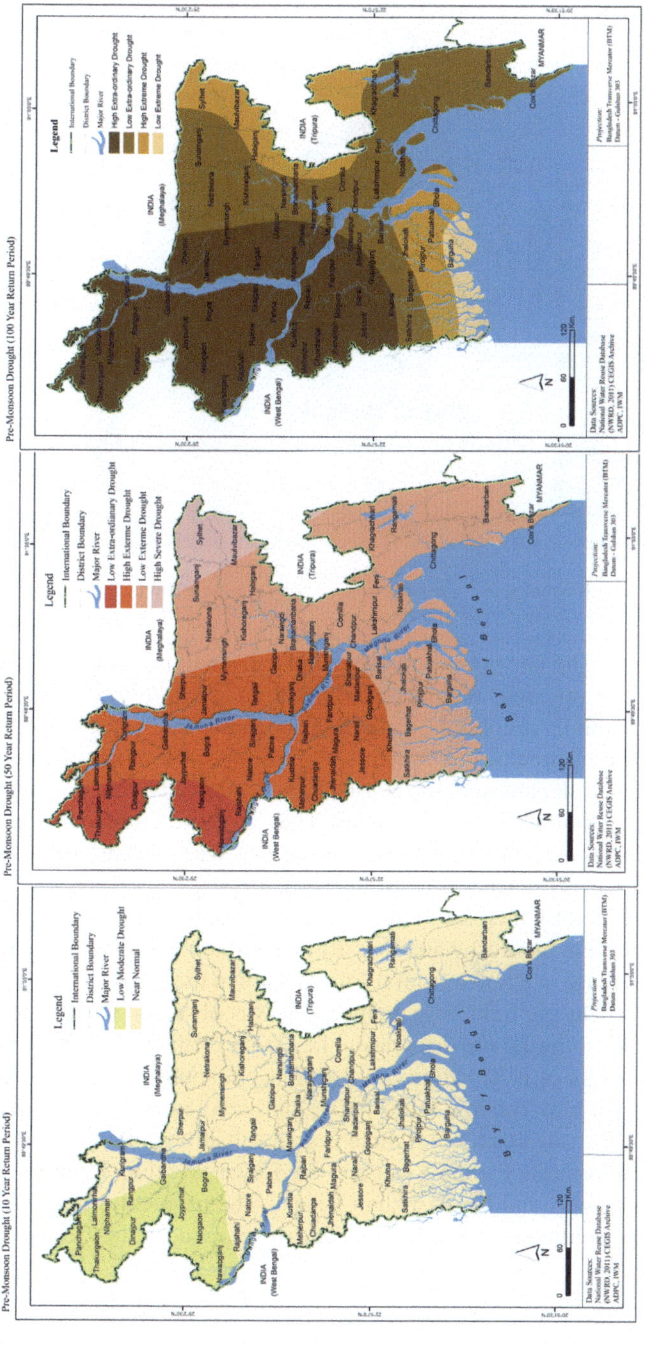

Fig. 6.9 Drought map of Bangladesh

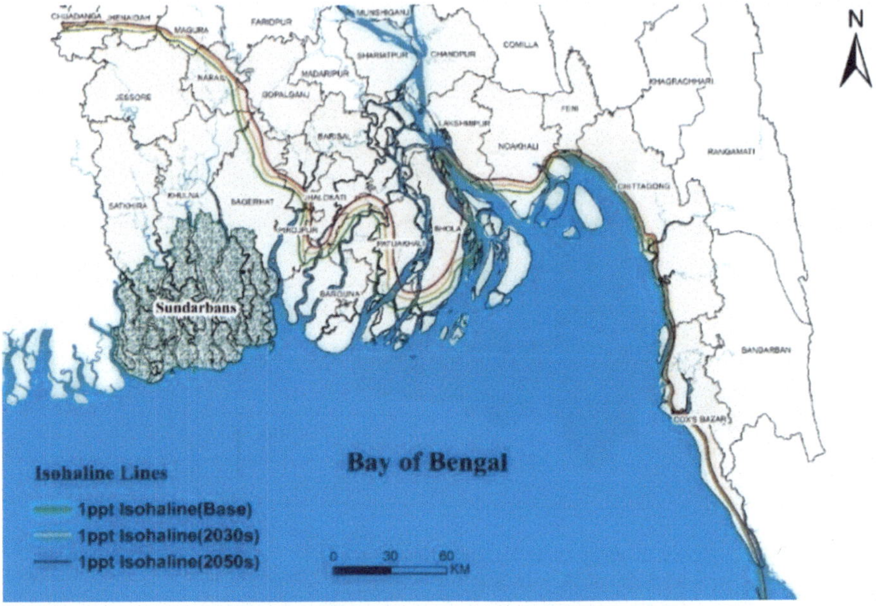

Fig. 6.10 Isohalines with equal salinity points. *Source* SIBDP Consortium

near and mid-term.[3] The one ppt salinity-affected areas will increase by 7.5% in the 2050s, while the five ppt areas will increase by 9% under SSP5-8.5[4] (Fig. 6.10).

The river water salinity in coastal Bangladesh depends on the volume of freshwater discharges from the upstream river systems, the salinity of the Bay of Bengal near the coast, the circulation pattern of the coastal waters induced by the ocean currents, and the tidal propagation to the river systems. Thus, average salinity concentrations of the rivers in the coastal area are higher in the dry season than in the monsoon because of the lack of freshwater flow from upstream.

Salinity level generally increases almost linearly from October to late May with the gradual reduction in the freshwater flow from upstream. Observations of salinity show a salinity increase in Khulna from 0.7 to 16.8 ppt in the Rupsa River from 1962 to 2011 (50 years). During field measurements, it was observed that the maximum river salinity occurred during the dry season, i.e., from October 2010 to May 2011.

However, the levels of river salinity in the adjacent south-central and southeast coastal zones remain low. The low salinity (0–2 ppt) level in the south-central zone, i.e., in much of the Barisal area for the whole year, results from the significant volume of freshwater flow from the Padma River and the Lower Meghna River through the Arialkhan, the Buriswar, and the Bishkhali Rivers. The problem of river salinity is most severe in the southwest zone. Salinity intrusion reduces the freshwater area,

[3] CEGIS Technical Note: *Impacts of projected Sea Level Rise on coastal embankments.* 2021, Dhaka, Bangladesh.

[4] Ministry of Environment, Forest, and Climate Change (2022).

decreasing agricultural production in many parts of the coastal zone, especially the Khulna region and the extreme south of the Patuakhali region and locally in the Noakhali and Chattogram regions. An increase in salinity damages the freshwater fish habitat and adversely impacts the Sundarbans ecosystems.[5]

6.1.5 Cyclones and Storm Surges

The entire southern boundary of Bangladesh is exposed to the Bay of Bengal and eventually to the Indian Ocean. Nearly every year, cyclones hit the coast and cause damage to the lives and properties of the coastal population. The cyclone that hit in 1970 at greater Barisal caused nearly 300,000 deaths. However, the one with almost the same severity in terms of wind speed hit in 2007 (the Sidr) took the lives of nearly 3406 people. This is credited briefly to improved cyclone warnings, more cyclone shelters, awareness building, improved communication systems (road communication and telecommunication), and highly developed disaster risk management capabilities.

In 2020, Cyclone Amphan affected more than a million people in 26 districts and caused 26 deaths. It damaged 55,667 houses, 149,000 ha of agricultural lands, 1,80,500 hatcheries, 150 km of embankments, 200 bridges and culverts, and 100 km of roads, causing a total loss of BDT 11 billion. In future, a 50 cm sea-level rise combined with a Sidr equivalent cyclonic storm surge could potentially inundate large parts (11–12%) of the coastal area of Bangladesh.[6] Previously, Cyclone Sidr (2007) created a similar kind of devastation in the coastal areas of Bangladesh.

After Sidr, an extensive program called the Comprehensive Disaster Management Program (CDMP) was launched to address all main hazards and to develop ways for disaster management in Bangladesh. In CDMP phase II from 2013 to 2018, the Storm Surge Study was conducted, in which hazard maps were developed for all 19 coastal districts up to the upazila/union level.

Recent major cyclones include Roanu (2016), Mora (2017), Fani (2019), Bulbul (2019), Amphan (2020), and Yaas (2021) (Fig. 6.1). Due to frequent disasters, crop production is much lower, and the poverty rate is much higher in the coastal zone.[7] Figure 6.11 shows the cyclonic storm tracks over the years.

6.1.6 Erosion and Accretion–Riverbank and Coastal

The rivers are very dynamic in nature. Figure 6.12 shows erosion-prone areas of the major rivers. It is estimated, on average, that some 6000 ha of major river banks are

[5] Alam (2018b).

[6] Ministry of Environment, Forest, and Climate Change (2022).

[7] Alam (2018b).

Fig. 6.11 Bangladesh cyclonic storm tracks. *Source* SIBDP Consortium

eroded annually, leading to the displacement of about 50,000 people. The net erosion and accretion for 1973–2020 (47 years) is estimated to be 310,000 ha.

The erosion rate increased to an average of 4820 ha/year from 1980 to 1989 and decreased to 1500 ha/year from 2000 to 2010. The rapid drop-in riverbank erosion rate was partly attributed to natural processes and partly to human interventions, starting with the construction of major riverbank protection schemes during the mid-1990s.[8] Figure 6.13 is showing the erosion and accretion in the major rivers. The chart indicates that erosion along the Brahmaputra/Jamuna Rivers was significantly higher from 1973 to 2021. Along the Padma and Lower Megna up to the estuary, erosion is higher than the accretion. Erosion accretion along the Ganges River during 1973–2020. Net erosion is higher than the accretion. The erosion and accretion along the Brahmaputra/Jamuna River during 1973–2020 are shown in Fig. 6.14. From this figure and map, it is evident that net erosion is much higher than the accretion.

Erosion accretion along the Ganges River during 1973–2020 is shown in Fig. 6.15. Net erosion is a little higher than the accretion. The BWDB information showed that along the common border 110 km of the Ganges River, about 77 km of bank protection or revetment work had been implemented up to 2021, so this part mainly erosion stopped. In the part of the Ganges River inside Bangladesh up to Goalondo, there are few erosion-vulnerable reaches.

The Meghna estuary is a delta-building estuary, the size of which is ever-increasing. On the other hand, the north part of the estuary is eroding. About 40% of Sandwip Island in the east was eroded, and there were considerable losses in the north of Hatiya, north-east of Bhola, and the south-west of the former Ramgati island. Beyond the estuary, the rates of coastal changes are small or undetectable. On the south-western coast, there have been small amounts of erosion locally, with amounts generally increasing westward towards the Hooghly estuary in India.

Preventing erosion in the Meghna estuary is difficult due to its highly dynamic nature. The lower Meghna River is eroding the left bank along the Ramgati and Kamalnagar Upazila under the Laxmipur District, and the erosion rate is in the range of 11–27 ha/km/yr over a length of 31 km. Bhola Island has been experiencing erosion of over 80 km, and about 5640 ha of land was devoured by the river from 2003 to 2014.[9]

In the lower part of the Lower Meghna and estuary, the net erosion was slightly higher than the erosion; see Fig. 6.16, during 1973–2020.

6.1.7 Groundwater Issues

The shallow groundwater in the southwestern coastal region is generally too saline for domestic or irrigation use due to connate (inborn) salts or estuarine flooding. However, sufficient flushing of saline water has taken place in isolated pockets to

[8] Alam (2018b).

[9] Alam (2018b).

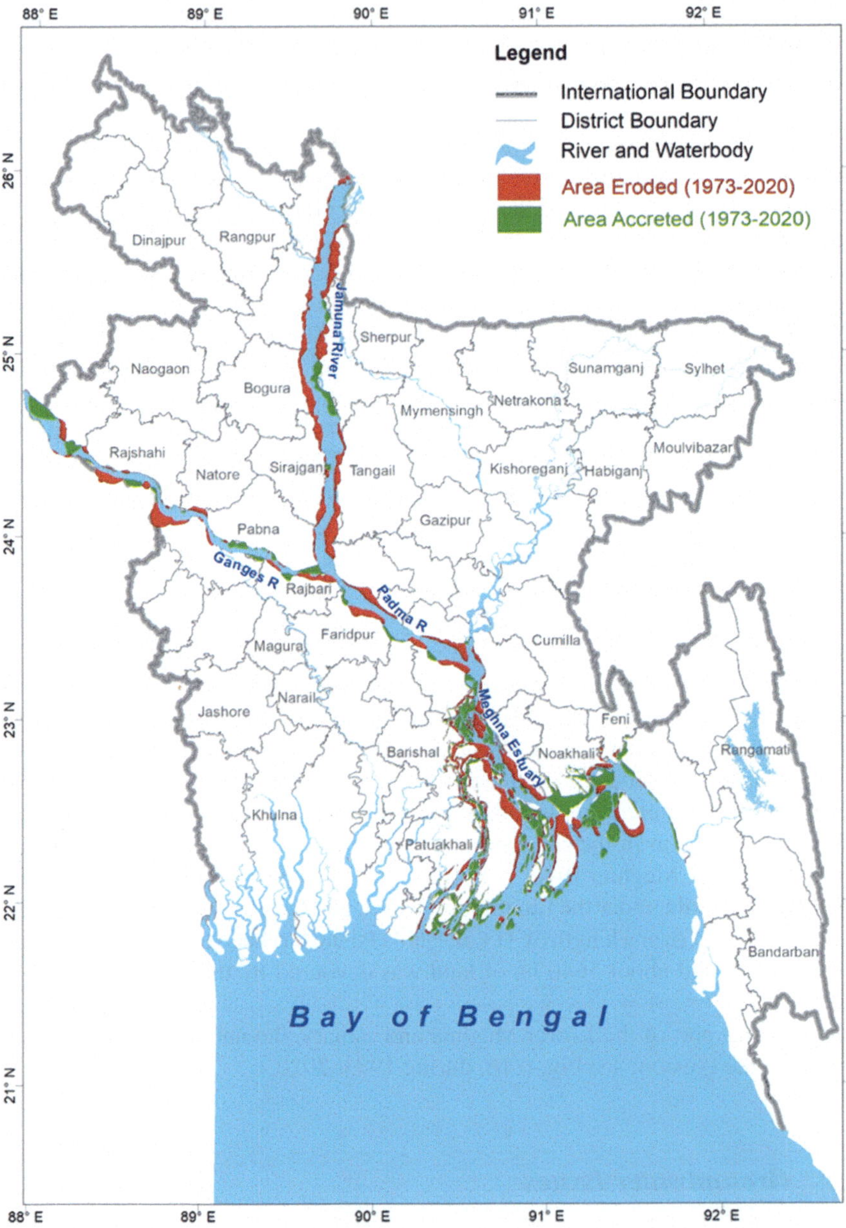

Fig. 6.12 Map showing the erosion and accretion in the major rivers, the Ganges, the Brahmaputra/ Jamuna and the Padma and the Lower Meghna Rivers and the coast of Bangladesh. *Source* SIBDP Consortium

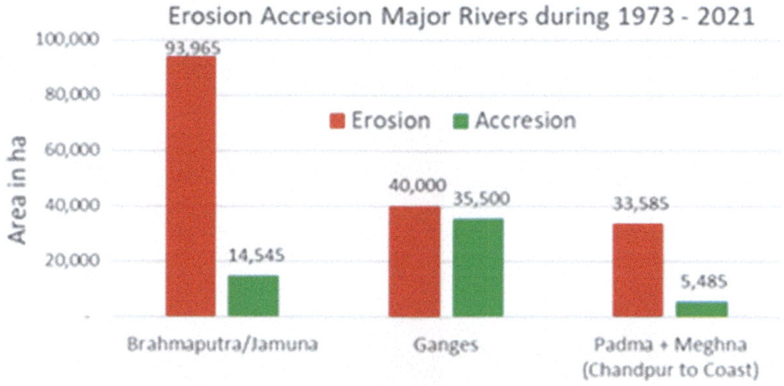

Fig. 6.13 Erosion and accretion in major rivers. *Source* SIBDP Consortium

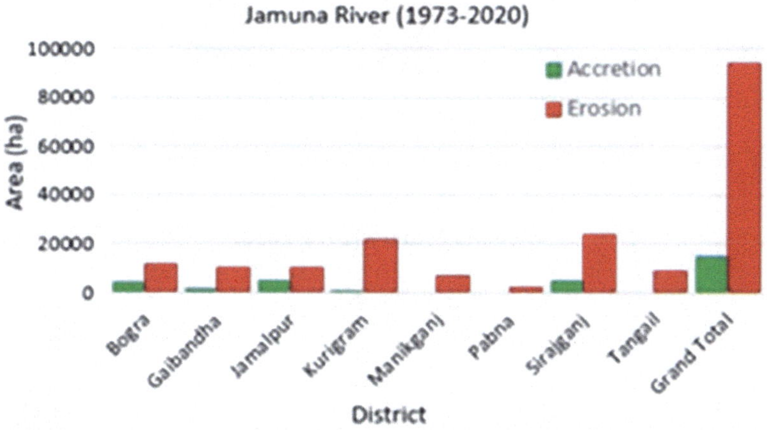

Fig. 6.14 Erosion and accretion in the Brahmaputra/Jamuna River. *Source* SIBDP Consortium

enable a limited domestic use of freshwater in the shallow aquifer. In the deep aquifer pattern, salinity distribution is more uniform on a regional basis, as is the continuity of the aquifer. The change from potable to very saline water is sharp and occurs over a relatively short distance. Groundwater quality monitoring studies by the Bangladesh Water Development Board (2006), the Department of Public Health Engineering (1999), and the hydrogeological investigation by the British Geological Survey (1999) revealed that groundwater is generally fresh outside the coastal zone. However, 25% of the population was exposed to contamination exceeding Bangladesh Standards in 2001 and about 14 districts are affected by salinity in the coastal belt, with an affected population of about 12 million people, in accordance with the BWDB study, 2013.[10]

[10] Alam (2018a).

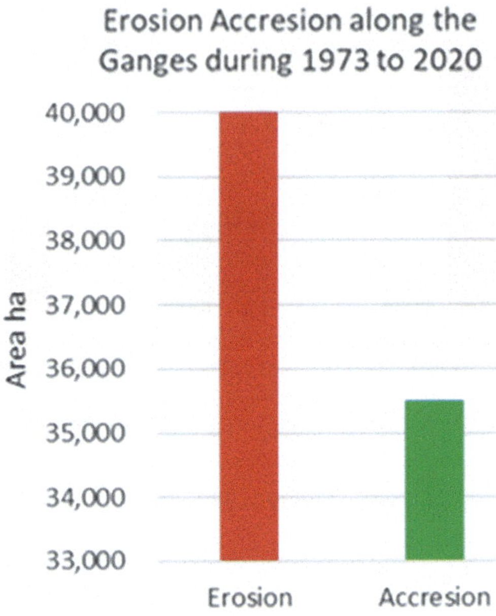

Fig. 6.15 Erosion/accretion along the Ganges. *Source* SIBDP Consortium

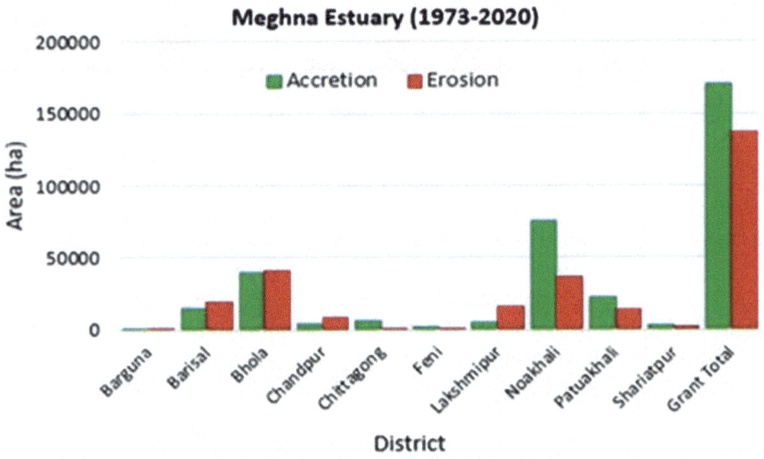

Fig. 6.16 Accretion and erosion in the Meghna estuary. *Source* SIBDP

The result of the intensive groundwater extraction has been a gradual (in the case of the Dhaka area, rapid) decline of both pre- and post-monsoon groundwater tables. In intensively irrigated areas, groundwater level fluctuates between 5 to 15 m below ground level, in some places even up to 23 m during the dry peak irrigation season. In Dhaka city, groundwater levels have fallen to 80 m below ground level in the central

Table 6.3 Tube well development in Dhaka City

Year	1970	2000	2005	2014
Nos. of Tubewell	49	300	423	690
Withdrawal Mm^3/day	0.18	1.2	1.5	2.5
Water Table (m)	0.5–10.5	19–41.5	19–54	24–71

Source SIBDP Consortium

part, as 85% of the water supply depends on groundwater. The apparent reason is the explosive growth of the number of tube wells for drinking water supply. Table 6.3 indicates the trend of tube well development in Dhaka from 1970 to 2014.

6.1.8 Land Subsidence

The lower deltaic area of Bangladesh is located on two active troughs, Faridpur Trough and Hatiya Trough. Although most of the Bengal Basin is slowly subsiding, the troughs are subsiding more rapidly.[11]

6.2 Socio-economic Delta Features and Issues

6.2.1 Population

Socio-economic features are relevant to developments in the delta area. The population of Bangladesh is 169.83 million (BBS census 2022a, 2022b), with a density of 1119 persons per sq. km, the highest in the world; Fig. 6.17. Out of this, 84.08 million are male, and 85.65 are female. The chart shows the increasing trend of the population since 1974 (76.4 million) in census years and 2022 (169.83 million). The increasing rate has been reduced in recent decades.

During first decade of this century, the population growth rate was 1.37%. The population is on the increase. The population is projected to rise to 202.9 by the year 2050 and then it will decrease as projected by the UN. It is expected that the growth rate will be zero in 2060. As per projection, the population of Bangladesh will be 173.55 million with a negative growth rate of -2.84% by the year 2100.

[11] Ibid.

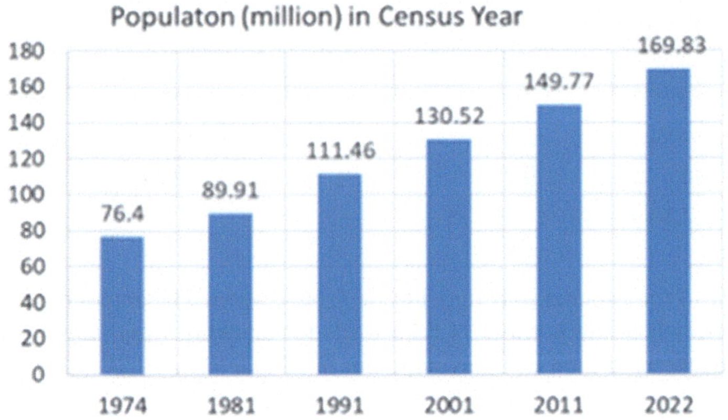

Fig. 6.17 Bangladesh's population growth. *Source* BBS

6.2.2 *Food Production*

Food production in Bangladesh has been constantly increasing, surpassing the population since independence in 1971 (Alam et al., 2018d). From 1971, the population increased 2.3 times, and food production increased more than 3.5 times. The chart in Fig. 6.18 shows the increment of rice production from 2005–06 to 2021–22.

In addition, the production of fish, meat, milk, fruit, vegetables, potatoes, etc., also significantly increased. Previously, food security meant rice and pulses (in Bangla Dal Vat); recently, food security meant nutrition, a balanced diet with rice, vegetables, meat, fruit, etc. (Alam et al., 2018d). The following chart in Fig. 6.19 shows the increment of agricultural food production since 1971.

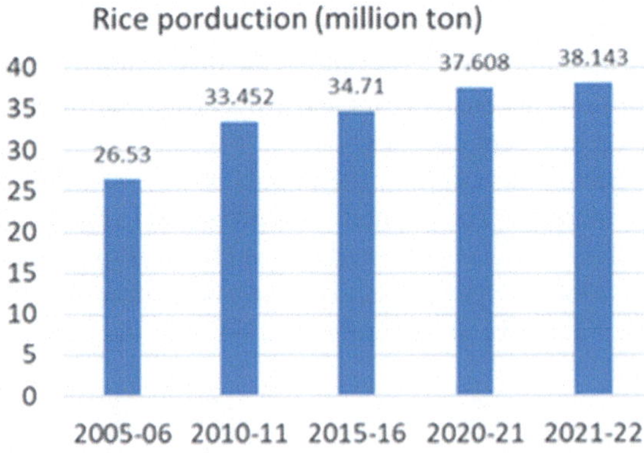

Fig. 6.18 Rice production in Bangladesh. *Source* DAE and DLS data

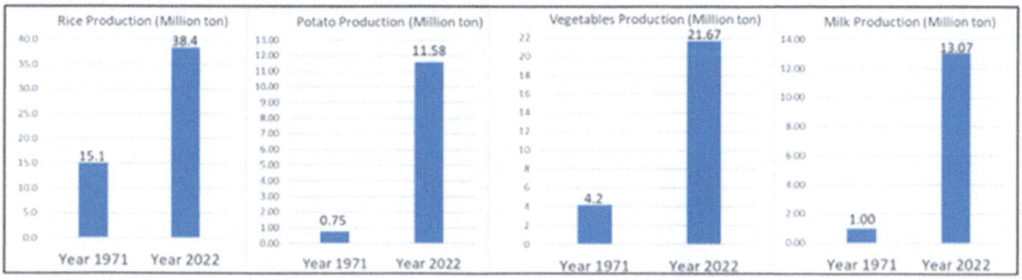

Fig. 6.19 Increment of food production since 1971. *Source* DAE and DLS

6.2.3 Standard of Living

The Household Income and Expenditure Survey (HIES) 2022 (BBS) data shows that the household's average monthly income was Taka 11,479 in 2010 and Taka 15,988 in 2016 and it stood Taka 32,422 in 2022. Income has increased 2.82 times during the 2010 and 2022 period; see Fig. 6.20.

The average household size is 4.30 in the rural areas and 4.18 in the urban areas. The average household size is bigger than that of urban areas. Average monthly per capita income increased to Taka 7614 in 2022, while this was Taka 6091 in 2016 and Taka 2553 in 2010. The household-level access to electricity increased to 99.3% in 2022 from 75.9% in 2016 and 55.3% in 2010; see Fig. 6.21.

Similarly, 92.3% of HHs have access to improved toilet facilities, and 96.1% have access to improved drinking water sources.

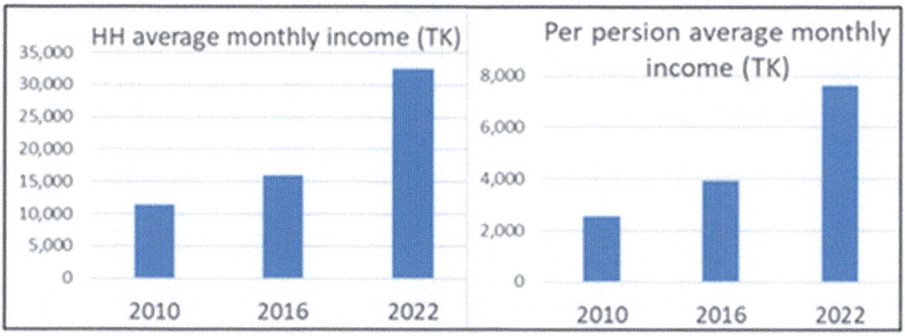

Fig. 6.20 Household and per person average monthly income. *Source* Survey HIES 2022; BBS

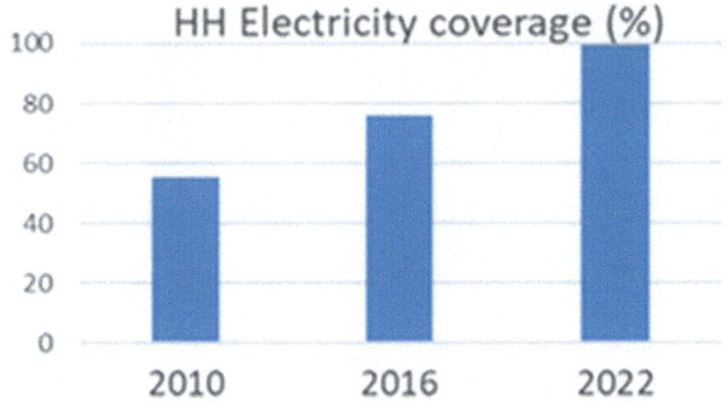

Fig. 6.21 Household-level electricity coverage. *Source* HIES 2022; BBS

6.2.4 Health

Concerning food consumption, the HIES 2022 (BBS data) illustrates that the share of food and non-food consumption expenditures in the households has changed. Non-food expenditures are increasing gradually. The percentage of food consumption expenditure is 45.8%, and non-food consumption expenditure is 54.2% in 2022, compared to 47.7% for food and 52.3% for non-food in 2016.

The life expectancy rate from 2010 up to 2022 is gradually increasing, except from 2020 to 2021 (Fig. 6.22). That might be due to the pandemic 'COVID-19' situation. The chart also revealed that life expectancy is higher for females than males. Female life expectancy was 68.8 years in 2010, which increased to 74.2 years in 2022, an increase of 5.4 years. Life expectancy for males was 66.6 years in 2010, which increased to 70.4 years in 2022, an increase of 3.8 years.

During this period, 2010–2022 (12 years), the rate of increase for female life expectancy (increased by 5.4 years) is higher than that for males (increased by

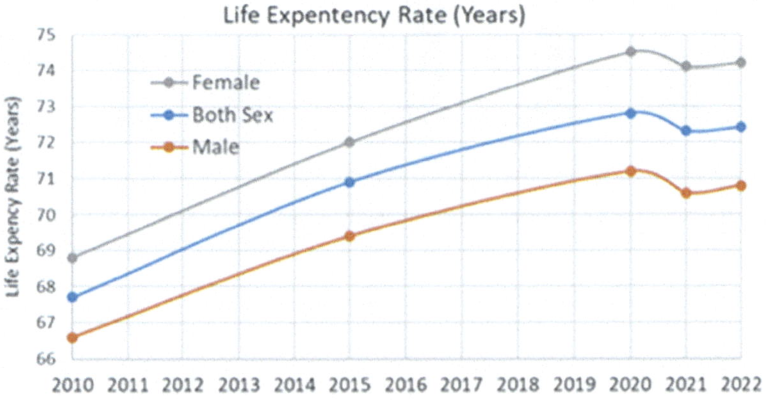

Fig. 6.22 Life expectancy rate. *Source* HIES 2022; BBS

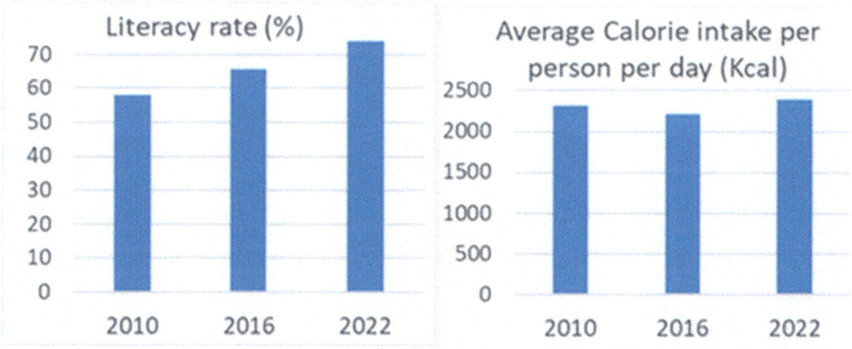

Fig. 6.23 Literacy rate and average calorie intake per person per day. BBS (2022a, 2022b)

3.8 years). It is assumed that the increment is due to better health access, better food security (or nutrition), water supply sanitation, and improved living conditions.

The literacy rate (persons of 7 years and over) rose significantly to 74.0% in 2022 from 65.6% in 2016 and 57.9% in 2010; see Fig. 6.23.

The calorie intake per person per day was 2393.0 kcal in 2022, 2210.4 kcal in 2016, and 2318.3 kcal in 2010.

6.2.5 Poverty

The poverty rate was as shown in Fig. 6.24 48.9% and the extreme poverty rate was 34.3% in 2000. During the period from 2000 to 2022, the poverty rate has reduced from 48.9 to 18.7% and the extreme poverty rate reduced from 34.3 to 5.6% (BBS 2022a, 2022b). In the rural areas the poverty rate is 26.4%, which is higher in comparison with the urban poverty rate of 18.9%. The extreme poverty rate in the urban areas is 7.6%, and in the rural areas is 14.9%.

The spatial distribution of poverty is important for national-level planning and development efforts. Considering the division-wise, the poverty rate is highest in the Rangpur division at 47.2% and lowest in the Dhaka division at 16.0%, (Fig. 6.25). Considering district-wise, the poverty rate is highest in Kurigram District at 70.4% and lower in Gazipur District at 6.9%. Extreme poverty is also highest in Kurigram 53.9% and least in Dhaka 1.7%.

The reason for this disparity of spatial distribution of poor and extreme poverty in the northern district may be attributed to the frequency of high natural disasters like floods, erosion, low rainfall, drought, extreme heat and cold, etc. Agriculture is the prime livelihood of this region, and natural and climatic disasters or calamities have a high adverse impact on income, livelihood, and poverty.

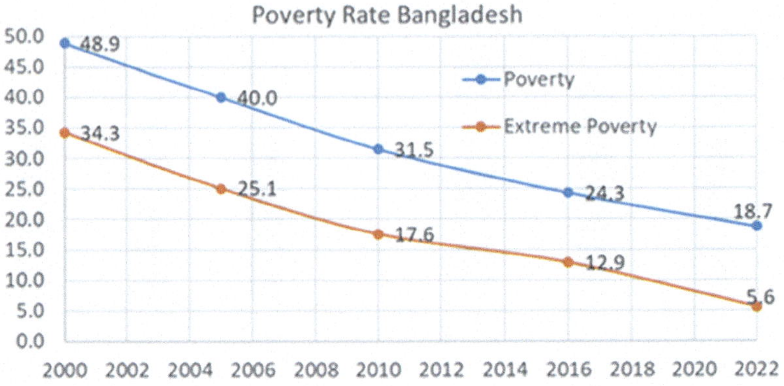

Fig. 6.24 Poverty rate in Bangladesh 2000–2022. *Source* SIBDP Consortium

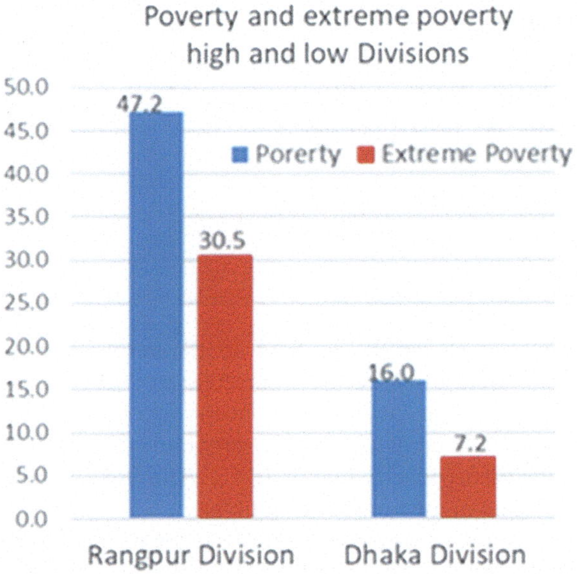

Fig. 6.25 Poverty and extreme poverty in Rangpur and Dhaka division. *Source* SIBDP Consortium

The higher-level goal of the BDP 2100 is to eliminate extreme poverty by 2030. The BDP 2100 program and projects essentially address this goal of achieving a poverty-free nation by 2030 and making a prosperous delta by 2041.

Concerning urban and rural population, the HIES survey (BBS) found that 68.34% of the population (116.07 million) is living in rural areas, and 31.6% of the population (53.76 million) is living in urban areas (BBS census 2022a, 2022b). Population is

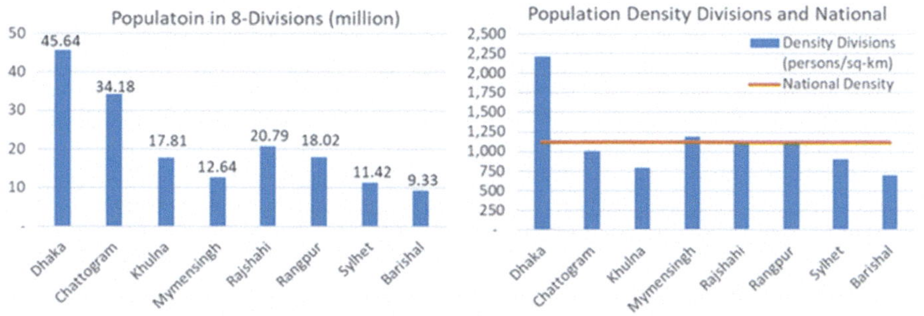

Fig. 6.26 Division-wise population. *Source* SIBDP Consortium; (HIES Survey BBS 2022a, 2022b)

increasing rapidly in urban areas. With accelerated development activities, urbanization is increasing rapidly; the urban area is also increasing along with the population density.

Considering the rapid rate of increasing urban population, the government has emphasized development in the rural areas with the betterment of education facilities, health, power supply, water and sanitation, communication, etc., in the rural areas to decelerate the population migration from rural to urban areas.

The country-wise, that the highest number of people living in Dhaka division (26.88%), followed by Chattogram division (20.13%) and Rajshahi division (12.24%) and the lowest population living in Barisal division (9.99%). The population density is also highest in the Dhaka division at 2216 persons per sq. km, which is almost double of the national population density of 1119 persons per sq. km. The population density in Mymensingh division is 1194 persons per sq. km, which is higher than Rajshahi division, 1145 persons per sq. km. The lowest density is in Barishal division 705 persons per sq. km (see Fig. 6.26).

6.2.6 GDP Contribution from Sectors

With a huge population living in a small country with land of 147,570 km^2, for a very long period, the contribution from the agricultural sector dominated the GDP. In recent decades, industry and service sector development accelerated, and thus, the GDP is dominated by industry and service sectors Alam et al. (2018e). Sectoral shares of GDP (BBS) are in the chart below. However, the agricultural sector provides most of the workforce or jobs. The sectoral contribution to GDP from agriculture and services sectors has gradually reduced, while the contribution from industry sector has gradually increased since 2015–16 (Fig. 6.27).

Although agricultural (crop, fish, poultry, livestock, fruit, etc.) production is consistently increasing, the size and output from industrial growth are higher than that of the agriculture; thus, the percentage of the agricultural sector's contribution in the total GDP is reducing, while the contribution of the service sector remains almost

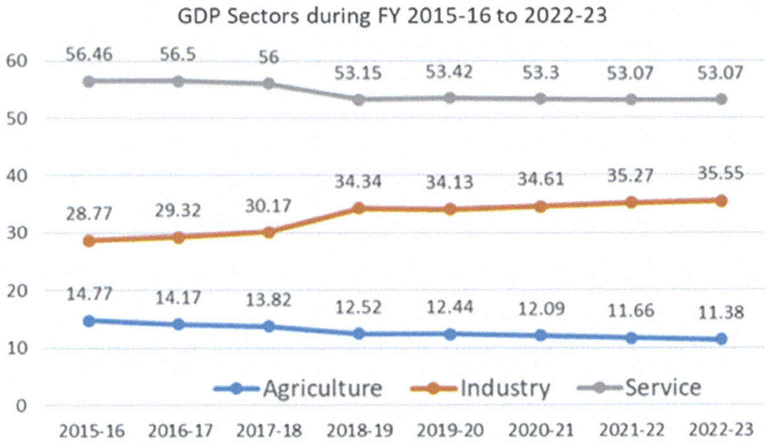

Fig. 6.27 Sectoral distribution of GDP. *Source* BBS (2022a, 2022b)

same. In the 2019–20 the contribution of agriculture sector to GDP was 12.44% which has reduced by 1.06–11.38%, and that of industry sector was 34.13% has increased by 1.42–35.55% in 2022–23, (see Fig. 6.28).

Regarding employment opportunities, the agriculture sector provides most of the employment to 40.62% of the population, followed by manufacturing sector 14.43%, business and tourism sector 14.34%, transport and communication sector 10.5%, goods and private services sector 6.08%, health, education, public services & defense 6.08%, construction sector 5.58%, economic business and services 1.97% in 2021, power, gas & water supply 0.02%, and mineral resources 0.02%. (BBS data), (Fig. 6.29).

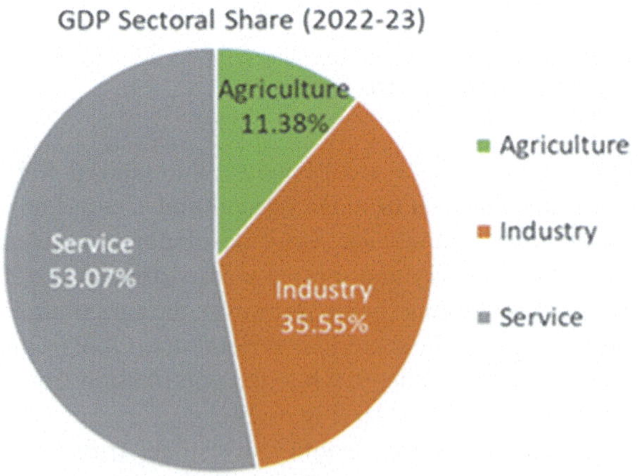

Fig. 6.28 GDP sectoral distribution 2022–2023. *Source* BBS (2022a, 2022b)

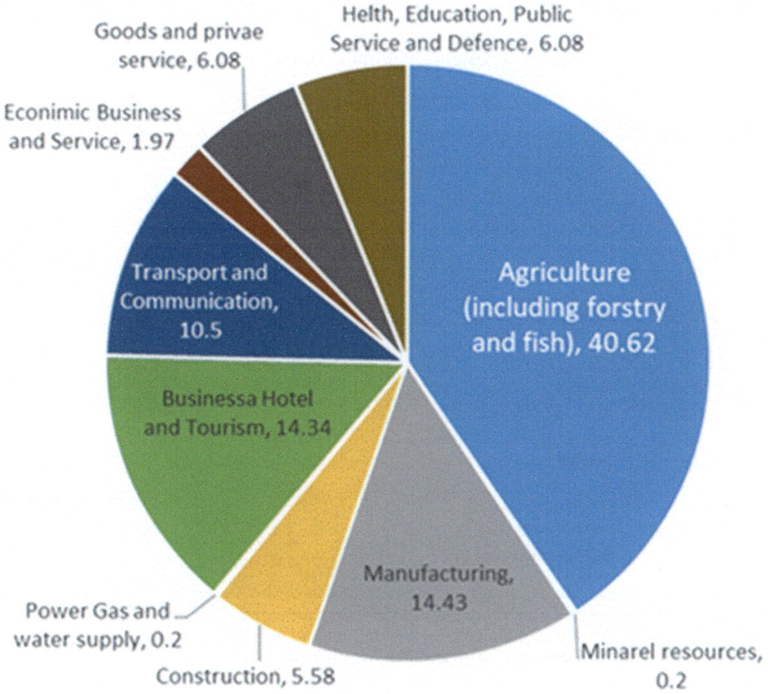

Fig. 6.29 Employment distribution by sector. *Source* BBS (2022a, 2022b)

6.3 Transboundary Issues

Bangladesh, Bhutan, China, India, and Nepal share some of the most intricate and complex river systems in the world. The formidable network of waterways includes the three major rivers the Ganges/Padma, the Brahmaputra/Jamuna, and the Meghna. When combined, these rivers constitute the worlds' second largest riverine drainage basins: the Ganges–Brahmaputra–Meghna basins (GBM); (Fig. 6.30).

The 64% catchment of these rivers lies in India, 18% in China, 8% in Nepal, 7% in Bangladesh, and 3% in Bhutan. In the case of the Ganges, 79% catchment lies in India, 14% in Nepal, 4% in Bangladesh, and 3% in China. In the case of the Brahmaputra, 50% of the catchment lies in China, 36% in India, and 7% in Bhutan and Bangladesh. In the case of the Meghna, 57% of the catchment lies in India and 43% in Bangladesh. The average annual flow of the Ganges is 373 km^3, of the Brahmaputra 650 km^3, and the Meghna 180 km^3.

The total population of the GBM basins is 630 million, the share of India is 477 million (75.8%), Bangladesh 128 million (20.3%), Nepal 22 million (3.5%), Bhutan 1.26 million (0.2%), and China 1.26 million (0.2%).

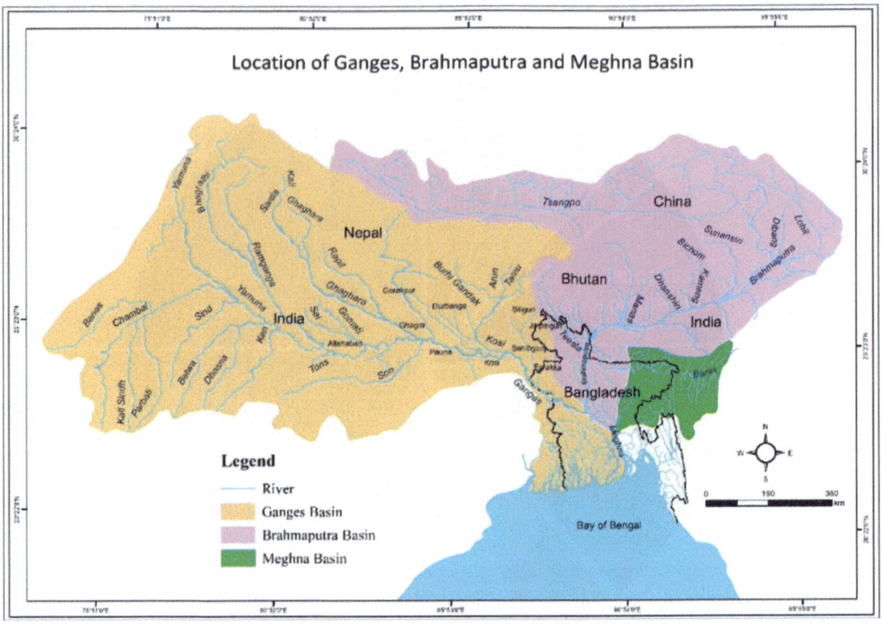

Fig. 6.30 The Ganges, the Brahmaputra, and the Meghna basins

Bangladesh shares 54 common/border rivers with India. Being a lower riparian country, water resource management is complex and dependent on upstream developments. Continuous withdrawal of waters from these rivers upstream of international boundary has caused severe hydro-morphological as well as ecological impact within Bangladesh, especially in the dry season. As these river basins are also vulnerable to flooding, the frequency and intensity of floods are expected to increase due to climate change. An example of it is the Farakka barrage in West Bengal located just few kilometers away from international boundary causing severe reduction of dry season flows as well as changes in the flood regime. The Ganges-dependent area is suffering from shortage of irrigation water in dry season, increased sedimentation on the river beds including the Gorai, impeding navigation including the operation of Mongla Port, increased salinity as well as reduction of fish production, etc.

For the last few decades, India has been constructing barrages and dams on tributaries of the Ganges and Brahmaputra rivers while also diverting water from these rivers for irrigation purposes. India is likely to continue the same path as irrigation plays an increasing role in India's economic growth and sufficient food for its rapidly expanding and urbanizing population. Of particular concern for Bangladesh are the Indian proposals to construct 16 barrages on the Ganges River and the country's plan to divert water from the Ganges and the Brahmaputra rivers towards the south of India. In addition, India is planning to construct the Tipaimukh Dam in the northeastern part of the country. All these projects will impact the water availability in Bangladesh as well as the ecological condition of the rivers. These developments are expected to affect fisheries and agriculture activities within Bangladesh adversely.[12] The proposed Indian river-linking project is expected to impact the dry season flows notably.

Among the major rivers, the Ganges has been found to be the most negatively impacted river, followed by the Brahmaputra and, to a lesser extent, the Meghna. Other most affected regional rivers are the Teesta, the Dudhkumar, the Dharla, the Mahananda, the Manu, the Khowai, the Gumti, and the Muhuri. For example, a reduction of the dry season flows in the Teesta in recent years resulted in an alarming situation due to water withdrawal in the upstream of the Teesta in the Indian part. During the low flow season, the mean monthly minimum flow has decreased from 139 to 50 cumec (1967–2006).

Resolving water issues between Bangladesh and India is vital for water resources development in Bangladesh. The Indo-Bangladesh Joint Rivers Commission was established in 1972 to maintain liaison between Bangladesh and India in order to ensure the most effective joint efforts in maximizing the benefits from common river systems to both the countries. Joint Rivers Commission, Bangladesh, is since its inception working with India on sharing waters of common rivers, acquisition of flood related data from upstream stations of different rivers in India, resolving problems of river training works of the commonborder rivers and other related issues. The Joint Rivers Commission has so far held 38 meetings alternately in Dhaka and New Delhi. After a prolonged negotiation spreading over several decades, Bangladesh and India signed the Ganges Water Sharing Treaty in 1996 for a period of thirty years. The treaty regulates the dry season flow from January to May by three different clauses. In case, that water availability at Farakka is less than 70,000 cusecs (cubic feet per second), the water will be equally shared by India and Bangladesh. If the water availability is between 70,000 and 75,000 cusecs, Bangladesh will get 35,000 cusec and rest will be used by India. In the case that 75,000 cusec is available at Farakka, India will receive 40,000 cusec and Bangladesh will get the remaining flow. In the event that flow at Farakka falls below 50,000 cusecs in any 10-day period, the two governments will enter into immediate consultation to make adjustment on an emergency basis, in accordance with the principle of equity, fair play, and no harm to either party. The treaty expires in 2026. The water sharing agreement of the Teesta and other medium rivers are being negotiated, but so far, no agreement could be reached.

China is constructing hydropower dams upstream of the Brahmaputra to use the river's potential hydropower capacity. At present, relatively smaller dams are being constructed to create a suitable situation for constructing the large Great Bend hydropower dam. This dam is estimated to generate 40,000 megawatts after its completion. According to the Chinese Government, these projects would not significantly impact the river discharge downstream, as no water would be extracted. However, there might be ecological consequences due to change of the river regime, that are yet unknown.

India and Bangladesh continue to remain concerned over water diversion from the upper reaches of the Brahmaputra to the northern China. Although such a large

---

[12] Alam (2018f).

infrastructural project is regarded as unlikely in the short term, dams constructed in China in the Great Bend of the Brahmaputra show China has so far not disclosed the diversion project.

India and China state very clearly that the countries will only negotiate bilaterally with other riparian countries within the Ganges–Brahmaputra–Meghna basins. However, even though this approach has been prevalent for decades, a shift can be noticed in India's attitude towards a more multi-lateral approach with regard to the country's transboundary rivers. The state visits between India and Bangladesh in 2010 and 2011 paved the way for the Joint Working Group, an initiative involving Nepal, Bhutan, India, and Bangladesh to jointly manage the Ganges and the Brahmaputra rivers. So far, this initiative is in a premature stage but seems as a promising development with regard to water diplomacy practices in the Ganges–Brahmaputra–Meghna basins.

Bangladesh receives flood-related data from China on the Brahmaputra and from few tributaries of the Ganges from Nepal. In addition, the countries hold frequent meetings at the government level to discuss water issues. Bangladesh took up the issue of diversion of water from the Brahmaputra based on media reports through its embassy in Beijing. Besides, Bangladesh has expressed its concern regarding dams in China on the Brahmaputra. There is also a diplomatic relation with regard to water cooperation with Nepal, however, limited with Bhutan and almost absent with Myanmar. Bangladesh shares three rivers with Myanmar.

6.4 Issues, Challenges, and Formation of Hotspot Areas

The Bangladesh Delta includes all administrative districts that face various natural hazards owing to the deltaic formation of Bangladesh and the related interface with the vast river networks, the Bay of Bengal and climate change. Hence integrated water resources management in the context of its interaction with climate change, environment, ecology, biodiversity, agriculture, and land management is an integral part of the BDP 2100. For water resource planning purposes, Bangladesh has been divided into eight hydrological regions; (see Fig. 6.31), which are the Northwest (NW), Northeast (NE), North-central (NC), Southeast (SE), South-central (SC), Southwest (SW), Eastern Hills (EH), and the major rivers and estuaries (RE).

Using the eight hydrological zones as the starting point, the BDP 2100 sharpens the focus on the magnitude of the natural hazard vulnerabilities facing each of the hydrological regions. This has led to a modified grouping of districts and areas facing similar risks of natural hazards. These groups are called 'Hotspots' that simply define a broad grouping of districts and areas facing similar natural hazard risks. Hotspots are prototypical areas where similar hydrological and climate change vulnerability characteristics and problems converge (such as sea-level rise, river erosion, intensity of flooding, water shortages, siltation constraints, etc.). The aggregate hotspot grouping provides a convenient analytical tool to summarize broad socio-economic and common risk profiles. The six Hotspots are: (i) Haor and Flash Flood Areas, (ii)

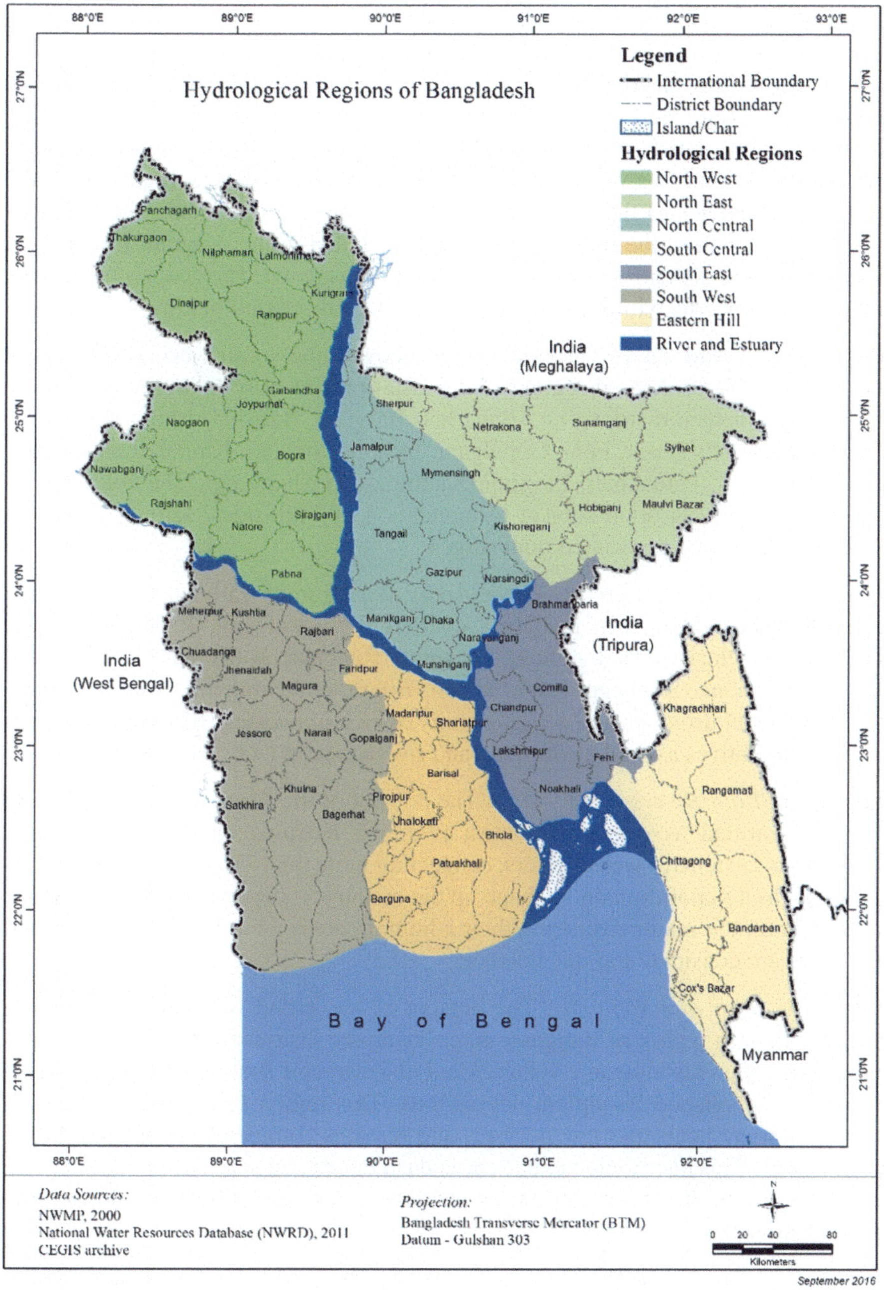

Fig. 6.31 Hydrological regions of Bangladesh

Coastal Zone, (iii) Chattogram Hill Tracts (CHT), (iv) Urban Areas, (v) Barind and Drought-Prone Areas and (vi) River Systems and Estuaries, (Fig. 6.32).

The area falling beyond these hotspots are termed cross-cutting. The hotspot areas are 'problems and issues'-oriented, as each hotspot comprises areas having common water-related issues and problems, and similar development potentials. The proposed strategies or programs have been designed in accordance with the geographical extents of the different delta hotspots. It is to be mentioned here that not only areas within the hotspots have been included in the program formulation process, but their dependent areas have also been considered. The geographic extents and generic concerns within each hotspot have been discussed below.

Hotspot 1: Coastal Zone. The coastal zone covers the south-west, south-central and south-east hydrological regions and coastal part of hilly hydrological region, comprising nineteen districts, i.e., Bagerhat, Barguna, Barishal, Bhola, Chandpur, Chattogram, Cox's Bazar, Feni, Gopalganj, Jashore, Jhalkati, Khulna, Lakshmipur, Narail, Noakhali, Patuakhali, Pirojpur, Satkhira, and Shariatpur. The coastal zone in its natural state is often subject to inundation by high tides, salinity intrusion, cyclonic storms, and associated tidal surges.

Hotspot 2: Barind and Drought-Prone Areas. Barind tracts falls on the north-west region hydrological region, including districts of Nawabganj, Rajshahi, Natore, Naogaon, Panchgara, Thakurgaon, Dinajpur, Rangpur, Kurigram, Lalmonirhat, Gaibandha, Pabna, Sirajganj, Jaipurhatand, Bogra, and Sirajganj. The area experiences frequent droughts and water scarcity, which immensely hamper dry season irrigation initiatives as well as flooding and river erosion; (see also: FAO, 2011).

Hotspot 3: Major Rivers and Estuary. The hotspot of the rivers and their adjoining areas are entitled as one of the most principal hotspots of Bangladesh delta. Bangladesh's extensive web of rivers, connecting more than 750 rivers and their tributaries act as a major domain to shape up the country's abundant natural resources. Among these, the major rivers and their adjoining areas, i.e., the distributaries and char lands are considered in this hotspot.

Hotspot 4: Haor and Wetlands. The hotspot of haors and wetlands is located in the north-eastern region of Bangladesh, covering the districts of Sunamganj, Habiganj, Netrakona, Kishoreganj, Sylhet, Maulvibazaar, and Brahmanbaria. The haors are large bowl-shaped floodplain depressions. This region dries up in dry season except for some beels and low pockets, and remains completely inundated during the monsoon. This haor region experiences flash floods and shortage of water during the pre-monsoon, floods during the monsoon and drainage congestion during pre- and post-monsoon.

Hotspot 5: Urban Areas. The hotspot of urban centers comprises the divisional headquarters of the country, i.e., Dhaka, Chattogram, Khulna, Rajshahi, Sylhet, Rangpur, Mymensingh, Kumilla and Barisal, with emphasis on city corporation areas including Narayanganj and Gazipur, and their influence zones. These urban areas frequently

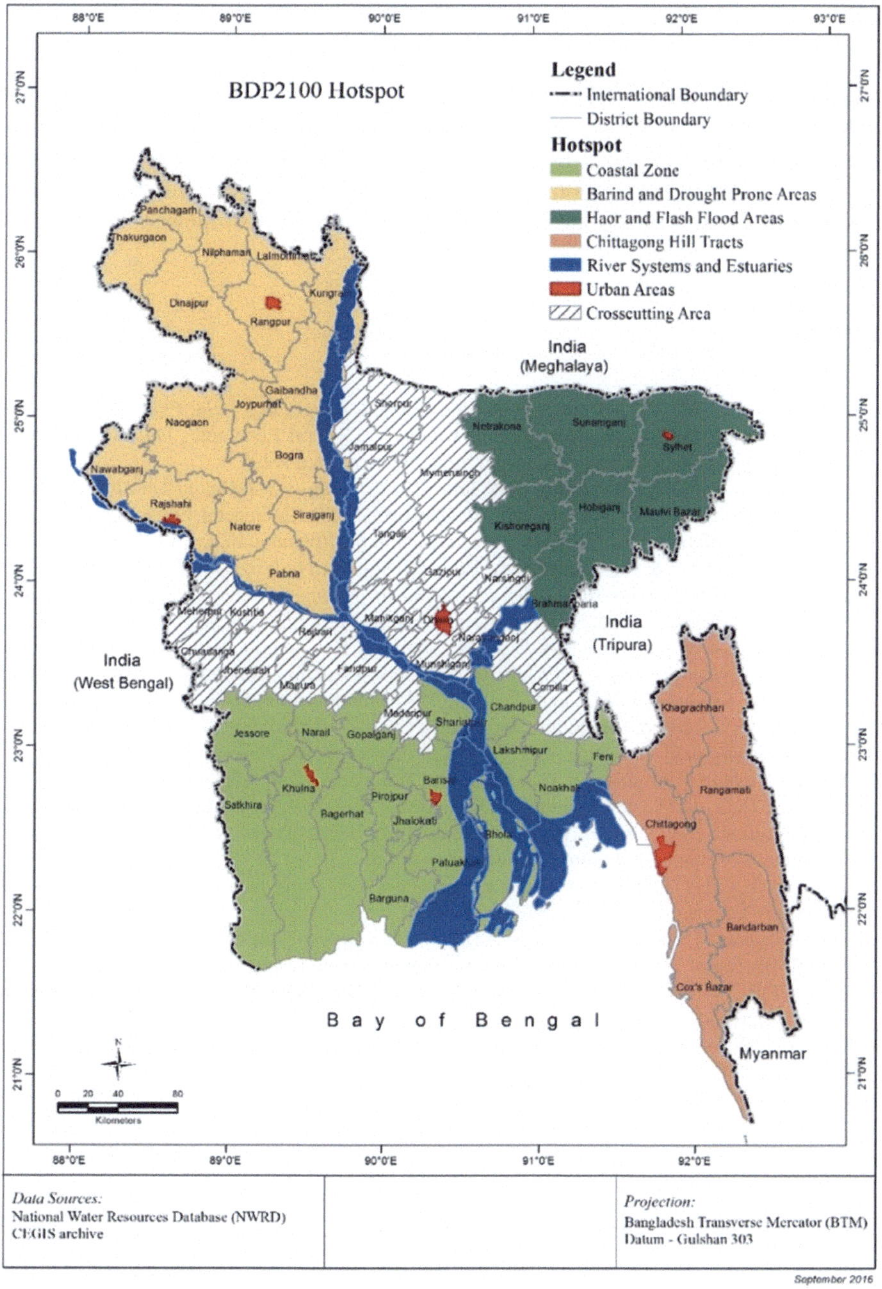

Fig. 6.32 Identified Hotspot areas. *Source* BanDuDeltAS, 2017

experience floods, drainage congestion, water logging, water shortage, environmental degradation as well as high incidences of water, air, soil, noise, and thermal pollution.

Hotspot 6: Chattogram Hill Tracts. The hotspot of Chattogram Hill Tracts is confined within the districts of Khagrachari, Rangamati, and Bandarban. The region experiences cyclonic rainfall, flash flood, and landslides during the monsoon and water shortage during the dry season, both for domestic and agriculture purposes. The region has the potential for development of small or mini-hydropower.

References

Alam, S., De Heer, J. M., & Choudhury, G. A. Eds. (2018a). *Bangladesh Delta Plan 2100: Baseline Studies on Water Resources Management. Volume 1.* General Economics Division, Planning Commission, Government of Bangladesh. https://plancomm.gov.bd/

Alam, S., De Heer, J. M., & Choudhury, G. A. Eds. (2018b). *Bangladesh Delta Plan 2100: Baseline Studies on Disaster and Environment Management. Volume 2.* General Economics Division, Planning Commission, Government of Bangladesh. https://plancomm.gov.bd/

Alam, S., De Heer, J.M., & Choudhury, G.A. Eds. (2018c). *Bangladesh Delta Plan 2100: Baseline Studies on Land Use and Infrastructure Development. Volume 3.* General Economics Division, Planning Commission, Government of Bangladesh. https://plancomm.gov.bd/

Alam, S., De Heer, J. M., & Choudhury, G. A. Eds. (2018d), *Bangladesh Delta Plan 2100: Baseline Studies on Agriculture and Food Security. Volume 4.* General Economics Division, Planning Commission, Government of Bangladesh. https://plancomm.gov.bd/

Alam, S., De Heer, J. M., & Choudhury, G. A. Eds. (2018e), *Bangladesh Delta Plan 2100: Baseline Studies on Socio Economic Aspects of Bangladesh Delta. Volume 5.* General Economics Division, Planning Commission, Government of Bangladesh. https://plancomm.gov.bd/

Alam, S., De Heer, J. M., & Choudhury, G. A. Eds. (2018f). *Baseline Studies on Governance and Institutional Development, Volume 6.* General Economics Division, Planning Commission, Government of Bangladesh. https://plancomm.gov.bd/

Aquastat, FAO. (2011). *FAO's information system on water and agriculture.* FAO, Food and Agriculture Organization of the United Nations.

Bangladesh Bureau of Statistics (BBS). (2022a). *Household Income and Expenditure Survey (HIES).*

Bangladesh Bureau of Statistics (BBS). (2022b). *Population and Housing Census (Census).*

FFWC, BWDB: Flood Forecasting and Warning Centre, Bangladesh Water Development Board. www.ffwc.gov.bd

Ministry of Environment Forest, and Climate Change (2022). *National Adaptation Plan of Bangladesh (2023–2050).*

Chapter 7
Climate Vulnerabilities and Impacts

Abstract Bangladesh has been ranked as the seventh most climate vulnerable country in the world in terms of risks from natural hazards according to the Global Climate Risk Index 2021. Being a disaster-prone low-lying delta, the country experiences temperature rise with impact on weather patterns, sea-level rise and erratic heavy rainfall in addition to normal regular floods, droughts, cyclones, storm surges, salinity intrusion, river erosion, transboundary water diversion, waterlogging. These cause substantial loss and damage to agricultural and industrial production, disruption of communication, education, and health services, damages of infrastructure and housing resulting in colossal economic loss almost every year. This situation will according to BDP 2100 further worsen due to impact of climate change. In this section, the impact of climate change is discussed in addition to the 'normal' issues and challenges and ambitions with respect to food security and socio-economic development brought in the perspective of a need-based long-term delta plan.

Keywords Climate · Climate change · Climate vulnerability · Climate risk · Climate migration · Impact · Weather pattern · Disaster · Hazard · Sea-level rise · Weather pattern · Rainfall · Erratic rainfall · Monsoon · Temperature · Temperature rise · Heat stress · Gender · Health · Waterlogging · Flood · Cyclone · Storm surge · Loss and damage

7.1 Temperature Rise

Bangladesh is a sub-tropical country with warm and humid climate. The average temperature ranges from 150 to 340C. Long-term mean of country-average monthly minimum temperature ranges from 12.50C in winter (January) to above 250C in summer. The monthly maximum temperature ranges from 250C in winter, with peak in summer (April) 33.50C.[1]

---

[1] See for a good introduction and overview Brammer, 2014. Also: Carbon Brief, 2018, and Jakariya and Islam, 2021.

J. M. de Heer and G. A. Choudhury, *Strategies for Adaptation to Climate Change in a Transformative Approach*, Springer Climate,
https://doi.org/10.1007/978-3-031-88884-7_7

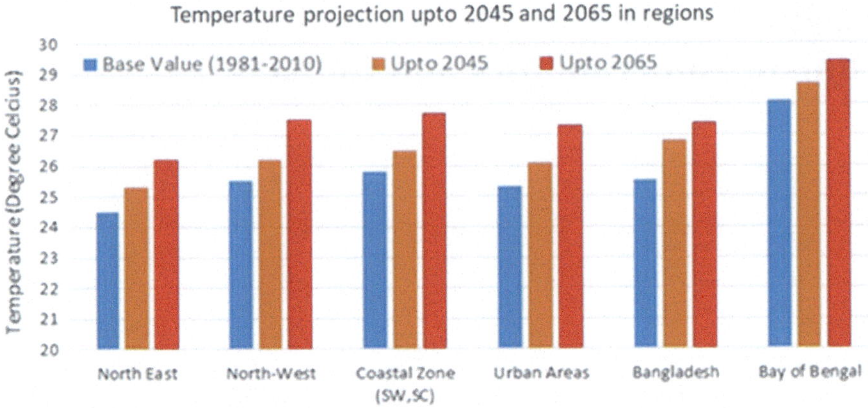

Fig. 7.1 Temperature projection up to 2045 and 2065 in different regions due to climate change. 2022 *Source*: IPCC; NAP,

The analysis of average surface air temperature of BMD data from 1961 to 2019 shows a trend of rising temperature during the last three decades. The average annual temperature is rising sharply during 1991–2019 compared to 1961–1990. For instance, temperature variations were minimal 0.0067 °C per year during 1961–1990. However, the variations rose sharply to 0.03 °C per year during 1991 to 2019. In recent decades, the mean temperature has been rising very rapidly from 1991 to 2000, the mean temperature by 0.390C, which further increased by 0.530C during 2001–2010 and 1.060C during 2011–2019.[2] The minimum temperature for the winter has risen by 0.450C and for the monsoon by 0.520C meaning the winters and summers arc becoming warmer. The Berkeley Earth dataset suggests an average temperature rise of 1.03 °C in Dhaka city over the period 1900–1917 to 2000–2017.

Future temperature projections show a warming range of 0.44–0.69 °C in the near term (2030s, up to 2045) and range of 1.3–2 °C in the mid-term (2050s, up to 2065) based on IPCC Assessment Report 6[3]; see Fig. 7.1.

The base value (average of 1981–2010) and future projections of temperature under two scenarios of GHG emissions in selected regions are shown in table below. The projected temperature under both scenarios will be minimum in the north-east and maximum in the Bay of Bengal which is harmonic with the base value. In the coastal zone, the temperature in base (1981–2010) value and projected scenarios are higher than other inland areas. In the urban areas, the projected temperature will be 7.9% increase from the base value in high greenhouse gas (GHG) emission scenario. In the north-west, the projected temperature will increase 7.84% from the base value in high GHG emission scenario; see Table 7.1.

[2] SIBDP Consortium.

[3] Ministry of Environment, Forest, and Climate Change (2022).

Table 7.1 Temperature ranges, up to 2045 and 2065. *Source*: SIBDP

Region	Base value (°C) (1981–2010)	Future projected temperature (°C)	
		Upto 2045*	Upto 2065**
North-East	24.5	25.3	26.2
North-West	25.5	26.2	27.5
Coastal zone	25.8	26.5	27.7
Urban areas	25.3	26.1	27.3
Bangladesh	25.5	26.8	27.4
Bay of Bengal	28.1	28.7	29.4

* IPCC SSPI-2.6 Represents low future GHG scenario
** IPCC SSPI-8.5 Represents very high future GHG scenario

7.2 Rainfall

Bangladesh climate is largely influenced by the South Asian monsoon. About 80% of the annual rainfall occurs during the monsoon (June–September) and little in winter (November–February). Most places receive more than 1525 mm of rain a year, and areas near the hills in the north-east receive 5080 mm. With climate change, there are projections of shifts in monsoon patterns, leading to changes in the timing, duration, and intensity of the monsoon season. For instance, rainfall is decreasing in the winter (December–February) and pre-monsoon (March–May) at 1.3 mm per year and 0.5 mm per year, respectively. However, rainfall is increasing for both the post-monsoon (October–November) and monsoon (June–September) at 0.05 mm per year and 4.5 mm per year, respectively. This indicates that winters are becoming dryer and monsoons are becoming wetter.

In recent years, Bangladesh has also experienced several extreme rainfall events, for instance, 433 mm of rainfall in 24 h in Rangpur in 2020, a recorded highest in the last 60 years. Analysis of extreme rainfall indicates that consecutive dry days (CDD) significant increasing trend all over the country (Ezaz et al., 2022). These alterations are currently affecting agricultural cycles and water resource. The future projections of rainfall for different climate stress areas of Bangladesh up to 2045 and 2065 are based on downscaled climate data.[4] Due to climate change, the rainfall will vary between 0.1–1.4 and 2.4–34.5 percent in in the 2030s and in the 2050s, respectively. The rainfall will be higher in the north-eastern hills as well as in the Chattogram Hill Tracts, while the remaining areas will experience lesser rainfall. In the 2030s, yearly rainfall will be slightly higher or similar, while in the 2050s, rainfall will increase all over country. On the contrary, the winter precipitation will decrease throughout the country, except in the coastal and Chattogram Hill Tracts regions. In contrast, the Bay of Bengal and the Chattogram Hill Tracts will get less pre-monsoon rainfall during the pre-monsoon. The rest of the country will experience more rainfall during the monsoon and post-monsoon. The frequency of heavy rainfall events is projected

[4] Ibid.

Table 7.2 Ranges of rainfall projection up to 2045 and 2065. *Source*: SIBDP

Region	Base value (rainfall mm) (1981–2010)	Future projected rainfall (mm)	
		Upto 2045*	Upto 2065**
North-East	3653	3683	3783
North-West	1560	1631	1643
Coastal zone	2129	2138	2215
Urban areas	2759	2785	2882
Bangladesh	2579	2582	2669
Bay of Bengal	1650	1353	1707

* IPCC SSPI-2.6 Represents low future GHG scenario
** IPCC SSPI-8.5 Represents very high future GHG scenario

to rise while that of light rainfall events will fall, inferring a shift towards a lower number of wet days with an increase in the intensity of rainfall on days with rain, entailing an increased risk of flash floods (Caesar & Janes, 2018).

Table 7.2 shows the base value and future projections of rainfall in selected regions of Bangladesh (NAP). The base value (1981–2010) and predicted future rainfall will be higher in Bangladesh in both low and high GHG emission scenarios. According to the projected future rainfall the Barind and drought-prone area showing an increase of 5.32% from the base value (1981–2010) up to 2065 under high GHG emission scenario. The urban areas projected 4.46% increase rainfall up to 2065 from the base under high GHG emission scenario. Comparatively less increase of projected future rainfall up to 2065 is 3.56% in the north-eastern haor area of Bangladesh under high GHG emission scenario.[5]

The base value of rainfall and future projections under two scenarios is shown in the chart in Figs. 7.2 and 7.3 for the selected areas of Bangladesh. In the coastal zone, the increment of predicted rainfall is indicating 4.04% up to 2065 under high GHG emission.

7.3 Sea-Level Rise

Sea-level rise is a secondary effect of global warming caused by the volumetric expansion of seawater and the addition of liquid water to the sea due to melting of the polar and mountains glaciers. For low-lying countries like Bangladesh, the coastal zones are highly vulnerable to sea-level rise (see, e.g., Anwar et al., 2022). The coastal zone is 39.000 square kilometer, out of which 6% of land is 1 m below, 34% between 1 and 2 m, 22% between 2 and 3 m and 38% is above 3-m mean sea level. Both the projected and historical sea-level rise will exacerbate due to climate change. Even a

[5] See in Table 7.9 mentioned IPCC Reports and IPCC 2022, 2023.

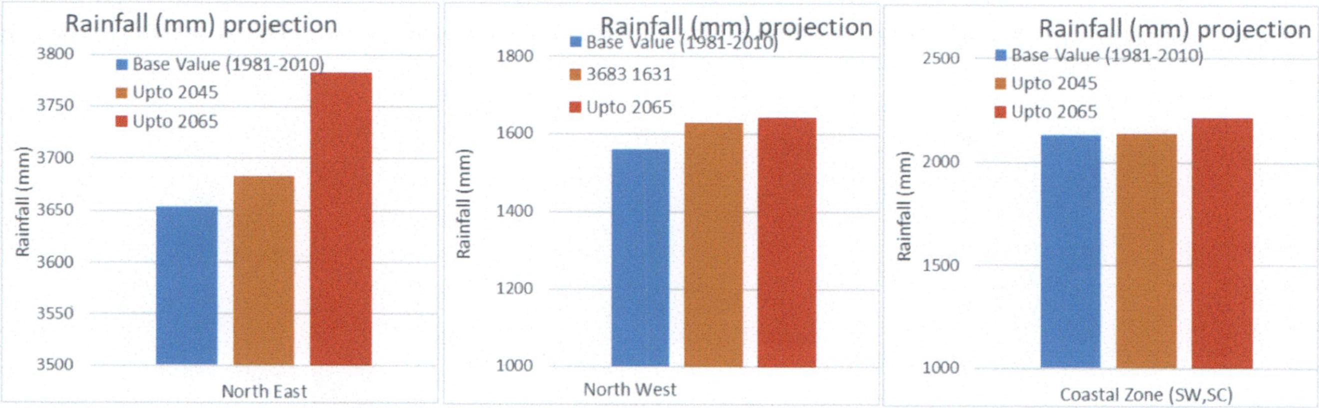

Fig. 7.2 Rainfall projection up to 2045 and 2065 due to climate change in North-East, North-West, and coastal regions. 2022 *Source*: IPCC; NAP

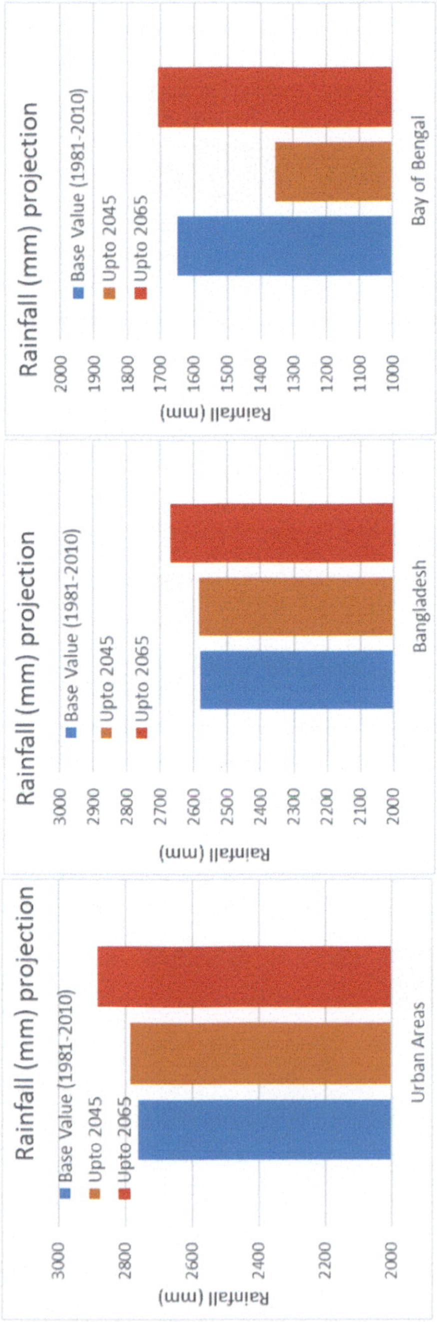

Fig. 7.3 Rainfall projection up to 2045 and 2065 due to climate change in urban areas, Bangladesh, and Bay of Bengal. 2022 *Source*: IPCC; NAP

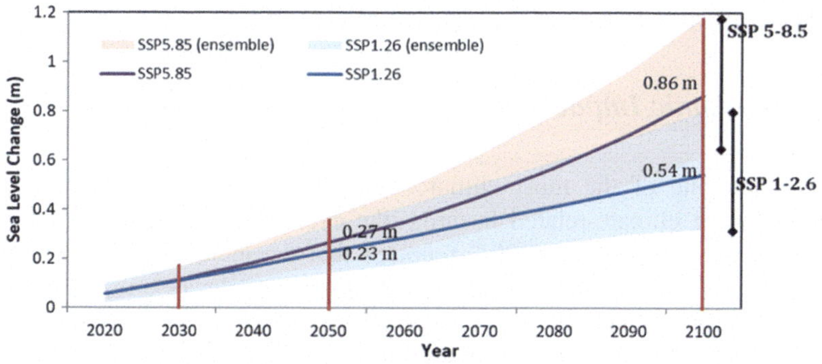

Fig. 7.4 Sea-level rise projections at the Bangladesh coast. 2022 Source: IPCC and NAP

small sea-level rise will significantly affect the coastal areas of Bangladesh, as about 62% of the coastal land is less than three meters below mean sea level.

The BDP 2100 Baseline Study[6] and, e.g., SMRC (2003) have reported increasing trends in Relative Sea-Level Rise (RSLR) using tidal observation data of 1978–1998. The RSLR was found to be 4 mm/year at Hiron point (western coast)., 6 mm/ year at Char Changa and (Meghna estuary) 7.8 at Cox's Bazar (East Coast) during the period of 1977–1998. The study of CEGIS (2011) has shown that the sea-level rise is 5.5 mm/year for the western coast at Hiron Point station, 7.5 mm/year at Moheshkhali, 5.1 mm/year at Cox's Bazar and 7 mm/year at Sandwip for the central region at Sandwip (baseline study).[7] This can contaminate water supplies, affecting agriculture, drinking water availability, and ecosystems. This higher tides and water levels will limit the drainage capacity of the water infrastructures and increase the risk of flooding. Low-lying areas behind the coastal polders in the south-central region will be inundated mostly. This will damage crops, agriculture-based livelihood, fisheries, and livestock and impact the coastal population and their way of life. Salinity intrusion will be further worsened.

Future projections show the following. According to IPCC AR6, the SLR projections in the Bay of Bengal may be 23 cm to 27 cm by 2030 and may be 54 cm to 86 cm by 2100; see graph in Fig. 7.4. This will retard the monsoon as well as flood water flows to the Bay of Bengal, prolonging the flood duration. The infrastructure, livelihoods, and coastal communities will all be significantly impacted. (Habiba et al., 2014). A study by the CEGIS (2021) found that almost 18% of the coastal area may be inundated due to sea-level rise as forecasted, based on the SSPS-8.5, considering the current coastal polder setup by the medium term (2050).

Updating will be frequently needed, to see actual developments and to incorporate sea-level legacy data.[8]

---

[6] Alam et al. (2018).

[7] Alam et al., (2018).

[8] Climate Analytics. (2018); Mengel et al. (2018).

7.4 Impact of Climate Change on the Delta

7.4.1 Economic Impact

Bangladesh is one of the most climate vulnerable countries as the country is constantly facing climate-related hazards like cyclone. After studying 122 years of cyclone data, Singh et al., (2001) reported that the incident of cyclones in the North Indian Ocean has increased at an alarming rate. Because of these cyclone incidents, Bangladesh has become more prone to the incident of the severe cyclone. For Bangladesh, UNDRR (2021) estimated considerable losses to disasters at around $36 billion, between 2005 and 2050. Although these losses are primarily focused on the agricultural sector, they are increasingly impacting Bangladesh's urban areas. Ahmed and Suphachalasai (2014) suggest that by mid-century, climate change is likely to cost Bangladesh a further 2% of GDP on top of its baseline losses to climate hazards. This figure could potentially rise to 9% of GDP by the end of the century if mitigation action is not increased. Particularly, the country can face serious downside risks to food security, growth momentum, and poverty reduction efforts. The BDP 2100 indicates the possible impact of climate change on overall rice production, predicting a declining rate of 17 percent by 2050, impeding food and nutrition security. It is crucial for Bangladesh to understand these risks and properly adapt to the changes influenced by climate change. The following section discusses about the impact of climate change on different sectors of Bangladesh.[9]

7.4.2 Impact on Crops

Between 1976 and 2019, Bangladesh experienced an average rise in temperature of 0.5 °C. Though the increase in maximum temperatures during this period has been steady month to month, the increase in maximum temperature across the country has not been constant. For instance, during this period, maximum temperatures increased by 0.9 °C in the eastern parts (Sylhet and Chattogram divisions) compared to a 0.5 °C rise in central parts covering Dhaka and neighboring districts. However, this discrepancy has not been limited to maximum temperature only. At present, summers are becoming longer, winters are getting warmer, and monsoons are becoming unpredictable beyond the norm. It is generally expected that Bangladesh will experience an increase in temperature of about 1.5 or more °C by 2050.

This temperature increase will adversely impact the agriculture sector since crops are sensitive to changes in temperature. As plants maintain all physiological processes within temperature thresholds, sudden temperature changes hamper important activities like flowering and germination. In the context of Bangladesh, climate change projections outline an increase in daily temperature with more hotter days during the

[9] See Alam et al., (2018), See also Patz et al., 2020.

pre-monsoon and fewer rainy days. This discrepancy is projected to have a severe effect on the overall production of the crops. For instance, according to BDP 2100 studies, Aus production would decline by 27% under a moderate climate change scenario while wheat production would fall to 61%. Under a severe climate change scenario (with 60 percent moisture stress), the yield of the dry season or Boro rice might decrease by 55–62%. Likewise, moisture stress might force farmers to reduce Boro cultivation.

Moreover, the impact of these climate change scenarios would alter the overall cropping pattern by, e.g., a shift in crop choices among farmers, implying changes in the future rice-cropping pattern. A temperature increase could compel farmers to choose irrigation-based rice as Boro, Aus, and other crops in favor of the rain-fed Aman rice crop. Though irrigation-based Boro, Aus, and other crops would be the best options for the farmers, these varieties are also not free from the effect of climate change, considering that Aus rice seems to be the most vulnerable to temperature increase. As a result of this increasing temperature and CO_2 levels, it could result in a decline in rice yield and threaten food security. An apparent fertilization effect due to the doubling of atmospheric concentration of CO_2 may result in an overall 20% increase in rice production; however, a 60% moisture stress on top of other effects might cause a decline in Boro yield as high as 32%. It is feared that moisture stress would be more intense during the dry season, which might force farmers to reduce the area under Boro cultivation. The associated shortfall in food grain production would severely threaten food security (see also Aggarwal et al., 2010). These rising temperatures, in combination with uncertain rainfall, may create phenological changes in plants that result in lower yields. Similarly, higher temperatures beyond certain thresholds affect plants and their outputs. This includes spikelet sterility in rice, reversal of vernalization (cooling of the seed during germination) in wheat, the reduced formation of tubers in potatoes, and the loss of pollen viability in maize. Increased temperatures can also speed up phenological development, reducing the crop's grain-filling period and lowering yield. Likewise, infestations of pests and diseases may also intensify with rising temperatures. A recent study has found that leaf folder infestation in rice fields might increase by 80–136% in different regions of Bangladesh by 2050. This concern will be more prominent in the north-central, northwest, and northeast regions.

Furthermore, another area of concern is weed infestation of crops. Climate change also influences the dynamics of the crop–weed competition. Besides weed growth, climate change also significantly influences herbicide efficacy, affecting the herbicidal mode of action. Considering this, climate change is projected to favorably influence the weeds of wheat crops, which are vital to food security.

Overall, the decline in paddy yield due to climate change will negatively impact Bangladesh's real GDP. Given the high prevalence of poverty, food security in the country is closely linked to 'rice security'. A fall in rice production would increase its price, which would then place upward pressure on the overall consumer price index of the economy (i.e. resulting in a decrease in real household consumption). This would put additional pressure to ensure the country's long-term food security, see e.g. Malhi et al. (2021).

7.4.3 Impact on Fisheries and Aquaculture

The fisheries sector plays a vital role in the economy of Bangladesh and alone accounts for 3.52 percent of GDP (26.37 percent of agricultural GDP), 1.39 percent of export earnings, and more than 60 percent of the animal protein supply in people's diets. Besides, 12 percent of the population works in fishing directly or indirectly. However, this fisheries sector, especially aquaculture, where more than 50 percent of fish is produced and is susceptible to climatic signals and hazards. Aquaculture is especially vulnerable to a combination of climatic factors, including global warming, temperature fluctuation, rainfall variation, flood, drought, and salinity change.

While coastal fish production is more vulnerable than inland aquaculture, upcoming climate change will critically affect overall aquaculture production. This can be seen among the shrimp farmers and fry catchers in Bangladesh. These poor farmers are recurrently affected by extreme climatic conditions like cyclones and storm surges, severely damaging the entire coastal aquaculture. Studies conducted on these shrimp farmers have concluded that weather has changed abruptly during the past five years and that high temperature is most detrimental to shrimp farming. This damaging effect is observed in the form of lowered growth rate, increased susceptibility to diseases, including deformation, and affected production. Besides that, low production in shrimp farming is also caused by variation and rainfall intensity, which most farmers perceive as climate change impacts.

In general, climate change has both direct and indirect impacts on fish stocks. The immediate effects alter the growth, physiology, reproductive capacity, distribution of fish, and mortality. Conversely, indirect effects change the structure and composition of the fish food chain and marine ecosystems. Besides, there are many other factors, such as fishing, biological interactions, and non-climatic environmental factors (Rahman et al., 2021).

Commonly, climate change triggers high temperatures and other climatic conditions to which particular fish species are highly sensitive (Hossain et al., 2021). Further research indicates the degradation of pond water quality when the temperature exceeds 25 °C. This is mainly because temperatures above 22 °C can lead to less feed intake by fish, while more than 32 °C can cause slow growth and mortality. Another study has found erratic rainfall as another major climatic factor that has a positive relationship with reduced fish. Such a phenomenon was observed in the Hakaluki Haor, affecting fish catch significantly. In addition to reduced fish, the combined effect of extreme temperatures and erratic rainfall directly impact fish growth, mortality, physiology, reproductive systems, feeding behavior, production, and migration in inland and marine waters. Considering the contribution of the fisheries sector to the overall economy of Bangladesh, this situation acts as a barrier in the way of achieving food security (see for Haor area Aziz et al., 2021).

Another major concern for the fisheries sub-sector is the impact of sea-level rise. According to projections, from 2012 to 2050, the freshwater river area of the southwest coastal zone of Bangladesh is expected to decrease from 40.8% to 17.1–19.7% under different sea-level rise scenarios. Consequently, this occurrence can displace

0.9–2.1 million people by direct inundation in southern Bangladesh by 2050. Besides that, projected SLR, related aquatic salinization, and alteration in fish species composition brought about by SLR could impact the availability of freshwater fish with adverse implications for poor communities. (Barua et al., 2020).

It is also found that climate change is indirectly responsible for the lower income of fishermen. Currently, fishery-dependent communities of Bangladesh are facing added vulnerability regarding fewer stable livelihoods, the availability of fish catch, and safety risks due to changing climate conditions. For instance, in the coastal areas, fishers' income and livelihood are highly seasonal, and there is also an acute scarcity of alternative livelihoods. High seasonality, combined with a lack of alternative livelihood, weakens the adaptive capacity of fishers against the adverse impacts of climate change. On top of that, it has been reported that most fishers in the world are uneducated and poverty-stricken, which makes them unable to take other occupations/vocations as adaptive measures.

7.4.4 Impact on Livestock

Livestock, a significant and affordable source of protein in Bangladesh, is at high risk due to the adverse impact of climate change. Regarding the effects, they can be both direct and indirect. The direct impact of climate change on livestock health includes temperature related to frequent disease incidence and death. The indirect effect trails more intricate pathways and involves the climate influences on pathogen density and distribution and multiplication of vectors as well as vector-borne diseases and soil-, food-, and water-borne conditions. Despite many factors, livestock health is mainly affected by increased temperature and precipitation variation. The rest of the factors influencing livestock health are highly complex as they involve environmental forces such as social, ecological, and economic interests and individual and/or community behavior (Forastiere, 2010). Livestock health, specially the cattle, also impacted by the increase soil salinity.

The livestock sector, including meat and dairy production, is highly vulnerable to heat stress, flooding, droughts, strong winds, and extreme weather events. Such hazards have the potential to change the quality of meat and milk. DLS reported that meat and dairy production negatively impacted due to soil and water salinity increase in the coastal area. For instance, temperature and humidity impact milk production and fat content, while wind speed and hours of sunshine per day affect protein content. Livestock is further susceptible to changes in water availability for drinking or bathing. This is mainly seen in drought-prone and coastal regions, where disease outbreaks under extreme heat stress and salinity are prevalent. The DLS reported that cattle growth is much higher in the non-saline zone, in the northern Bangladesh than the coastal areas with river and soil salinity. Causality of livestock due to cyclone and storm surge and high flood, as shown in Fig. 7.5, is high.

Fig. 7.5 Flooded farm land. *Photo* courtesy of Md. Amanat Ullah

Overall, climate change impacts on livestock include health issues such as heatstroke; disease outbreaks; regenerative changes; reduced milk, meat, and egg production. Other adverse impacts are a water availability crisis, a fodder crisis, livestock deaths dues to disasters, reduced protein intake, and malnutrition. All these, in combination, act as threats to the livestock and poultry sector and increase food and nutrition insecurities. (also: Ali et al., 2020).

7.4.5 Impact on Forestry

Forests provide materials like timber, pulp, pole, fuel wood, food, and medicine; habitat for wildlife; and a primary base for biodiversity. They also provide oxygen, control or reduce the intensity of the cyclones and tidal surges in the coastal areas of Bangladesh, and influence rainfall and water yield in the river systems. Of the total area of Bangladesh, agricultural land makes up 65% of its geographic surface, forest lands (2.52 million hectares) account for almost 17%, and urban areas are 8%. Water and other land use account for the remaining 10%. Many of the anticipated adverse effects of climate change, such as sea-level rise, higher temperatures, and an increase in cyclone intensity, will damage the forest resources of the country, put pressure on many climate-sensitive species, and cause increased erosion and deterioration of soil quality in many upland forested areas.

The world's largest mangrove forest, the Sundarbans, is highly vulnerable to climate change. Sea-level rise will increase saltwater intrusion and negatively affect the forest, leading to considerable mangrove loss. Loucks et al. (2010) predict a 96% decline in tiger habitat in Bangladesh's Sundarbans mangroves with a 28 cm sea-level rise if sedimentation does not increase surface elevations. Considering that the salinity regime inside the forest will significantly change as a consequence of climate change, it has been argued that increased salinity would have discernible adverse impacts on forest regeneration and succession.

Besides that, climate change would bring adverse economic impacts. For example, in the Sundarbans, unproductive livelihoods are common among residents of both India and Bangladesh. Observations depict that local livelihoods are rapidly becoming unproductive due to the loss of fish and increasing salination, which makes agriculture increasingly difficult. This scenario is already prevalent and expected to be exacerbated by climate change impacts (high confidence). In addition, the coral reef of St. Martin's Island, which contributes 33.6 million USD yr–1 to the local economy, is also in danger. Along with other anthropogenic activities, climate change has been identified as a threat to these habitats.

Community or social forestry could be a good option at local level for Climate Impact Adaptation. The community or social forestry also providing so income and livelihood impalement of the rural population.

7.4.6 Impact on Livelihood

Climate-induced natural disasters often come with intangible and non-economic losses such as loss of identity, belonging, quality of life, and social value. High river discharges due to torrential rains cause erosion due to the strong currents of sand hungry rivers. Many people lose their land and properties and leave as landless poor people to the cities; see Fig. 7.6.

Under 0.44 and 2 m mean SLR, direct inundation is anticipated to drive migration of around 0.73–2.1 million people by 2100 (Davis et al., 2018). Alarmingly climate change will trigger permanent migration as many affected people will fail to return to their respective homes once such calamity is over. For instance, many households in villages affected by Cyclone Aila in Dacope and Koyra Upazilas of Khulna district in Bangladesh migrated permanently to other areas after the cyclone due to long-term flooding (e.g. 2–3 years) following the cyclone.

7.4.7 Impact on Health

Climate change is impacting people's medical expenditure as health hazards are intensifying in different regions of Bangladesh delta. Increased frequency of periods of prolonged high heat is a major threat to human health and living standards in

Fig. 7.6 Heavy erosion at village along the Jamuna river. *Photo* J.M. de Heer

Bangladesh, particularly in urban environments and for outdoor laborers. A World Bank study of 15,000 people in 3600 households in Bangladesh has illustrated the influence of climate variability on physical and mental health. According to this study, respiratory illnesses generally increase with rising temperatures and humidity. For instance, a 1 °C increase in temperature rises the chances of contracting a respiratory infection by 5.7 percentage points. Similarly, a 1 percent increase in humidity raises the likelihood of getting a respiratory infection by 1.5 percentage points. Water-borne diseases, such as diarrhea and dysentery, and vector-borne diseases, such as malaria and dengue, are also climate sensitive.

Worldwide, public health researchers have connected the climate crisis to substantial human health risks for over two decades through direct and indirect mechanisms (Patz et al., 2000, 2014). Concerning the direct mechanisms, climate change intensifies human health risks from exposure to extreme heat, prolonged drought (an occurrence linked to air pollution and mental distress), and intense storms (related to flooding and injuries). Chakraborty et al., 2019). Conversely, regarding the indirect mechanism, warmer ocean temperatures trigger stronger coastal storms (Marsooli et al., 2019), while warmer ambient air temperatures endorse the formation of ground-level ozone smog air pollution and allow insect vectors to carry diseases like West Nile Virus and Lyme Disease to new geographies (Crimmins et al., 2016).

7.4.8 Impact on Infrastructure

The impact of climate change will further aggravate historical sea-level rise and increase the vulnerability of the low-lying coastal areas of Bangladesh. Climate change and the resultant floods and cyclones will have a significant negative impact on capital stock in construction and infrastructure in Bangladesh. Study shows that at a 1 m SLR, a significant part of dryland in Bangladesh will be permanently inundated, and this increase in sea-level rise will substantially affect coastal communities, infrastructure, and livelihood. In this regard, the estimation by CEGIS (2021) for variable sea-level rise is quite alarming. This estimation shows that by the mid-term, around 18 percent of the coastal area might be inundated due to sea-level rise projections based on the SSP5-8.5 scenario.

Climate change and resulting floods and cyclones will significantly impact Bangladesh's infrastructure. In the absence of any estimates of the projected loss on infrastructure, it is assumed in the 'BGD-Din-CGE model' that due to climate change's impact, the construction sector's capital stock would be depleted by 0.05% annually until 2100. Simulation shows that the infrastructure shock would have a negative impact on all sectors of the economy. In this regard, climate-induced SLR is expected to impact power infrastructure, in severe cases necessitating power plant relocation. In Bangladesh, approximately one-third of power plants may need to be relocated by 2030 to avoid inundations caused by SLR (SSP2, RCP8.5). It is estimated that while power plants in the northern region will probably see a decrease in output because of droughts, an additional 30% of power plants are likely to be affected by increased salinity of cooling water and increased frequency of flooding. Furthermore, in many cases, low-carbon technology diffusion might make certain infrastructures redundant, leading to stranded assets.

Already Bangladesh faces yearly GDP losses of around 1.3 percent due to climatic events, which may rise to 2 percent per year by 2050 under extreme climate change conditions. Part of this loss will be borne by the industrial and services sectors, and the magnitude of these adverse effects will intensify beyond the end of the century. A negative shock on infrastructure due to climate change would have severe negative impacts on the overall economy of Bangladesh. The sectors closely linked to the construction sector, such as cement and metals, would experience large falls in production. Since the poor status of infrastructural development is already a severe binding constraint to realizing growth targets, the upcoming impact of climate change will further exacerbate the existing situation.

7.4.9 Impact on Energy

Projected climate change of Bangladesh can affect both energy generation (especially hydropower and thermal) and energy demand during hot summer seasons (especially during peak hours). Changes in climate patterns are currently affecting agricultural practices and crop yields in Bangladesh, impacting energy demand for irrigation and other agricultural activities. For instance, an increase in drought conditions is increasing energy demand as the pumping load is rising with falling water tables during the summer season. Currently, the energy demand–supply gap is worsening during irrigation periods. This can produce a negative feedback loop where poor farmers are economically hit hard due to reduced harvest and energy supply shortages.

Further, regular cyclones and floods are causing power supply failures and infrastructure damage, especially in the coastal areas of Bangladesh. These disruptions in energy infrastructure are leading to supply shortages and power outages. Changes in precipitation patterns can affect the availability of water resources for hydroelectric power generation. Reduced water flows could result in decreased hydroelectric output.

7.4.10 Gender and Climate Change

Drought, floods, extreme weather events, and food and water insecurity affect women and men differently, making the poorest more vulnerable. Because of salinity issues, the availability of freshwater is limited in some of the coastal areas of Bangladesh like Satkhira. In Bangladesh, women usually collect fresh water and in areas like Satkhira women need to walk long distances to collect drinking water for family and household activities. This aggravates their health condition, makes household responsibilities more stressful, and decreases time availability to invest in income-earning activities.

The degree to which people are affected by climate change impact is partly a function of their social status, gender, poverty, power, and access to and control over resources. In Bangladesh, women have lesser economic, political, and social empowerment. According to the 2019 index, Bangladesh has a Gender Inequality Index value of 0.537, and ranks 133 out of 162 countries. In addition to such low ranking, women's limited access to deltaic resources, limited mobility to explore opportunities, and limited voice in shaping decisions make them highly dependent and vulnerable to climate change. Therefore, many women are less able to cope with the increasing climate vulnerabilities and are more exposed to the adverse effects of the changing climate.

7.4.11 Climate Migration

Bangladesh ranked as the seventh climate vulnerable country in the Global Climate Risk Index (UNDP, 2021). The country already suffers regularly from natural hazards: floods, droughts, cyclonic storm surges, bank erosion, sea-level rise, salinity, waterlogging and drainage congestion, leading to loss of life and livelihood, land, and homes, ensuing to forced displacement of people. The intensity and frequency of these hazards will increase due to climate change, further aggravating the migration situation.[10]

In case of sea-level rise, around 17% of the country's coastal lands will be submerged, displacing could be as large as about 20 million people by 2050. A large number of people lose their land and houses due to river erosion every year and many of them are turned into landless. Intensity of erosion is expected to increase in the future due to climate change. Most of the displaced become migrants and move to urban slums.

The people living in the chars within the Ganges, the Padma, the Brahmaputra/Jamuna, and the Meghna floodplains periodically migrate to the nearby safer areas as an adaptation strategy during flood seasons. However, the idea of 'migration as an adaptation strategy becomes questionable while most of the climate migrants cannot find suitable accommodation having a livable environment and services. The migrants usually move to shanty houses in already overpopulated urban areas and slums having poor water supply, sanitation, and health facilities. The migrants move to urban areas and slums in the hope to improve their economic condition. On the contrary, the rural–urban migration often pushes the migrants deeper into poverty.[11]

7.5 Need-Based Long-Term and Holistic Delta Plan

Bangladesh is considered one of the most vulnerable countries to the impact of climate change. According to the 6th IPCC, it is estimated that Sea-Level Risk (SLR) in southern Bangladesh could displace 0.9–2.1 million people by direct inundation by 2050. Water salinization brought about by SLR is expected to impact the availability of freshwater fish in southwest coastal Bangladesh with adverse implications for poor communities.

Different organizations have prepared and adopted many initiatives and plans for the water sector and agricultural development in the country since 1960 as shown in Chap. 3. The IECO Master Plan for the water resources sector, National Water Plan-I followed by National Water Plan-II, The Flood Action Plan (after the devastating flood in 1987 and 1988), etc., have been adopted in the country. The FAP studies were very elaborate, and one of the outcomes of the study is the Bangladesh Water and Flood Management Strategy, based on which the National Water Policy and National Water Management Plan were formulated.

The National Water Policy provides a useful basis for the Delta Plan Institutional Framework, as it outlines an integrated approach towards sustainable water resources management. The National Water Management Plan is meant as an operationalization of the NWPo. It is a comprehensive water resources plan involving 13 ministries and more than 30 agencies. Within the existing institutional context, however, the NWMP shows many implementation issues, which are essential to analyze and to consider as 'lessons learned'.

---

[10] See also, UNFCCC, 2012.

[11] Bangladesh Delta Plan 2100, 2018.

Along with different water management policies, Bangladesh has also focused on creating policy guidelines for addressing climate change over the years. Bangladesh has also taken various proactive initiatives, with its own resources and international cooperation, to adapt to climate change. The Ministry of Environment and Forest formulated in 2005 the National Adaptation Program of Action and in 2009 the Bangladesh Climate Change Strategy and Action Plan to address climate change's adverse impacts (see Fig. 7.7), including variability and extreme events, and promote sustainable development.

Other government ministries and departments also devised their own plans and strategies. The Master Plan for Haor areas was prepared for Haor areas located in the country's north-eastern region by Bangladesh Haor and Wetland Development Board in 2010. The Ministry of Agriculture and FAO prepared the Master Plan for Agricultural Development in the southern region of Bangladesh. The National Plan for Disaster Management was prepared by the Ministry of Food and Disaster Management in 2008 to reduce the risk to people, especially poor and disadvantaged, from the effects of natural, environmental, and human-induced hazards.

The Local Government Engineering Department has recently prepared the District Development Plan and Upazila Development Plan. Furthermore, some strategic plans, such as the Five Year Plans and Perspective Plan, have been formulated by the General Economics Division of the Ministry of Planning, which steers the country's development activities.

The Ministry of Local Government and Rural Development has prepared the National Policy for Safe Water Supply and Sanitation to provide safe drinking water

Fig. 7.7 River bank erosion threatening a village. *Photo* J.M. de Heer

and sanitation. Sanitation has received international attention. In September 2000, the United Nations General Assembly endorsed 8 Millennium Development Goals (MDGs). Bangladesh has already met the MDG target of fifty percent of the population with access to improved sanitation (MICS, 2013). However, the target set in 2000 of 100% sanitation coverage in accordance with the National Sanitation Policy is still to be achieved.

Following the Millennium Development Goals (MDGs) of 2015, the United Nations has taken the Sustainable Development Goals (SDGs), an intergovernmentally agreed set of targets relating to international development. To achieve these goals, Bangladesh needs to invest heavily in infrastructure development and technological innovations to make the nation safe, resilient, and sustainable.

The good thing that came out of these various phases of water resources planning is the new planning approach that has undergone many stages and refinement, responding to the socio-economic needs of the people. The existing strategies and plans have advanced the country in water resource development and climatic risk reduction. However, none of these plans were implemented totally and also not promptly. All these sectoral plans are short-term oriented and independently pursued by the formulating ministries or departments. The looming requirement is to maintain food sufficiency in the face of the increasing population and the challenges of climate change. Bangladesh needs a long-term vision and planning to meet the higher water demand. Given the delta issues and challenges, the implementation gap has to be bridged. The challenge lies in better integrating and mainstreaming these sectoral and independent plans into long-term coherent strategies, considering climate change and future demand.

A comprehensive, adaptive, and integrated approach to address the issues and challenges including climate change is important for three reasons. First, the sectoral plans typically tend to be short-term oriented and independently pursued by the formulating ministries or departments. However, goals and targets are at the national level, and climate change and natural disaster risks present major downside risks and uncertainties that require long-term strategies and multi-sectoral coordinated policy management under uncertainty. For example, the national challenge to maintain food sufficiency in the face of increasing population and decreasing agricultural land, as well as the threat posed by climate change, requires coordinated policy actions involving MoA, MoEFCC, Ministry of Land (MoL), Ministry of Fisheries and Livestock (MoFL), Ministry of Water Resources (MoWR), Ministry of Local Government, Regional Development and Cooperatives (MoLGRD&C), Ministry of Finance (MoF), and Ministry of Planning (MoP).

Second, due to the large uncertainties concerning climate change and socio-economic development, planning is being enriched with adaptive strategy making in several deltas in the world. Rather than providing linear recipes, robust and flexible strategies and measures have been taken, with strong institutions and a good knowledge base that allows policy makers and stakeholders to anticipate and decide on the most appropriate investments. Learning from these international experiences, BDP 2100 has been similarly developed in light of possible paths and is designed to be changed over time as new information becomes available or policy priorities change.

So, instead of only focusing on short-term 'trial and error' actions and projects, the idea is to keep the long-term vision in mind while prioritizing short-term 'no-regret' actions.

Finally, sound management of water safety, food security, and socio-economic development issues requires a balanced combination of investments, policies, and institutions. Earlier sectoral plans were mostly investment-centric. Policies and institutions for sustainable management of water resources have been generally overlooked. The First Five-Year Plan emphasized policies and institutions for sustainable management of water resources. These included decentralized management of water resources through the involvement of beneficiaries, the importance of O&M, the need for cost recovery (beneficiary pays principle), and a balanced combination of flood control with irrigation schemes. These sound water management principles were not well integrated into practice, and a national-level plan was needed to incorporate these issues.

The Bangladesh Delta Plan addresses all these issues by developing long-term strategies for each hotspot. The Delta Plan also focused on multi-sectoral coordinated policy management under uncertainty through the Adaptive Delta Management approach. The Bangladesh Delta Plan 2100 is different from previous water management plans in several ways:

- Long-term perspective: The Bangladesh Delta Plan 2100 is a long-term plan that covers the period from 2010 to 2100, whereas previous plans have been shorter-term and focused on specific issues or projects.
- Holistic approach: The plan takes a holistic, multi-sectoral approach to addressing the challenges facing the delta region, including water management, agriculture, infrastructure development, and climate change adaptation. It herewith meant to facilitate food security and socio-economic development.
- Climate change adaptation: The plan strongly focuses on climate change adaptation, which is a key consideration for the delta region given the projected impacts of sea-level rise, increased flooding and cyclones.
- Integrated and adaptive: The plan is designed to be integrated and adaptive, considering the complex interactions between different sectors and the dynamic nature of the delta region.
- Involvement of stakeholders: The plan involves a wide range of stakeholders, including government agencies, the private sector, civil society organizations, and international partners, in its development and implementation
- Emphasis on research and monitoring as well as reflective evaluation and updating: the plan strongly emphasizes research and monitoring to improve understanding of the delta region and track progress in implementing and renewing the plan.
- Emphasis on community participation: The plan encourages community participation in various stages of the planning process, implementation and monitoring of the plan, which is crucial for the success of the plan.

The formulation of the Bangladesh Delta Plan 2100 is hosted by the GED, an agency responsible for matters relating to long-, medium-, and short-term national plan preparation and formulation of economic policies. Thus, the BDP takes into consideration the ongoing and future initiatives and investment portfolios and analyses them in light of the long-term delta vision, goals, scenario, and preferred strategies.

References

Aggarwal, P. K., Kumar S. N., & Pathak H. (2010). *Impacts of climate change on growth and yield of rice and wheat in the Upper Ganga Basin.* WWF Report.

Ahmed, M., & Suphachalasai, S. (2014). *Assessing the costs of climate change and adaptation in South Asia.* Asian Development Bank. https://www.adb.org/publications/assessing-costs-climate-change-and-adaptation-south-asia

Alam, S., De Heer, J. M., & Choudhury, G. A., (Eds.). (2018). *Bangladesh delta plan 2100: Baseline studies on water resources management* (Vol. 1). General Economics Division, Planning Commission, Government of Bangladesh. https://plancomm.gov.bd/

Alam, S., De Heer, J. M., Choudhury, G. A., (Eds.). (2018). *Bangladesh delta plan 2100: Baseline studies on disaster and environment management* (Vol. 2). General Economics Division, Planning Commission, Government of Bangladesh. https://plancomm.gov.bd/

Ali, M. Z., Carlile, G., & Giasuddin, M. (2020). Impact of global climate change on livestock health: Bangladesh perspective. *Open Veterinary Journal, 10*, 178–188. https://doi.org/10.4314/ovj.v10i2.7

Anwar, M. S., Rahman, K., Bhuiyan, M. A. E., and Saha, R. (2022). Assessment of sea level and morphological changes along the eastern coast of Bangladesh. *Journal of Marine Science and Engineering, 10*(4), 527.

Aziz, M. S. B., Hasan, N. A., Mondol, M. M. R., Alam, M. M., & Haque, M. M. (2021). Decline in fish species diversity due to climatic and anthropogenic factors in Hakaluki Haor, an ecologically critical wetland in northeast Bangladesh. *Heliyon, 7*(1), e05861.

Barua, P., Rahman, S. H., Barua, S., & Rahman, I. M. (2020). Climate change vulnerability and responses of fisherfolk communities in the South-Eastern coast of Bangladesh. *Water Conservation and Management, 4*(1), 20–31.

Brammer, H. (2014). *Climate change, sea-level rise and development in Bangladesh.* UPL, Dhaka.

Caesar, J., & Janes, T. (2018). Regional climate change over South Asia. (2018) In R. J. Nicholls, C. W. Hutton, W. N. Adger, S. E. Hanson, M. Munsur Rahman, M. Salehin. (Eds.). *Ecosystem services for well-being in deltas* (pp. 207–221).

Carbon Brief. (2018). *Mapped: How every part of the world has warmed—and could continue to.* Infographics-multiple authors. URL: https://www.carbonbrief.org/mapped-how-every-part-of-the-world-has-warmed-and-could-continue-to-warm.

CEGIS and Department of Environment. (2011). *Final report on programs containing measure to facilitate adaption to climate change of the second national communication project of Bangladesh.* Dhaka, Bangladesh.

CEGIS. (2021). *Technical note: Impacts of projected sea level rise on coastal embankments.* Dhaka, Bangladesh.

Chakraborty, J., Grineski, S. E., & Collins, T. W. (2019). Hurricane Harvey and people with disabilities: Disproportionate exposure to flooding in Houston, Texas. *Social Science and Medicine, V, 226*, 176–181.

Climate Analytics. (2018). *Sea level legacy: 20cm more rise by 2300 for each 5-year delay in peaking emissions*, https://climateanalytics.org

Crimmins, A., Balbus, J., & Gamble, J.L. et al. (2016). *The impacts of climate change on human health in the United States: a scientific assessment.* U.S. Global Change Research Program.

Davis, K. F., Bhattachan, A., D'Odorico, P., & Suweis, S. (2018). A universal model for predicting human migration under climate change: examining future sea level rise in Bangladesh. *Environmental Research Letters, 13*(6), 064030.

Ezaz, G. T., Zhang, K., Li, X., Shalehy, M. H., & Hossain, M. A. (2022). Spatiotemporal changes of precipitation extremes in Bangladesh during 1987–2017 and their connections with climate changes, climate oscillations, and monsoon dynamics. *Global and Planetary Change, 208*, 103712.

Forastiere, F. (2010). Climate change and health: A challenge for epidemiology and public health. *International Journal of Public Health, 55*, 83–84.

Habiba, U., Abedin, M. A., Shaw, R., & Hassan, A. W. R. (2014). Salinity-induced livelihood stress in coastal region of Bangladesh. In *Water insecurity: A social dilemma* (Vol. 13, pp. 139–165). Emerald Group Publishing Limited.

Hossain, P. R., Amjath-Babu, T. S., Krupnik, T. J., Braun, M., Mohammed, E. Y., & Phillips, M. (2021). Developing climate information services for aquaculture in Bangladesh: A decision framework for managing temperature and rainfall variability-induced risks. *Frontiers in Sustainable Food Systems, 5*, 677069.

Hussain, S.G. (2011). Assessing impacts of climate change on cereal production and food security in Bangladesh. In *Climate change and food security in South Asia* (pp. 459–476).

IPCC. (2022). Climate change 2022: Impacts, adaptation and vulnerability. Contribution of working group II to the sixth assessment report of the intergovernmental panel on climate change. In H.-O. Pörtner, D. C. Roberts, M. Tignor, E.S. Poloczanska, K. Mintenbeck, A. Alegría, M. Craig, S. Langsdorf, S. Löschke, V. Möller, A. Okem & B. Rama (Eds.), (pp. 3056). Cambridge, UK, NY: Cambridge University Press. https://doi.org/10.1017/9781009325844

IPCC (2023). Climate Change 2023: Synthesis report. Contribution of working groups I, II and III to the sixth assessment report of the intergovernmental panel on climate change. In H. Lee & J. Romero (Ed.). Geneva, Switzerland.

Jakariya, M., & Islam, M. R. (Eds.). (2021). *Climate Change in Bangladesh. A cross disciplinary framework.*

Loucks et al. (2010). Sea level rise and tigers: Predicted impact to the Bangladesh's Sundarbans mangroves. In: *Climate Change* 98, pp. 291–298.

Marsooli, R., Lin, N., Emanuel, K., & Feng, K. (2019). Climate change exacerbates hurricane flood hazards along US Atlantic and Gulf coasts in spatially varying patterns. *Nature Communications, 10*, 3785. https://doi.org/10.1038/s41467-019-11755-z

Malhi, G. S., Kaur, M., & Kaushik, P. (2021). Impact of climate change on agriculture and its mitigation strategies: A review. *Sustainability, 13*(3), 1318.

Mengel, M., Nauels, A., Rogeli, J & Schleussner, C. F. (2018). Committed sea-level rise under the Paris agreement and the legacy of delayed mitigation action. In: *Nature Communications* (Vol. 9, Article number: 601).

MICS. (2013). Unicef Multiple Indicator Cluster Survey 2012–2013 Report.

Ministry of Environment, Forest, and Climate Change. (2022). *National Adaptation Plan of Bangladesh* (2023–2050).

Patz et al. (2000). Effects of environmental change on emerging parasitic diseases. In: *International Journal for Parasitology*, Vol. 30/12–13.

Patz et al. (2014). Climate Change: Challenges and Opportunities for Global Health. JAMA.

Patz, J. A., Stull, V.J., & Limaye, V. S. (2020) A low-carbon future could improve global health and achieve economic benefits. *JAMA 2.* https://doi.org/10.1001/jama.2020.1313

Rahman, M. L., Shahjahan, M., & Ahmed, N. (2021). Tilapia farming in Bangladesh: Adaptation to climate change. *Sustainability, 13*(14), 7657.

Singh, O. P., Khan, T. M. A., & Rahman, M. S. (2001). Has the frequency of intense tropical cyclones increased in the north Indian Ocean? *Current Science*, 575–580.

SMRC. (2003). *The vulnerability assessment of the SAARC coastal region due to sea level rise: Bangladesh case.* SAARC Meteorological Research Centre, Dhaka.

UNFCCC. (2012). *Climate Displacement in Bangladesh.*

UNDP. (2021). Global Climate Risk Index.

UNDRR. (2021). Annual Report.

Part III
Adaptation and Transformation Practices

Strategy Making Experiences in Bangladesh

Chapter 8
Strategy Development Under Uncertainty

Abstract The indicated issues and challenges are considered in the perspective of development of Bangladesh. In this way, the Bangladesh Delta Plan is expected to take up a role in contributing to national development goals. Given the climate challenges, a long-term strategic vision, mission, and goals are formulated. However uncertainty about the way climate change and transboundary issues will develop and what the impact will be dominates the future perspective and decision making on the right measures for goal achievement. The country is extremely vulnerable to (climate change-induced) disasters due to its natural conditions and limited protection. The major rivers are, e.g., already highly erosion prone, which may increase with larger discharges due to climate change-related heavy rains and melting water from the Himalayas. Every year, around 50 ~ 60 thousand river erosion victims migrate towards the cities and chars and adapt to an altered way of living. There are other natural disasters as well, affecting lives and livelihood. A way to deal with uncertainties over time is creating plausible narratives of possible futures. These narratives are normally indicated as scenarios. To be more precise, they are external scenarios, meaning that the narratives are built on external drivers of change which cannot or hardly be influenced. Examples are climate, transboundary water issues, natural hazards, population growth, etc. The idea is that under each scenario the realization of the vision and goals is explored and that strategies and measures are identified accordingly. Strategies and measures which play an important role in multiple scenarios for realization of the vision and goals are considered to be robust with a high probability of being selected as preferred strategies and measures. This chapter presents the way uncertainty has been approached, in order to make adaptive decision making possible on effective strategies and measures.

Keywords Strategy · Strategy development · Strategy formulation · Uncertainty · Uncertainty management · Risk · Risk reduction · Visioning · Back casting · Adaptation by development · Scenario · External development · External driver · Driver uncertainty matrix · Narrative of possible future · Robust · No regret measure · River discharge · Melting ice · Water variability · Economy

J. M. de Heer and G. A. Choudhury, *Strategies for Adaptation to Climate Change in a Transformative Approach*, Springer Climate,
https://doi.org/10.1007/978-3-031-88884-7_8

8.1 Adaptation to Climate Change in Broader Bangladeshi Considerations

8.1.1 A Delta in Development

Bangladesh is located in the downstream of the largest delta in the world, lying at the confluence of three of the worlds' major rivers the Ganges, Brahmaputra, and Meghna. The country only entails 7% of the integrated basin area of these rivers, the remaining portion lying within upstream countries India, China, and Nepal. These river systems carry between 1 and 1.6 trillion cubic meters of freshwater and 0.75–1 billion tons of sediment annually, the status of which is dynamic and considered by the people as lifeline of the Bangladesh delta. This preserves a decisive significance and challenge, with around 93% of the total basins actually lying outside Bangladesh. Over the centuries, the natural formed delta has also become highly modified due to human-induced activities interfering with the natural water dynamics. The delta is densely populated with about 172 mln people (2023; increasing by about 1% per year) and growing economy with GDP of $ 454 billion (FY 2022–2023).[1] A major concern in the country is that the usual dynamics of the delta systems and various areas, combined with natural hazards and climate change are threatening the large population and further socio-economic development.

The lower riparian geographic location and flat terrain of Bangladesh forms concerns, for which Bangladesh is largely dependent on the upper riparian countries for water availability and use. The country faces water scarcity in the north–west drought-prone areas during the dry season, whereas a significant portion of the country is flooded during monsoon. Such seasonal water variances affect the water-fed occupations of the country (irrigation, fishing, etc.), and hence shape up the socio-economic conditions and development. In addition, physical and chemical pollution to both the surface and groundwater sources is a serious problem, which hampers the domestic use of water. In the urban areas, lowering of the groundwater table due to uncontrolled water abstraction is a prominent concern which impacts the long-term water security and leads to soil subsidence. Furthermore, the existing wetlands and surface water sources are often subjected to uncontrolled encroachments due to rapid anthropogenic expansion.

One of the most prominent issues of the delta is climate variability, and the uncertainties in predicting long-term climate change impacts. The country is extremely vulnerable to (climate change-induced) disasters due to its natural conditions and disaster management difficulties. The major rivers are highly erosion-prone. Every year, around 50 ~ 60 thousand river erosion victims migrate towards the cities and chars and adapt to an altered way of living. There are other natural disasters as well, affecting lives and livelihood. Around 20% of the country is inundated under average flooding situations (around 60% during extreme flood events). In the haor and wetland regions, flash floods are more common, which bring in huge damage to

[1] Ministry of Finance, Chap 2-GDP, Savings and Investment (2023).

agricultural crops and other assets. Moreover, in the coastal regions of the country, a lot of flat and unprotected places are severely impacted due to cyclones and storm surge events. There are also climate change-induced issues such as sea-level rise and saltwater intrusion problems. The salinity frontier of the country is moving upstream and around 40% area of Bangladesh is currently susceptible to dry season saltwater intrusion. In many coastal places, saltwater intrusion even contaminates the subsurface layers, where safe drinking water is very scarce.

The dry season water shortage often brings in drought and other water scarcity issues. Many of the rivers directly dependent on the flow of Ganges and Jamuna, have died and de-linked from its tributaries. On the other hand, in the coastal region, rivers are continuously being silted up, leading to reduced conveyance and drainage congestion issues. In the greater coastal areas (Noakhali, Khulna), drainage congestion and waterlogging have emerged as the most prominent issues in recent years.

The water sector of the country is also affected by demographic and economic growth. A high rate of urbanization has stressed the land use-water use interaction, and thus affected freshwater resources reserve. Low capacities in managing the available water resources, coupled by the existing operations and maintenance limitations in many places of the country have also resulted in poor sustainability of resources.

As with many deltas, the Bangladesh delta has many challenges, a lot of which are closely related to its position downstream of major, sediment-laden rivers and an intense pressure on the scarce land and water resources. Although the challenges are many and important and both man-made and natural, the country is equally characterized by its resilience. There is traditionally a strong ability to adapt to complicated and changing conditions and to profit from the abundant resources available in the delta. Building a climate-resilient society is perhaps the biggest challenge of the country in this dynamic delta. The country needs to overcome its hurdles in tackling the consequences of frequently occurring natural disasters. Maintaining the coastal polders and sustaining food security are real challenges in the coastal areas, whereas ensuring community-level climate sustainability is also important. Bangladesh wants to strengthen its capacity for long-term climate change adaptation, accepting uncertainties while using scenario development, and hence implement flexible strategies for ensuring sustainable development.

Management of saltwater intrusion, flood risk mitigation, climate proofing, dry season flow augmentation, erosion protection and river navigability restoration, removal of drainage congestion and water logging, and restoring surface and groundwater quality in the sensitive hotspots are some of the salient challenges of Bangladesh. Overall, sustainable use of water resources is important, especially in the north–west and north central regions where water has gradually emerged as a scarcer resource. Moreover, maintaining connectivity by restoring the river system, revitalizing regulated freshwater flow in the water courses form a big challenge.

Also the quality of water has challenges in many places. The urban areas suffer from water pollution and need integrated water treatment and sustainable water supply facilities. Spatial planning implications for sources allocation and equitable utilization of resources are another endeavor for water resources development. The implementation of the Integrated Water Resources Management (IWRM) concept in the development projects is required, in relation to strategic choices in other sectors. Besides the national, multi-sectoral problems, choices and priorities, resolving conflicts and disputes regarding community-level water management and use is also of high significance in the rural areas.

8.1.2 *Considering Earlier Strategies*

The Bangladesh government initiated a number of sectoral master plans in the past, dealing with different development strategies to ensure Integrated Natural Resources Management, see Sect. 3.1. The sectoral plans and policy documents before or during the late 1990s (IECO Master Plan, National Water Plan, Flood Action Plan Studies National Water Policy, National Water Management Plan) do not, or indirectly, address contemporary issues and challenges of Bangladesh, such as climate change, sea-level rise, transboundary water management, etc. The policies and plans formulated later on (Coastal Zone Policy, Bangladesh Climate Change Strategy and Action Plan, Haor Master Plan of Haor Areas, Master Plan for Agricultural Development in the Southern Region of Bangladesh) have expanded their domain to cover such aspects, however the temporal extent of such plans is confined within short- to medium-term (20 to 30 years), limiting the prospects to properly address long-term developments and their implications in an adaptive way (Alam et al., 2018a). The advantage of the earlier studies, considerations and plans is that suitable strategies and measures which have not been realized can be considered and possibly utilized in the BDP 2100 process.

8.2 Visioning: Adaptation by Development and Transformation

A Bangladesh Delta Plan would offer a long-term perspective with an adaptive planning concept, considering the spatial and temporal extents of all the aforementioned issues. It is seen as an adaptive, holistic, and long-term strategic plan to help steer the country's sustainable development pursuit to the next century. The delta plan clusters the past and existing planning experiences of the country during its formulation process, and formulates flexible strategies in the form of programs and projects to be implemented in the different sectoral planning domains. The use of scenarios show plausible and possible futures, based upon which the robustness of alternative strategies is analyzed.

By developing and adopting this plan, government undertakes to tackle the challenges mentioned above in an integrated and coherent manner and take concrete short- and long-term steps towards developing a safe, resilient, and prosperous delta, the backbone of which is formed by a robust water system. In addition, in view of the major uncertainties concerning the future, Bangladesh aims to develop its capacity to adapt flexibly and productively to future climate and socio-economic changes as well as international developments in the basins.

A widely accepted and approved long-term delta vision and a related set of goals are of utmost importance, to guide the formulation of short-, medium-, and-long term strategies. The long-term broader strategies are aimed at fulfilling the delta vision, the medium-term strategies are targeted towards achieving different immediate goals and targets in compliance with the country's vision for 2041, and the short-term strategies guide a set of 'no regret' measures to help the country achieve comprehensive benefits within the existing five-year planning horizons up to 2030. So, and that is quite a difference with other delta plans, the overall ambition is development of the country towards a prosperous Bangladesh delta, without frequent physical and socio-economic setbacks. Ambition of BDP 2100 in relation to the outlined broader policy context is to achieve a sustainable, integrated, and adaptive delta, water management and governance, promoting regional cooperation and managing risks of climate change.

The Delta Vision builds on key strategic documents such as Vision 2021, Perspective Plan 2011–2021, National Sustainable Development Plan and Social Protection Plan and related strategic and sectoral plans. This has been developed while Vision 2041 was under process. Within the framework of the national vision, the Delta Vision is essential by indicating opportunities and choices regarding its water and land resources in close relationship with the people and environment.

The future economic development of Bangladesh, as well as the deltaic challenges and problems are likely to be influenced by socio-ecological uncertainties such as, environmental degradation, climate change-induced sea-level rise and salinization, frequent drought and floods, land subsidence, population growth, urbanization, industrialization, etc. As such, an integrated approach to foster future land and water management, in relation to water safety and food security is needed; see Fig. 7.5. In order to optimize the interventions and investments to prepare for climate change as well as for other uncertainties, elaborating an adequate water governance framework and investment plan belong to the strategy process. The existing water governance setup of Bangladesh needs strengthening and integration by emphasizing the future challenges to achieve water-related safety, and increased agricultural production to enable positive arrangements for long-term sustainable growth.

8.2.1 Delta Vision

Given the socio-economic diversification of Bangladesh and the expected impacts of climate change, thinking about the future appeals to the imagination and leads to formulating the aspirations in an integrated and long-term vision. The comprehensive delta vision is defined as: 'Achieve long term economic growth and environmental sustainability while effectively coping with climate change, natural disasters and other delta issues through robust, adaptive and integrated strategies, and equitable water governance'.

This delta vision is supported by a set of contemporary goals, tangible and intangible indicators and phase-wise targets to be achieved. The vision also provides guidance for preparation of measures and investment plan and hence development of a portfolio of possible 'no regret' projects to achieve short-term targets. To support the sustainable development of the water system of the Bangladesh delta as outlined in the Delta Vision, and to tackle the key issues outlined before, the Delta Plan aims to achieve six core goals within the timeframe of the plan. The core goals are each linked to a set of decision support indicators (DSIs) indicators, differentiated in time and space.

8.2.2 Delta Goals

The goals to achieve the delta vision have been identified based on three distinctive subjects or vision elements namely, water safety and security/availability, food security, and socio-economic development. A sustainable environment is closely related to these three and is also a contributing factor to the goals. Comprehensive review of national-level policies, plans, or strategies provided the principles of the identified goals; see Fig. 8.1.

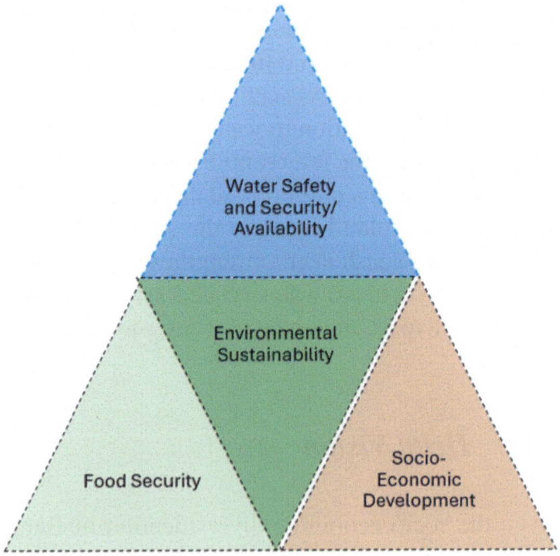

Fig. 8.1 Main vision elements for goal formulation on achieving the delta vision. *Source* BanDuDeltAS

8.2.2.1 Goal 1: Ensure Safety from Floods and Climate Change-Related Disasters

The first goal of the Delta Plan focuses on managing the risks of floods and other climate change-related disasters in the Delta at a well-defined and acceptable level. This level is set to facilitate (i) a safe living environment for all, in urban and rural environments; (ii) reliable water system conditions for long-term economic development; and (iii) the performance of key societal and economic functions. The risks of floods and climate change-induced disasters are managed according to knowledge base and spatially differentiated safety levels and include the combined approach of climate proofing, early warning and prediction, prevention, evacuation, and disaster relief. The goal considers ensuring safety of lives and livelihood against flood, flash flood, drought, salinity intrusion, river bank erosion, cyclones, storm surge, etc., through introducing improved adaptation and mitigation options. Using of technological innovations act as a focal point to achieving the goal.

8.2.2.2 Goal 2: Enhance Water Security and Efficiency of Water Usages

Maintaining a balance between the economic developments, expanding population and the need to secure water for multifaceted uses are the core of the second goal of the Delta plan. The goal aims to ensure reliable and adequate provision of freshwater to support equitable and sustainable economic development, environmental sustainability and livelihood security. This includes: (i) sufficient and timely provision of safe surface and groundwater for drinking, agriculture, fisheries, environment, navigation, industry, etc.; and (ii) controlling pollution, ensuring water quality and providing sanitation at acceptable levels in relation to defined standards as well as environmental, health, agricultural, and industrial needs.

8.2.2.3 Goal 3: Ensure Sustainable and Integrated River Systems and Estuaries Management

This goal considers sustainable management of river systems and estuaries, considered as one of the cornerstones of the Bangladesh Delta. The river systems of the country offer key development ingredients for the economy, society, and the environment. Sustainable river and estuaries management as a whole, contributes to (i) long-term solutions for navigation; (ii) sufficient sediment supply to floodplains; (iii) land reclamation in support of community development and economic growth; and (iv) an inter-connected major and minor river system for environmental and economic development. This goal seeks to foster sustainable river system management by coping with its morphological dynamics and results in cost-effective, environment-friendly, and socially viable options for reclamation and developments of lands.

8.2.2.4 Goal 4: Conserve and Preserve Wetlands and Ecosystems and Promote Their Wise Use

The goal realizes the term 'conservation' as a more generic target to achieve to safeguard and maintain the wetlands, whereas 'preservation' as a more converging approach, to protect some sensitive ecosystem areas. Some of the components to be safeguarded under the goal are mangrove forest, water, and riverine ecosystem, coastal greenbelt, seasonal and perennial wetlands, tidal freshwater frontier, etc. Conservation and preservation of wetlands is also emphasized by the Government of Bangladesh, in its recent national-level policy documents.

8.2.2.5 Goal 5: Develop Effective Institutions and Equitable Governance for In-Country and Transboundary Water Resources Management

The goal aims to improve the existing water governance for tackling both the endo and exogenous challenges of water resources management. At the same time, the community-level functioning of local institutions would also be rejuvenated. Key elements include: (i) ensuring adequate financing for investment, operation and maintenance; (ii) capacity building and institutional strengthening; (iii) stakeholder and private sector participation at all levels; (iv) minimizing the gaps between planning and implementation; (v) supporting integrated water resource management at regional and local levels, taking into account the interests of all prominent water users; (vi) knowledge development for integrated decision making though modelling, data and information sharing and management; and (vii) transboundary water negotiation.

8.2.2.6 Goal 6: Achieve Optimal and Integrated Use of Land and Water Resources

The primary quest of the goal is to ensure efficient integration and proper interaction between land and water use aspects. The goal would mainstream the inter-connected resources functions (drainage, navigation, etc.) as well. The key topics covered by the goal are land use, water use, multi-modal transportation infrastructures, navigation network, port connectivity, utilities, etc. Establishment of connectivity of water courses, and inducing integration between different modes of resources interactions are highly encouraged under this goal.

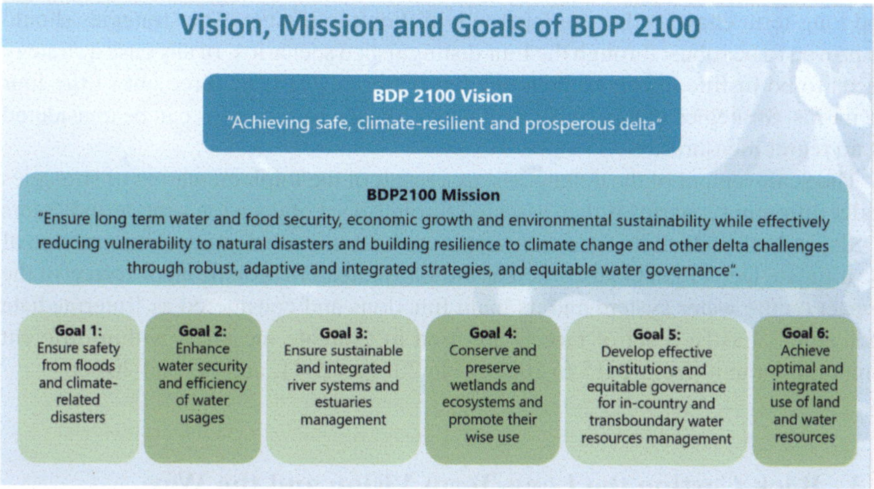

Fig. 8.2 BDP 2100 vision, mission, and goals. *Source*: BDP 2100, Volume 1: Strategy, pp. 180–183 (2018)

8.2.3 Overview of Delta Vision, Mission, and Goals

Considering the long-term challenges presented by climate change and natural hazards, Bangladesh has adopted a long-term delta vision of achieving a safe, climate-resilient, and prosperous delta. For achieving this vision by the end of the twenty-first century, Bangladesh defined 3 higher level goals on: elimination of extreme poverty by 2030, achieving upper middle income status by 2030, and being a prosperous country beyond 2041. Six dedicated delta goals have been formulated additionally; see Fig. 8.2. BDP 2100 has taken the climate-change issue as an exogenous variable in developing the macro-economic framework of the plan. Here lies the unique characteristic of BDP 2100. The principle of BDP 2100 is adaptation by development while living with nature.[2]

Indicators are identified to operationalize these goals in such a way that the present state of the Bangladesh delta can be monitored and quantified based on the baseline studies. This enables selection of strategies to come from the present state to the goals to be achieved. An Assessment Framework is formulated to assess and evaluate goal achievement by the strategies. The framework is prepared with a set of indicators of the water system (state indicators) and policy priorities (decision support indicators), used to evaluate the strategies in terms of achieving the formulated delta goals.

Scenarios (see Sect. 8.5) are descriptions of future situations based on plausible socio-economic and physical development paths. Scenarios are used to evaluate the robustness of selected strategies and design a preferred strategy that includes no-regret measures, and a number of alternative pathways that can deal with medium-

[2] See also: Shaw et al. (2022).

and long-term changes that are inherently difficult to predict. The strategies should be proven to be robust through the four distinguished scenarios. In any case a strategy needs to be justified in more than one but preferably at least three out of the four scenarios. Strategies that perform well under a range of scenarios can be considered as no regret measures.

The achievement of the delta goals, as a result of the implementation of strategies under different scenarios is then assessed through a set of decision support indicators (DSIs) and sub-indicators. Decision support indicators provide impact information of the ultimate Delta Plan Goals, whereas the sub-indicators provide an overview of the impact on the water system and its main functions and considered as 'Intermediate Indicators'. Lastly, a set of targets are used to evaluate the goals within different temporal domain (from 2015 to 2021, from 2021 to 2041, and beyond 2041).

8.3 Back Casting the Long-Term Vision and the Way Forward

8.3.1 Back Casting for Vision Realization Under Conditions of Uncertainty

Although Bangladesh has adopted several initiatives for ensuring climate change adaptation and resilience, policy makers are still struggling to deal with uncertain impacts of climate change. This happens because of two key issues. First, many parameters of future climate factors are uncertain. For instance, salinity intrusion is a key challenge for Bangladesh, but policy makers do not have a precise idea regarding the total area that will be affected by salinity in 2040 or 2050. A similar situation is also prevailing in future rainfall patterns, temperature, and cyclone patterns. While the Bangladesh delta is fighting to address the increasing impact of climate change, policy makers do not have any precise idea of the future climatic outlook, hence are uncertain about strategies, measures, and investments.

Second, along with the future climate, future socio-economic changes are also uncertain in Bangladesh. The river system plays a crucial role in socio-economic development. However, the river system is dynamic, with the frequent shifting of banks. Along the Padma River, the average erosion and accretion rates are even 1472.056 and 1610.152 ha/year.[3] Often large rivers change their course with rapid erosion and accretion. For instance, the Old Brahmaputra River changed its initial course and moved in a separate direction from the main Brahmaputra River. With such large-scale erosion and accretion and movements, it often becomes difficult to fund sustainable solutions and design and execute interventions accordingly.

Most of the areas of Bangladesh are densely populated, with a majority of the people depending on agriculture and agriculture-related activities. But with

[3] Ophra et al. (2018).

rapid industrialization, this situation is expected to change in the future. Besides, the country is also moving from traditional agriculture to commercial agriculture. Bangladesh is rapidly experiencing changes in land use patterns in different areas with the introduction of shrimp cultivation and different cash crops like watermelon. The water requirement and use patterns for these crops are different. For instance, rice cultivation depends on the season-long availability of irrigation water, which is not the case for watermelon. Having too much water in the cultivation area can be counterproductive for watermelon farmers.

These changes will trigger a significant socio-economic change and modify people's development priorities. In an uncertain environment, individuals try to be risk-averse and avoid innovation. With such realities, policy makers and planners must come forward and assure individuals and firms of the government's long-term commitment. Primarily if policy makers can provide a very long-term strategic plan (like up to 2100) for the country that can help individuals and firms plan their activities. For instance, if a farmer is unsure about the future flood protection strategies in his area, he will be reluctant to try a new crop variety. However, if a national-level long-term plan specifically discusses the flood protection strategies for his/her region, then that same farmer can plan individual initiatives and, e.g., invest money and time to learn the cultivation process of a new crop variety that is more suitable for his region. Having said that, it is also challenging for policy makers and planners to develop that long-term plan, again because of the uncertain behavior of nature. Bangladesh's policy makers and planners started this difficult journey of developing a long-term delta plan up to 2100 with the support of the Netherland Government. In Bangladesh Delta Plan, the policy makers used a transformative technology management approach to address different uncertainties, meaning that technologies from many disciplines or sectors have been used and managed in such a way that strategies and measures for adaptation could be defined. Strategies and measures are called transformative, when they enable adaptation of systems or institutions to the significant changes triggered by climate change.

In the future, climate change will severely impact the people living in the Bangladesh delta. According to the latest IPCC report, from 2012 to 2050, freshwater river area is expected to decrease from 40.8 to 17.1–19.7% under different sea-level rise scenarios in the southwest coastal zone of Bangladesh. In coastal Bangladesh, projected saline inundation to wetland ecosystem services will result in ecosystem services losses of raw materials and food provisioning, ranging from USD 0 to 20.0 million under RCP2.6 to RCP8.5 scenarios.[4] While the IPCC report provides a projection of future outlook, the actual outlook of climate change can be more severe, if countries fail to take required adaptation and mitigation.

For addressing these challenges, Bangladesh intended to prepare a scenario-based adaptive Delta Plan. With this, it shifted from a 'predict and act' approach to an 'explore, anticipate and reflect' adaptation approach that accepts that the future is deeply uncertain. Instead of responding to events like the cyclones from the past and developing a plan for addressing similar events, the Delta Plan approach asks: what

[4] IPCC (2022).

could happen in the future, and what can we do now to achieve our vision, being able to adapt to the future? For Bangladesh, this vision aimed at 2100 is 'Achieving a safe, climate resilient and prosperous delta'.

Back casting is useful here. Because with this vision of Bangladesh in 2100 in mind, one can move backward from the future vision to the present situation and then move ahead with adaptive measures step by step towards the vision. Thus back casting the strategic vision to recent time and near future brings forward inspiring points of attention and direction for action. Realization of the vision in an uncertain future means that the strategy process should be adaptive, open to intensify when needed, or easing when possible. To learn more about different possible futures, scenario development provides a way to formulate and discuss narratives of possible futures. Strategies which perform well for achieving the vision and goals under different scenarios are seen as robust and preferred strategies.

For achieving this long-term vision, the government made detailed programs to be implemented as short-term strategies and actions while monitoring the progress of these actions in response to the climate situation and socio-economic context.

8.4 Achieving the Vision Under Uncertain and Diverse Conditions

8.4.1 Scenario Approach

Scenarios are used here (see also sect. 2.5) to consider different possible futures in which major but uncertain developments are expected to play a significant role. Altogether, BDP 2100 developed four scenarios to consider realization of the long-term vision under different possible futures. The idea is than to look per scenario to issues and challenges and define strategies and measures to deal with them. Subsequently specific strategies have been proposed and were tested on robustness under each scenario for addressing different future climatic, transboundary, and economic possibilities.

The scenarios for the BDP are being developed in a participatory approach, which is crucial to ensure proper development, including the Bangladesh perspective, and create a common understanding and ownership at national level. The two-day scenario workshop that was held on February 11 and 12 at the GED provided the building blocks for the scenarios (note that most of the pictures included in this report, are from this meeting). The scenarios are being further substantiated, validated and transformed to consistent scenario narratives based on the baseline studies and further research and consultation with key experts. The scenarios are quantified on key indicators that are relevant for the impact assessment and further modeling.

Because an interactive diagnostic search and strategy process with stakeholder involvement is characteristic for the whole delta plan approach, a brief description of preparations and the scenario workshop is provided as follows. To prepare the scenarios, the following methodology has been applied:

1. Identify major driving forces of change (trends or developments) based on political, economic, social, technical, environmental, and legal factors.
2. Plot drivers based on impact and uncertainty in an impact-uncertainty matrix.
3. Take two most important drivers with highest impacts and uncertainties along XY-axis, these will form a four-scenario framework.
4. Develop four narratives based on the scenario framework, including additional relevant drivers, pressures, or states.
5. Define impacts per scenario, what may happen and what are opportunities and vulnerabilities.
6. Add to each scenario narratives with proposed strategies.
7. Test the performance of proposed strategies in the different scenarios following an assessment framework; strategies performing well under at least three scenarios are indicated as robust strategies, to be elaborated further into measures and project proposals.

Preparations for developing scenarios took place in the scenario team in terms of data collection, trend analysis, and reflecting on the impacts of developments and of the drivers behind them. An impact-uncertainty matrix was used to identify and discuss the main drivers that are considered to be the most relevant in relation to the future of Bangladesh. The impact/uncertainty matrix is used to identify what key drivers of change should be considered for the scenario development. The delta scenarios should include those drivers that will have the most impact in relation to the objectives of the BDP 2100 and that are also very uncertain.

A two-day long workshop on Scenario Development was organized by the General Economics Division (GED) of the Planning Commission of the Government of Bangladesh in collaboration with the scenario team of the consultants of the Bangladesh Delta Plan 2100 Formulation Project. Because the scenarios play an important role in the process and also because many sessions were organized in this way, the following highlights the main items of the workshops.

8.4.1.1 Scenario Building Process Experiences

The workshop was held at the NEC Conference Room of the GED from 11 to 12 February 2015, chaired by the Senior Secretary of the GED and Member of the Planning Commission. The status of this conference room where normally NEC-meetings take place of Prime Minister and Minsters as well as the high-level chairperson contributed to the importance and atmosphere of the workshop. Stakeholders from various ministries, government agencies, development partners, research organizations, and civil society were present to add value to the workshop. The objective of the workshop was to explain the steps in development of scenarios and involve relevant stakeholders in the scenario development process. Scenarios are storylines of plausible and possible futures, developed to quantitatively and qualitatively describe future developments. On Day 1, the workshop was inaugurated by the workshop chair who mentioned that the Delta Plan 2100 is a 'unique plan which can have a significant impact on the livelihood of the common people'.

The first session of the workshop, the subtleties of scenario development in the context of delta planning was brought to light by experts in the BDP 2100 Technical Assistance team. They explained the use and importance of scenarios as a tool in relation to the strategic vision and required and appropriate strategies. Scenarios are descriptions of possible futures, and they are the fulcrum of any adaptive plan. Designing possible future scenarios are all about the level of uncertainty of a multitude of factors (or drivers) which are used to construct different scenarios. It was explained that it is notoriously difficult to accurately predict the future, underscoring the need for thinking in a number of plausible futures. Since the future cannot be planned for exactly, the only feasible mechanism is to plan for multiple futures taking multiple scenarios into account. This offers the flexibility that is needed by institutions to be able to adapt to different levels of intensities on the ground, as realities change over time. The second and third sessions were interactive exercises where the more than 150 participants were divided into groups where they brainstormed to identify the main drivers of change in Bangladesh, clustered in six domains: demographics, economics, social, technical, environmental, and political. This session was interactive and lively, as stakeholders from various different backgrounds expressed their views regarding what they thought were the most important trends and developments. A host of various developments were identified, listed, and categorized. In the third session of the first day, also carried out as an interactive exercise in the groups, aimed to develop an appreciation for which of the developments identified are both highly relevant for Bangladesh future as indicated in the vision and were seen as very uncertain. This was accomplished through an impact-uncertainty matrix.

8.4.1.2 Considering the Content of Scenarios

Day 2 of the workshop brought the same stakeholders, now about 125 participants, to the table once again to discuss a possible scenario framework, proposed main items in the storytelling and impacts of major developments expected in each scenario. The session began with a recap of the sessions on the first day and the facilitators indicated that the most important developments that were identified and discussed again and again in their groups were technological progress, political unrest, natural disasters, climate change, and bilateral relations with neighboring countries with transboundary water sharing issues.

In the next session, the process of formulating scenario narratives was explained, where one thinks of uncertain developments, how they have impact, e.g., on society and economy, and use this impact analysis over time to develop stories describing plausible, consistent, and relevant developments over time for a particular future. A horizontal and vertical axis was brought to the screen to show that developments with high level of uncertainty and considerable impact can be combined in four different ways to look at and create four different plausible futures for Bangladesh.

The outcome was that variability of water conditions due to climate change and due to transboundary issues was most relevant development with highest uncertainty and impacts. Besides, high and low economic development is seen as a development with high uncertainty and high impact. With these major developments, four scenarios were identified, named productive, resilient, urbanized, and reactive Bangladesh; see Fig. 8.3.

The fifth session of the workshop was again a group exercise and the most interesting one, which called upon the participants to envisage scenarios for Bangladesh in the future, each group selecting one of the four scenarios introduced in the fourth session. To explain thinking in different scenarios, a few examples were presented. This was after verbal explanation also done by showing a video on different scenarios of the European Union in 2030, based on how united the member states could be. Secondly, the story of Bangladesh's development and growth over 40 years from 1971 to 2011 was presented with indications of important developments, to provide an example of what a scenario narrative for the future could potentially include. Where historical facts and developments of Bangladesh over 40 years were presented, participants in the working groups in the fifth and sixth sessions, imagined future scenarios. The group facilitators worked with the group members to illustrate these scenarios on flipcharts and a spokesperson from each group presented their scenarios to all the participants at the workshop.

Fig. 8.3 Plenary meeting of the scenario groups. *Photo* J. M. de Heer

The feedback from the workshop audience was very positive and it was indeed their interactive participation that made the workshop a success. There were many comments in the open discussion on the scenarios in relation to strategy making and measures to be decided upon. Moreover, the remarks included ensuring that the delta plan becomes a professional and practical exercise, but also takes into account ground realities and the capacities of those who will be implementing the plan, and about how to develop an investment plan which can be used to develop interventions through which the GED can engage various development agencies and donors in the implementation of the Delta Plan going forward. The Technical Assistance team commented that these concerns will be taken into account. At the end of the two-day workshop, the TA-Team thanked the participants for being active on both days and commented that such interesting outputs could not have been produced so quickly if such a diverse group of people did not participate enthusiastically. He indicated that the findings of the workshop will be related to a delta vision and will be enriched with data of the Baseline Studies, that the scenario narratives will be elaborated and that a validation session will take place. The session was formally closed by the chair, who invited everyone to keep participating in the formulation of the delta plan the way he has seen over the two-day workshop. He said, 'Increased participation will minimize the risk of wrong steps in the Delta Plan'.

8.4.2 *The Main Drivers for Building the Scenarios*

The BDP external scenario narratives are concentrated around two key drivers: (1) water variability based on transboundary developments and climate change and (2) economic development and related land use changes. A brief overview of the four scenarios after consultation and validation is provided in Table 8.1.

8.4.2.1 Perceived Importance of Transboundary Developments

One of the key issues with both high uncertainty and impact defined during the Scenario Workshop was the rate and level of upstream (surface water) abstractions resulting from the Indian River Linking Project (IRLP).[5] The largest anticipated development is the diversion of flow from the Brahmaputra to the Ganges and further basins within the country. Irrigation development is a key driver for this abstraction. Further upstream abstraction is also expected from regional transboundary rivers such as the Dudhkumar, Dharla, Mahananda, and Teesta[6] rivers in the North–Western part of the country and the Surma and Kushiyara rivers in the North–Eastern part of the country. The flows in these rivers are not regulated through any treaty at present. Low

[5] See BDP Baseline Reports on Regional Cooperation and Water Resources for further information.

[6] Low flows in the Teesta River, much like the Ganges, are already impacted due to upstream abstractions.

Table 8.1 Key characteristics of scenarios including upstream abstractions

Productive (Sufola)	Resilient (Sohonio)
• Very high economic growth • Moderate climate change • Strongly globalized and diversified economy with many shared interests between riparian countries • Low to moderate increase in upstream abstraction from low to medium in the medium term (2030) and high in the long term (2050–2100) • Extensive data sharing and joint (water) knowledge development • Trans-boundary agreements for main and regional rivers • Joint water development projects in upstream catchments	• High economic growth • High climate change • Moderately diversified economy, strong agro-sector, moderate shared interests between riparian countries • Moderate increase in upstream abstraction from low to medium in the medium term (2030) and high in the long term (2050–2100) • Extensive data sharing and joint (water) knowledge development • Trans-boundary agreements for main rivers • Joint water development projects in upstream catchments
Moderate (Sushoma)	**Active (Sokrio)**
• Low but stable economic growth • Moderate climate change • Closed economy with little diversified - shared interests between riparian countries • Moderate increase in upstream abstraction from low to medium in the medium term (2030) and high in the long term (2050–2100) • Little to medium data sharing and joint (water) knowledge development • Extension of Ganges treaty only	• Low and decreasing economic growth • High climate change • Closed economy with little diversified—shared interests between riparian countries • Moderate to high increase in upstream abstraction from medium in the medium term (2030) and high in the long term (2050–2100) • Little data sharing and joint (water) knowledge development • Extension of Ganges treaty only

Source BanDuDeltAS

flows in the Ganges are regulated in the 1996 Ganges Water Treaty between India and Bangladesh. In the treaty, Bangladesh is guaranteed a share of the expected minimum long-term average flow of approximately 900 m^3/s (50% of minimum of 50,000 cusecs).[7] Actual low flows since the ratification of the Ganges Treaty are however 750 m^3/s on average.[8] For the purpose of scenario development, this average will therefore be used rather than the long-term average mentioned in the 1996 Treaty.

The IRLP also includes diversion from the Ganges to other basins. However, for scenario development, it does not seem plausible that Bangladesh will accept a lower share than the present minimum, as laid down in the treaty. This amount will therefore remain constant as lower level in all scenarios. Project documents and articles suggest that the Indian Government has developed a vision of a total abstraction of

[7] Annual inflows for the three major rivers are 380 BCM for the Ganges, 537 BCM for the Brahmaputra, and 195 BCM for the Meghna.

[8] Haque Khan, IWM, presentation at GED, November 2014.

some 33 Billion m^3 (BCM) annually from the Brahmaputra (Himalayan) to other rivers (Peninsular) in the medium- to long-term. Assuming such diversions would take place evenly throughout the year, this would amount to approximately 2.75 BCM in the 'final' IRLP situation. Average low flows for the Brahmaputra entering Bangladesh are in the order of 2700 m^3/s or 7.12 BCM per month.[9] The abstraction therefore equals some 38% of the total average monthly low flows for the Brahmaputra River. For the purpose of scenario development, this percentage is taken as the upper limit of upstream abstraction. It goes without saying that this approximation is precisely that. Given the importance and potential impact, detailed scenario modeling and monitoring should take place as part of the follow up phase of the BDP 2100 formulation.

For the purpose of developing workable scenarios, it is furthermore assumed that the percentage-wise reduction in flows for other regional rivers is equal to the percentage-wise reduction for the Brahmaputra[10] for each scenario. Given the status of the IRLP there is little uncertainty on whether upstream abstractions will take place or not, the uncertainty concerns the level of abstraction (% of low flows—volume of water) and the rate at which the upstream developments are implemented over time, taking into account the relevant time horizons 2030, 2050, and 2100.

While development narratives regarding abstractions for each of the four scenarios, the following factors are also considered:

- Higher global economic growth leads to a higher upstream water demand.
- Higher economic growth in Bangladesh leads to a better negotiation position vis-a-vis its upstream neighbors.
- Increased climate change (in Bangladesh and riparian countries) leads to an accelerated rate of abstraction.

The key characteristics of the narrative regarding upstream abstractions for each scenario are summarized in Table 8.1.

8.4.2.2 Perception of Climate Change and Natural Disasters

Both climate change and natural disasters were discussed in all groups as a key driver with high impact and uncertainty. Climate change is driven by increased greenhouse gas emissions resulting in global warming. A warmer climate results in higher temperatures, increased rainfall intensity, more frequent and long dry spells, higher wind speeds, increased cyclone intensity, and sea-level rise. These changes in the climate systems result in more frequent and severe floods, reduced dry season river flows, and more intensive and frequent droughts. This has an impact on food security through changes in agricultural water use and reduced dry season water availability. Also cyclones and major river floods can have devastating impacts on food production

[9] Delta Atelier Coast, September, 2015 Report, GED.

[10] Follow-up studies were recommended for the smaller regional rivers to substantiate these assumptions for each region (e.g., for the North–West).

systems. Urban flood risks will increase due to a combination of more intensive rainfall, increased cyclone intensity, and higher peak river flows. In the coastal region the sea-level rise will add to the increased flood risks. Water quality is affected especially by increased salt water intrusion caused by a combination of sea-level rise and reduced dry season river flows. Domestic and industrial water supply will become more erratic due to increased droughts and in the coastal zone by salt water intrusion. Climate change will also affect biodiversity and environmental quality in several ways (see Chap. 4 for features of climate change in Bangladesh).

8.4.2.3 Significance of Economic Development, International Markets, and Technology Development

In many groups, drivers related to the economic development of Bangladesh were discussed in relation to international relations and trade and technology enhancements. Key issues that were discussed are:

- Globalization and development of trade.
- Human capital development (investments).
- Global market of manufactured products of Bangladesh.
- Level and driver of industrial development.
- Economic transformation.
- Investments in infrastructure.
- Digital and agricultural technology.
- Modernization of agricultural sector.
- Impact of cross-border trade.

8.4.2.4 Demographics, Urbanization, and Land Use Changes

Developments related to the rapid population growth and inherent urbanization trends were discussed in all groups. Concerns are raised on the impacts on agricultural land due to the continuous urbanization and the pattern to which urbanization is currently stretching the capacity of the major cities. Two extremes are depicted, in which on the one hand unplanned urbanization is further encroaching large cities and paralyzing city life and on the other hand, outside Dhaka other economic hubs are emerging as second-tier cities across the country. This will help to reduce the pressure on Dhaka as the one and only hub, and facilitate spreading of the urbanization trend. Main issues that were raised are:

- Population growth and composition.
- Land use changes.
- Level of unplanned urbanization.
- Decrease of agricultural land.
- Emergence of 'high input, high investment, high return' agriculture (flowers, strawberries, broccoli, etc.) when compared with low input low risk agriculture.

- Food security, livelihoods and poverty.
- Access to education.
- Change in consumption patterns of Bangladesh.
- Equality–differences between rich and poor.

8.4.2.5 Governance and Political Situation

Governance and political instability was raised in most of the groups. Political and institutional reforms are considered to be key in the further development of Bangladesh. Effective governance will shape the direction of Bangladesh to a large extent. The future situation is seen as volatile and highly uncertain.

8.4.2.6 Environment

A few groups raised the issue of water quality and environmental degradation, including degradation of biodiversity and wetlands. Salinization is seen as a main driver with regards to environmental degradation in the coastal zone, and industrial pollution is considered an important threat (especially around Dhaka and in other industrial areas). In a few cases agricultural runoff is also considered as a key determinant. Arsenic contamination of groundwater and the use of this water in agriculture are also points that need attention. One group discussed the possibility of a scenario in which the Sundarbans were completely destroyed due to salinity intrusion and unsustainable industrial activities.

8.4.3 Impact on Water Resources

Flood is a recurrent event in Bangladesh. Every year about 20–25% land is inundated, and, in extreme event, inundation could be as high as 60% or more. The river flood normally occurs during the monsoon and flash floods during pre-monsoon period. As the country is being industrialized, the economic loss of flood is climbing. Due to climate change, the economic loss will be even more.

The flood projections for the GBM basins are presented here based on downscaled climate data obtained from GBM model under the highest emission SSP5-8.5 Scenario in the 2050s.[11] The projections were presented for pre-monsoon, monsoon, post-monsoon, and dry seasons (winter); see Table 8.2. According to NAP (2022), the mean annual flows of the Ganges, the Brahmaputra and the Meghna Rivers will increase by 17–28%, 2–5%, and 1–4%, respectively. The projected inflow will increase significantly in the Ganges Basin during monsoon and post-monsoon in a

[11] Ministry of Environment and Forest and Climate Change (2022).

Table 8.2 Predicted flow in GBM-basins till 2050

Basins	Hydrological seasons			
	Dry season	Pre-monsoon season	Monsoon season	Post-monsoon season
Ganges	Upto 10% increase	10–30% increase	20–30% increase	20–30% increase
Brahmaputra	Upto 2% decrease or 10% increase	10–20% increase	Upto 10% increase	Upto 10% increase
Meghna	Upto 2% decrease or 10% increase	Upto 2% decrease or 10% increase	Upto 10% increase	Upto 10% increase

Source BDP 2100

range of 20–30%. Dry season or winter inflow will either slightly decrease (may be 2%) or increase up to 10% in both Brahmaputra and the Meghna basins.

The table indicates that the predicted flow may increase in the Ganges in the dry season by 10% and other three seasons by 30%. In the Brahmaputra, the predicted flow may decrease slightly (may be by 2%) or may increase by 10% in the dry season. In pre-monsoon, the flow is predicted to increase by 10–20% in the Brahmaputra basin. In the monsoon and pre-monsoon, the flow may increase by 10%. It is worth mentioning that the Brahmaputra has the highest contribution (65%) to annual inflow and has much impact on the flood and erosion. In the Meghna Basin, the upper hilly catchment is the source of flash floods in the northeast area of Bangladesh. It is a big concern as it damages the boro crops of the region. The monsoon and post-monsoon inflow may increase by 10%, which will retard the drainage of the north–east region by delaying the planting of boro paddy.

Waterlogging in urban areas have recently become an environmental and economic concern. The BBS 2022 survey revealed that the 31.6% population (53.76 million) is living in urban area (BBS census 2022), which is expected to grow faster in the future. More rainfall due not climate change will further worsen the situation.

Increased inflow from the upstream catchments coupled with increased sea-level rise will worsen the flood situation in the south-central and southeast regions. High tide level will retard flood flow and will prolong flood duration, heighten the flood depth as well as the flood extent.

Riverbank erosion is a major cause of poverty as the affected people lose their land, and, in many cases, have to shift the homesteads or become into homeless. According to CEGIS estimates,[12] the Brahmaputra/Jamuna, the Ganges, and the Padma rivers have eroded 99,200 ha, 44,000 ha, and 388 ha, respectively, between 1973 and 2020. In 2020 about 725 ha land eroded along Brahmaputra/Jamuna, 460 ha land along the Ganges and 780 ha along the Padma rivers. Currently, on an average, total annual erosion in Bangladesh is about 3000 ha and displacing about 25,000 people annually.

[12] See CEGIS annual erosion predictions at cegisbd.com.

It is anticipated that the impact of climate change (especially higher rainfall and higher inflow from the upstream catchments) would alter river flows and sediment transport modifying the dynamics of these rivers. As a result of climate change, erosion may worsen in the near future due to an increase in peak river flows and high-intensity rainfall that will transport more sediment from upstream catchments (Alam et al., 2018b).

Salinity is increasing. Bangladesh has a coastline of 710 km along the Bay of Bengal. The coastal zone comprises of is 39.000 km^2, protected from tidal flooding by 139 polders. However, the Meghna estuaries as well as many rivers discharge into the Bay. The salinity problems will affect these rivers in the coastal zone. With the gradual decrease in upstream freshwater inflows during dry season, the salinity level (surface water, groundwater, or soil) often rises practically linearly from October to late May. Historical salinity statistics show that the salinity increased in the Rupsa River in Khulna from 1962 to 2011 from 0.7 to 16.8 ppt. A significant amount of freshwater flows from the Padma River through the Arial Khan River and the Lower Meghna River in the south-central zone resulting in low salinity (0–2 ppt) throughout the year. River salinity will increase significantly due to sea-level rise triggered by climate change during the dry season, which will be made worse by a lack of fresh water in the major rivers. In the short-term (up to 2030) and medium-term (up to 2050), sea-level rise will cause salinity to move further inland. The distribution of surface water salinity in coastal zone because of climate change is shown in Fig. 8.4. Under IPCC SSP5-8.5 scenario, the 1 ppt salinity-affected areas may grow by 7.5% until 2050s, while the 5 ppt areas would grow by 9% intruding further inland. Increasing salinity concentration will be higher in the south central and southwestern area. Salinity increase will impact river water and other surface water, up to certain depth, ground water, soil salinity, etc.

8.5 Scenarios, Narratives of Different Possible Futures

8.5.1 The Bangladesh Delta Scenarios Following Main Drivers

8.5.1.1 Productive (Sufola), "Market-Driven Delta"

The Sufola scenario is characterized by moderate water conditions, growing population, and an increase of per capita GDP growth as a result of fast global economic growth and development of Bangladesh towards an industrialized and moderately diversified economy. Driven by strong global trade liberalization and a high international demand for manufacturing products, large foreign and domestic firms invest in the manufacturing and agro-processing industries, leading to higher labor and land productivity. Although the overall income level rises, the gap between rich and poor also increases. (Public) investment in infrastructure is necessary to enable transport

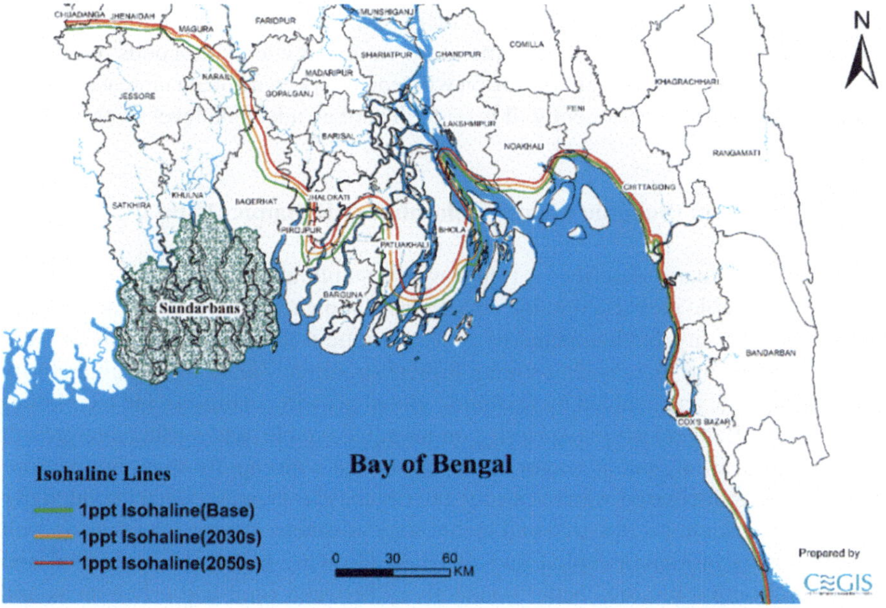

Fig. 8.4 Surface water salinity distribution in coastal Bangladesh due to climate change, the contour showing the 1 pp salinity line further inland up to 2030 and 2050. *Source* CEGIS Bay of Bengal Model

of goods and services to the markets, and other economic centers besides Dhaka and Chattogram emerge. The positive effects on the economy instigate further (private) investments in technology, human capital, and infrastructure, leading to many more high-skilled job opportunities. As the middle income population increases, food security objectives slowly shift to food variety, with an emphasis on a nutritious and varied offer. By 2050, Bangladesh has developed into a diversified market-driven economy in which the industry and service sector dominate, little regulation and in which large internationally operating firms dominate.

Towards 2100, Bangladesh has become a higher income and highly urban country, dominated by services and technology, driven by private investors and well-connected to international markets. Water demand for irrigated rice will reduce due to changes to less water demanding crops, varieties and agricultural practices. A secure water supply will become more important. Urban (domestic and industrial) water demand and water network requirements will increase due to rapid economic development and population increase. Water quality standards will increase significantly as a result of a growing and wealthier population with higher consumption standards. Competition between the agriculture, domestic, and urban water supply needs will increase, with the share of agriculture becoming less important, especially up to 2030 and 2050. Water and environmental quality will be under severe pressure in this period. The decrease of the population after 2050, accompanied by a higher standard of living and global environmental awareness will enable investment in clean technologies, a much more efficient use of water and an increased allocation to meet environmental needs. Ecosystem services will become more valued as economic goods. Increased living standards, assets, and globalization will lead to a proportionate increase in flood risks, for which the government will want to ensure an adequate level of protection.

8.5.1.2 Resilient (Sohonio Means 'Tolerable'), 'Dynamic Delta'

The Sohonio scenario describes a future with a stable increase of GDP per capita, driven by favorable global economic conditions leading to a transformation to a fully diversified economy, but hampered by more extreme water conditions, due to rapid climate change and increased upstream flow abstraction. Aware of its vulnerability to natural events, and enabled by favorable global growth conditions and technological changes, Bangladesh puts priority on an equal, stable, and sustainable growth and avails of a strong public sector equipped to deal with rapidly changing circumstances. Due to increased water scarcity, the urgency to increase water productivity is high and agricultural research and technology advances are considerable, as well as investments in human capital and services. Growing high-value industrial and agricultural output requires better connectivity between rural and urban areas and secure infrastructure. A substantial of the agricultural labor force is now employed in the agri-food processing sector, leading to high rural to urban migration. Also many people seek opportunities outside Bangladesh, which stabilizes the population growth in combination with declining fertility rates. By 2050, the country has developed in a resilient country with a stable economic growth per capita that is more equally distributed and well-equipped to adapt to natural hazards and climatic conditions.

Towards 2100, Bangladesh develops into a high-income country, which has mastered in technological adaptation and sustainability to cope with the harsh climatic conditions. Agro-technology and services drive the economy, as well as high international mobility of people, which has decreased the population significantly.

Water shortages for the agricultural sector due to climate and upstream abstraction for irrigation will be severe. This is (partly) mitigated by water saving technologies, the cultivation of high-value and low water demanding crops in the dry season and the application of sound zoning policies for agriculture, housing, industry, critical infrastructure, and environment. There will be a large increase in flood risks due to the increase of extreme events and continuing increase in asset value of industries and other valuable assets. Sea-level rise and increased cyclone intensity will further increase the flood risk in the coastal zone. Water quality will initially deteriorate due to water shortages, pressures from urban areas and industry as well as salinity intrusion in relation to (relative) sea-level rise. A growing economy and income level, especially of the middle class, in combination with a stronger enforcement of environmental standards will lead to improvements from the middle half of the century onwards.

8.5.1.3 Moderate (Sushoma), 'Delta Under Pressure'

The Sushoma scenario is characterized by a slower GDP per capita growth, due to moderate global economic growth with competition over labor-intensive production and a fast growing population in combination with low climate change and limited upstream developments. Due to slow economic growth, competition between Bangladesh and other developing economies over low-value production is high and protectionism grows. GDP per capita growth slows down, due to an increasing supply of low-skilled laborers and a slow pace of technological enhancements and inherent stagnating productivity increases. Thanks to moderate climate change and limited upstream developments, there is sufficient opportunity to develop and exploit traditional economic growth and industrialization. As output is not reinvested in much needed infrastructure, R&D, agriculture, education and human capital, inequality is rising. The large cities, Dhaka and Chattogram, grow rapidly and in an unplanned manner, due to a lack of spatial planning and enforcement, connectivity between urban and rural areas and urban facilities. Other urban and rural centers remain isolated. In this scenario, environmental degradation due to overexploitation of natural resources, agricultural, and industrial pollution is severe.

Towards 2100, as a result of slow global economic growth in decades, the rise of the middle income class has not sufficiently taken off and inequality has grown substantially, both within the large urban areas and between urban and rural areas.

Urban water demand will increase due to a higher population up to 2050, slowly declining up to the end of the century. A substantial part of the population will however not avail of improved water supply and sanitation, resulting in many diffuse and uncontrolled pollution sources and low water quality This will lead to increased (peri-)urban health problems and deteriorating soil and agricultural/livestock productivity. As climate change is relatively mild and upstream abstractions increase only gradually, transboundary low flows decrease slowly. Water demand however still increases, due to a higher rural population and water use, but less than in the Sokrio and Sohonio scenarios. The urban population is at high risk of floods and drainage congestion.

8.5.1.4 (Re)Active (Sokrio), 'Basic Needs First'

This scenario is characterized by a stagnating per capita GDP growth due to unfavorable global economic growth and an exponentially growing population, in combination with extreme water variability due to climate change and uncoordinated and growing upstream water abstraction. Stagnation of the global economy leads to a decrease of global demand in low-value manufacturing products, and a fierce competition between countries in the region. Due to limited international and education opportunities, migration is limited. As fertility rates decrease at a slow pace, the population continues to grow. There is a large number of climate refugees that flee Bangladesh in response to longer droughts, increased cyclones, and (relative) sea-level rise. The population, as a result of natural and man-induced disasters, continues to move to Dhaka and Chattogram, albeit at a slowly increasing pace. Migration to the cities is not attractive die to the low level of urban facilities and congestion. A sizeable part of the population remains in rural areas, dependent on subsistence rice farming. The increasing salinity and more extreme events due to climate change put a strain on agricultural productivity, leading to further increase of rural poverty and reduced food security. By 2100, population of Bangladesh continues to grow uncontrollably. In combination with harsh climatic and economic conditions, the struggle for food and natural resources is tremendous and Bangladesh is known for the largest number of climatic refugees worldwide.

Due to severe climatic conditions and high rural vulnerability, the increasing flood events have dramatic impacts on livelihoods. Flood frequency and intensity will increase due to more severe cyclones, higher peak river flows, and sea-level rise. This will affect both the rural and urban population. The low economic growth will limit the investment in large-scale adaptation and there is a focus on low-cost community-based adaptation and people struggle to survive. Adaptation will depend highly on foreign assistance. Water stress will be severe due to a large population that is dependent on rice cultivation, while water supply in the dry season will be seriously constrained due to high upstream abstractions and climate change.

Environmental degradation will be high and water quality will be severely stressed, both due to rural and urban pollution and salinity intrusion.

8.5.1.5 Overview Based on the Main Drivers

The BDP scenario narratives are concentrated around two key drivers: 1) future water conditions based on transboundary developments and climate change and 2) economic development and related land use changes. This approach follows the IPCC climate scenario approach and is both simple and effective and allows other key drives to be included using a consistent approach. Main drivers have been quantified for each scenario. The main characteristics of each scenario and accompanying urbanization patterns are shown in Fig. 8.5.

The aim of the Bangladesh Delta Plan scenarios is to identify key bottlenecks and appropriate measures per scenario. For each (set of) measures, the performance in the next decades is set against several key decision support indicators, both qualitative and quantitative. The outcomes of the decision support indicators will be different per strategy and per scenario. Within the overall framework of the BDP, the scenarios are inherently connected with the strategy identification, and form an integral part of the (strategy) assessment framework.

8.5.2 Implications for the National Planning Process

This scenario development process is essential for two reasons. First, through scenarios, policy makers can clearly articulate Bangladesh delta's future climatic

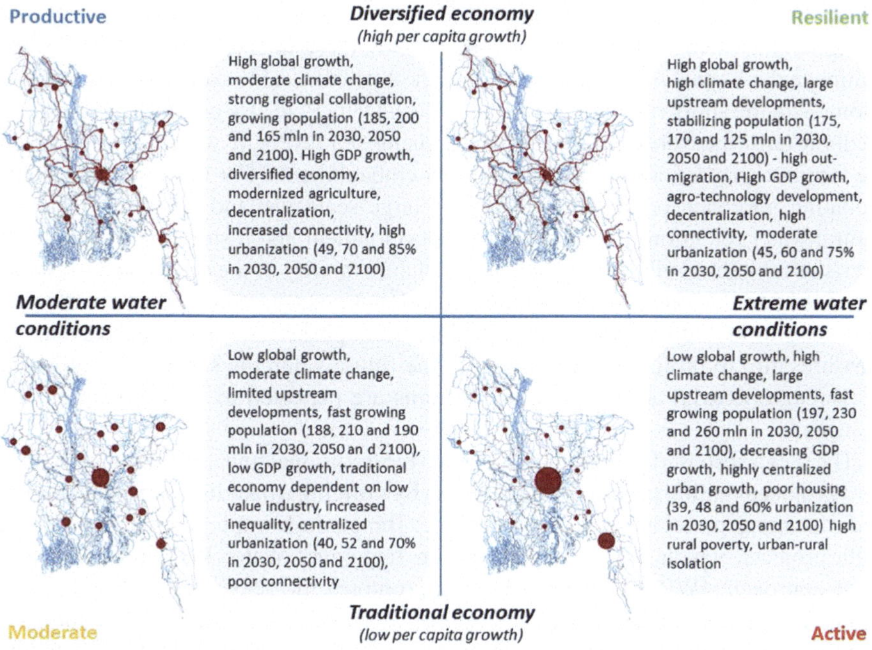

Fig. 8.5 Overview of the 4 scenarios. *Source* BanDuDeltAS 2016

and macro-economic possibilities. Second, policy makers can start preparing robust strategies for addressing different climatic scenarios. As said, strategies and measures are viewed as robust if they perform well under say three scenarios. However, one key aspect is that the positioning in the scenarios can gradually change as Bangladesh's climatic and macro-economic conditions alter over time. For instance, with the increasing impact of climate change and the economic impact of the COVID-19 pandemic and geo-political unrest, Bangladesh may currently move to the Sokrio (Active) delta scenario. Worsening climatic conditions require considerable investments while budget constraints can hamper these. Nevertheless, in the near future, the country can move into the Sohonio (Resilient) scenario. Therefore, it is essential to ensure that as the climatic and macro-economic conditions change, Bangladesh can shift flexibly with its strategies and measures from one scenario to another. Bangladesh Delta Plan ensured this flexibility through its adaptive setup in terms of alternative modes of operation. For each scenario, strategic adaptive pathways indicate options for what Bangladesh can take up in the short, medium, and long term. The adaptation pathways indicate when certain projects are expected to be necessary and appropriate in case earlier measures are no longer effective due to various circumstances. An example of Strategic Pathways focusing on the coastal region is given in the following. On the short run: Bangladesh Delta Plan includes the construction of new coastal embankments and raising existing embankments in the short-term phase to protect the coastal areas from flooding, cyclone impact, and storm surge. In fact, the delta plan also proposes widening embankments and strengthening them. In the short run (up to 2030), these interventions will be able to minimize the impact of climate change and protect vulnerable community members. Along with the short-run measures, Bangladesh Delta Plan also emphasized different medium-term measures because with the rising sea levels, it would be difficult in the future to protect coastal areas with only embankments. For instance, an accommodating approach could be followed by large-scale wetland creation to restore natural water retention. By proposing the development and restoration of wetlands, the risk of flooding will be reduced and the natural ability of the area to absorb water increased. On the long run, the Delta Plan includes creating new land and using new technologies including protection in the long-term phase. These seaward advancing measures aim to protect from flooding in the future when the sea level is expected to rise significantly, and the weather patterns are expected to undergo significant changes.[13]

The scenario-based planning transformed the traditional linear planning with standard project planning process in Bangladesh by bringing a paradigm shift in planning and managing projects. At the initial stage, Bangladesh implemented stand-alone delta projects, which later changed into an Integrated Water Resources Management approach. IWRM is a process that promotes the coordinated development and management of water, land, and related resources. This integrated planning is important as the interdependencies are particularly strong in the delta. Subsequently, Bangladesh followed integrated delta management, which considers the effects of interventions on a larger spatial scale and looks at the interrelations between different projects. Adaptive management as in BDP 2100 goes further, by considering uncertainties in drivers that influence water safety and water availability. These drivers include demography, economic development, and climate change. Water-related measures cannot influence these exogenous drivers, but these exogenous drivers can strongly influence water safety and water supply and demand. Examples are: population growth increasing needs for safety and water demand, climate change altering precipitation patterns and river discharges, and economic development also requiring water safety and influencing water demand and quality.

References

Ahmed, Y., Choudhury, G. A., & Ahmed, M. S. (2017). Strategy formulation and adaptation pathways generation for sustainable development of western floodplain of Ganges. *Journal of Water Resource and Protection, 9*, 663–691.

Alam, S., De Heer, J. M., & Choudhury, G. A. (Eds.) (2018a). *Baseline Studies on Governance and Institutional Development*. https://plancomm.gov.bd/

Alam, S., De Heer, J. M., & Choudhury, G. A. (Eds.) (2018b). *Bangladesh Delta Plan 2100: Baseline Studies on Water Resources Management. Volume 1*. General Economics Division, Planning Commission, Government of Bangladesh. https://plancomm.gov.bd/

[13] See also e.g. Ahmed et al. (2017).

Bangladesh Bureau of Statistics. (2022). *Census Report.*

IPCC. (2022): Climate Change. (2022). *Impacts, Adaptation and Vulnerability. Contribution of Working Group II to the Sixth Assessment Report of the Intergovernmental Panel on Climate Change.*

Ministry of Environment, Forest, and Climate Change. (2022). *National Adaptation Plan of Bangladesh (2023–2050).*

Ministry of Finance, Chapter 2-GDP, Savings and Investment 2023. www.mof.portal.gov.bd

NAP. (2022). Report on National Adaptation Plan. Government of Bangladesh.

Ophra, S. J., Begum, S., Islam, R., & Islam, M. (2018). Assessment of bank erosion and channel shifting of Padma River in Bangladesh using RS and GIS tech.

Shaw, R., Luo, Y., Cheong, T. S., Abdul Halim, S., Chaturvedi, S., Hashizume, M., Insarov, G. E., Ishikawa, Y., Jafari, M., Kitoh, A., Pulhin, J., Singh, C., Vasant, K., & Zhang, Z. (2022). Asia. In: Pörtner, H.-O., Roberts, D.C., Tignor, M., Poloczanska, E. S., Mintenbeck, K., Alegría, A., Craig, M., Langsdorf, S., Löschke, S., Möller, V., Okem, A., Rama, B. (eds.) *Climate Change 2022: Impacts, Adaptation and Vulnerability. Contribution of Working Group II to the Sixth Assessment Report of the Intergovernmental Panel on Climate Change.* Cambridge University Press.

Chapter 9
Adaptation by Three Types of Strategies

Abstract For ensuring climate change adaptation and fulfilling the delta vision throughout Bangladesh, three types of strategies and measures have been developed. This chapter discusses the strategies and measures coming forward as robust in terms of effectiveness under various scenarios. As flood risk and freshwater management is important for all the areas of Bangladesh, national-level strategies are introduced for addressing these two challenges. Subsequently, specific issue-related regional-level strategies are formed as different parts of the country have their own characteristics. BDP 2100 divides the entire country into six Hotspot areas and introduced strategies for each of the Hotspots. Finally, in order to further capture the socio-economic development goals, a number of thematic strategies are developed for addressing specific priority areas linked with the development ambitions, like developing agriculture and ensuring nutrition, drinking water supply, sanitation and waste management, land use planning, inland transportation and navigation, transboundary issues, renewable energy, earthquakes, and advancing the Blue Economy. Altogether, a large number of (sub) strategies came forward and even more measures offering a broad action repertoire in which choices by the government can be made.

Keywords National strategies · Hotspot strategies · Thematic strategies · Flood risk · Fresh water · Drought · River systems · Estuaries · Coastal zone · Haor · Flash flood · Urban · Hill tracts · Drinking water · Sanitation · Water quality · Waste management · Transboundary · Water diplomacy · Land use · Spatial planning · Water transport · Navigation · Renewable energy · Earthquake · Blue Economy · Protecting · Advancing · Accommodating · Retreat · Doing nothing · Building with nature · Ecosystems

J. M. de Heer and G. A. Choudhury, *Strategies for Adaptation to Climate Change in a Transformative Approach*, Springer Climate,
https://doi.org/10.1007/978-3-031-88884-7_9

9.1 National Strategies

During the development of strategies for achieving goals and mission, the clear expectation and sometimes message from the government were with respect to sea-level rise, that 'protection'[1] and 'accommodating'[2] and where possible 'advancing'[3] were strategic responses to look for, where possible combined with building with nature and or ecosystem-based adaptation. Responses as 'doing nothing', and 'retreat' (moving away) were during BDP 2100 formulation in relation to sea-level rise not elaborated.[4] Temporary 'retreat' is an often used strategy of disaster management when early warning leads to evacuation, e.g., during upcoming cyclones. In practice, community-based adaptation already shows small-scale protection works and also accommodating and retreat (e.g. from chars or continuously flooded salty areas) actions.

The BDP 2100 developed flood risk management and freshwater management strategies as national-level strategies. Such simultaneous development of flood risk management and freshwater management strategies is essential for protecting lives, property, ecosystems, and economic stability while ensuring water availability and promoting community resilience in the face of climate change and extreme weather events (Alam et al., 2018a)

9.1.1 Flood Risks

Three water flows of mighty rivers, the Ganges, Brahmaputra, and Meghna, meet in Bangladesh. With the effect of climate change, the upstream areas of these rivers face erratic monsoon rainfall, which leads to flooding; see, e.g., Fig. 9.1.

Bangladesh also experiences frequent tropical storms in the coastal zone. Of the total land, around 70% area is less than 1 m above sea level.[5] Therefore, even medium-scale storms cause considerable damage. All of these phenomena trigger frequent flood occurrences. An estimated 20–25% of the country becomes inundated due to river spilling and drainage congestion.

Flood risk management is, therefore, crucial to Bangladesh. By focusing on both the short and the long term, the BDP 2100 focused on ensuring protection against the destructive impacts of floods. BDP also emphasized enhancing the productive use of water. By focusing on both the short and long-term issues, the Bangladesh Delta Plan

[1] Protection (by embankments, dykes, dams, barriers) in an open (open connection with sea or major rivers) or closed way.

[2] Accommodating refers to allowing water in, living with water while mitigating effects and taking care of quick recovery; using floating or elevated structures).

[3] Advancing means building a buffer into the sea or: seaward development of well protected new land.

[4] See further: IPCC (1990, 2022) and e.g. Haasnoot et al., (2019).

[5] Ministry of Environment, Forest and Climate Change (2022).

Fig. 9.1 Flood in Turag river area. *Photo* Courtesy of Naymuzzaman Prince

2100 has developed flood risk management strategies by introducing three separate but closely linked operational strategies. These strategies include:

9.1.1.1 Protecting Economic Strongholds and Critical Infrastructure

The potential areas for rapid economic growth require a high flood protection standard to protect investments. The typical measures that can provide this level of protection include embankments, barriers, regulators, sluices, closures, bank revetments, groins, and efficient drainage systems. See for this strategy the overview in Fig. 9.2.[6]

Most of these measures are already in place, albeit at a basic level. In addition, adapted flood-proof locations and buildings are needed for key facilities such as hospitals, power stations, industrial plants, and major communication networks. In the Delta Plan, the indicative flood protection levels are considered and proposed as: 1/100–1/250 up to 2030[7]; 1/250–1/1000 up to 2050, and 1/1000–1/2500 by 2100. These protection levels are costly and form a great challenge. Adaptive design could offer possibilities to overcome this by raising the level of protection need based over time. The Bangladesh government is implementing different flood risk-related

[6] See all strategy overviews: BDP 2100 Volume I, Strategy document (2018).

[7] Meaning: 1 in 100 year to a 1 in 250 year event. This refers, e.g., to a 1 in 100-year flood, so to a flood height that has a long-term likelihood of occurring once in every 100 years. This is also called a recurrence interval or in risk terms a 1–2.5% chance of happening in any given year or, e.g., 0.01–0.025 Annual Exceedance Probability.

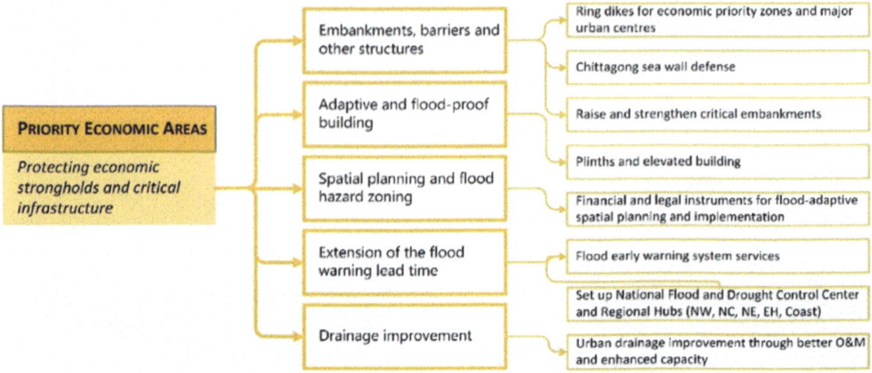

Fig. 9.2 Strategy flood risk management 1: protecting economic strongholds and critical infrastructure. *Source* BanDuDeltAS (2017)

projects. For specific purposes, design requirements are very high as for instance for the super dyke along the Mirsarai Economic Zone. The height of the super dyke is + 10 m above mean sea level and a width of 10 m. In the future, this dyke will also act as the peripheral road to provide connectivity in the economic zone.

9.1.1.2 Equipping the FCD Schemes for the Future

Creating a climate-proof Bangladesh implies that the measures are designed to meet current physical and socio-economic conditions and future developments. The country has already invested heavily in FCD development over the past 50 years; FCD schemes are in urgent need of maintenance and, in selected areas, remodeling to equip them for the future. Future designs need to consider climate and hydrological change projections, sea-level rise, land use, infrastructure, and urbanization.

9.1.1.3 Protecting Vulnerable Communities and Safeguarding Livelihoods

In addition to protecting economic strongholds, Bangladesh's policy makers focused on providing safety and supporting the economic resilience of vulnerable regions and communities in the country. This is in line with the principle of 'leaving no one behind'. This type of 'protection' is not aimed at avoiding floods and creating a flood-free Bangladesh for all. Given the current level of economic development, this is not feasible within the coming decades. Leaving no one behind merely means (i) mitigating the most undesired effects of large and extreme floods and (ii) enabling those affected by floods to recover quickly in the aftermath of the flood events.

9.1.2 *Freshwater*

Freshwater resources are exceptionally vital for Bangladesh's agriculture and transportation sector. Water quality is a growing concern for the country, as many rivers and wetlands are at serious environmental risk due to pollution and disconnection between wetlands and the river system. Bio-chemical pollution is a particular issue in Dhaka and Chattogram. For addressing freshwater management challenges, the Bangladesh Delta Plan 2100 has developed freshwater strategies through two distinct but interrelated strategies: Water availability and water quality.[8]

9.1.2.1 Ensure Water Availability by Balancing Supply and Demand

In areas where adequate water resources exist or can be developed, BDP 2100 emphasized providing additional irrigation water. It is essential for Bangladesh to gradually promote surface water use, especially in areas where surface water resources can substitute valuable groundwater supplies. At the same time, Bangladesh Delta Plan highlighted the importance of reducing the demand for freshwater in areas where additional water resources cannot be (economically) developed. This reduction of water demand is also applicable for areas where more urgent socio-economic priorities such as safe water supply, industry, ecology or navigation receive a higher priority. To reduce the demand for water, different interventions have been proposed, including promoting less water-consuming crops to more efficient irrigation and more effective management and pricing.

The Government of Bangladesh has initiated the implementation process of the freshwater management strategy. For instance, the GoB is realizing 'Expansion of Irrigation' in greater Rangpur district through best uses of surface water and conservation of rainwater projects. The Barind Multi-purpose Development Authority (BMDA) is executing this project to ensure efficient surface water usage. This five-year (2019–2024) project covers 35 Upazilas of Rangpur, Nilphamari, Kurigram, Lalmonirhat, and Gaibandha districts. This project will ensure greater agricultural surface water availability and improve the environment, ecology, and biodiversity.

For increasing freshwater flows in urban and rural rivers, BDP 2100 emphasized initiatives to improve water quality in the (peri-) urban areas and address salinity intrusion in the coastal zone. In this regard, Bangladesh also needs to focus on institutional development (capacity building, appropriate protection and management arrangements) and knowledge and innovation in support of long-term sustainable water resources planning.

[8] Bangladesh Delta Plan 2100 (2018).

9.1.2.2 Maintaining Water Quality for Health, Livelihoods, and Ecosystems

Water quality is a growing concern for the country, with many rivers and wetlands at serious environmental risk. Pollution of surface water is caused by three major sources of contamination: industry, agriculture, and domestic/municipal. Hazardous chemicals of both organic and inorganic compositions are discharged into the water bodies from the industries like textiles, tanneries, pulp and paper mills, fertilizer, and chemical plants. There are also domestic contaminations in rural areas with fecal coliforms. For addressing these challenges, Bangladesh Delta Plan highlighted the importance of protecting environmentally valuable and sensitive areas and the prevention of pollution by enhanced treatment. The environmental and urban drinking water and sanitation authorities need to play a key role in implementing these interventions.

9.2 Hotspot-Specific Strategies

Bangladesh Delta Plan divided the entire country into six hotspots (see Fig. 6.32 and Sect. 6.4) to address the unique challenges and vulnerabilities faced by different regions within the country. Different regions within the country have varying geographical and hydrological conditions, such as river dynamics, tidal patterns, and land formations. The development of hotspot-specific strategies under the Bangladesh Delta Plan acknowledges the diverse geographical, hydrological, socio-economic, and climate change-related challenges faced by the different regions. Thc BDP 2100 aims to promote effective and targeted actions, enhance stakeholder participation, and achieve sustainable and inclusive development in each hotspot by tailoring strategies to address these specific needs.

9.2.1 Barind and Drought-Prone Areas

In general, the Barind region is considered a drought-prone area, especially concerning agricultural drought. This region includes the Rajshahi and Nagaon areas. In many areas of this Hotspot, the soil has low water holding capacity (Rashid et al., 2017). The North-West part of this region annually receives 1400 mm of rain and over 80% of this rainfall occurs during monsoon (June to September). Along with drought, depleting groundwater tables is another major concern in this region. BDP 2100 proposed a set of preferred strategies and sub-strategies for the Barind Hotspot area. The main strategies are as follows.

9.2.1.1 Balancing Supply and Demand for Sustainable and Inclusive Growth

Ensuring water availability by balancing supply and demand for sustainable and inclusive growth lies at the heart of this hotspot. Increased droughts and water shortages for different socio-economic sectors are expected in the future and these constraints will hamper sustainable growth.[9] For ensuring sustainable growth for this hotspot, Bangladesh Delta Plan proposed:

- Supply management and additional irrigation: where adequate water resources exist or can be developed, from the Ganges and regional rivers, this sub-strategy is geared towards providing additional irrigation water for sustainable growth.
- Demand management and efficient water use. This sub-strategy is aimed at reducing the demand for freshwater, for those areas where additional water resources cannot be (economically) developed or where more urgent socio-economic priorities such as safe water supply, industry, ecology, or navigation receive a higher priority. For achieving this, Bangladesh Delta Plan proposes a mix of interventions, consisting of promoting less water-consuming crops to more efficient irrigation. This sub-strategy also emphasizes water productivity by more effective water management, smarter water use and a pricing policy.

9.2.1.2 Minimizing Losses Due to Floods and Drainage Congestion

Although the Barind region is representative for the drought-prone areas of the country, floods and drainage congestion are key issues here as well. An integrated approach also entails considering the impacts on both floods and droughts as well as the multiple functions and interests in the region.[10] Two sub-strategies are part of the strategy:

Minimizing losses due to floods: The lower Barind area, and the Atrai basin in particular, is typically affected by floods during the monsoon season when prolonged and intense rainfall coincides with high water levels in the main rivers, impeding drainage. Many FCD systems were designed over 20 years ago and are in urgent need of repair and modernization. A key aspect is taking into account multiple functions and interests and appreciating the positive impacts of flooding in the lower Barind area, including replenishment of beels and wetlands, fish migration and spawning, groundwater recharge and flushing of nutrients and pollutants from the water system.

Alleviate waterlogging in urban and rural areas.

[9] Bangladesh Delta Plan 2100 (2018).

[10] Ibid.

9.2.1.3 Ensuring Water Supply and Sanitation

This strategy centers on achieving basic water security and sustainable development goals: ensuring safe and reliable access to water supply and sanitation. There is a strong link with water quality through the development of improved sanitation and wastewater treatment systems in all district headquarters and major settlements, with cost recovery principles advocated for all water supply and sanitation services.[11]

The Government of Bangladesh has already initiated the implementation process of this strategy. For instance, GoB is implementing the 'Expansion of irrigation through conservation of canal water in the Barind region'. This project aims to provide a seamless supply of irrigation water to farmers across 3,490 hectares of croplands during the dry seasons.[12] The project will create, restore, and conserve surface water at a total investment of about € 20 mln..[13]

9.2.2 Major River Systems and Estuaries

The major rivers of Bangladesh, the Brahmaputra/ Jamuna, the Ganges, and the Meghna, are the backbone of the delta system. These three rivers influence the hydrological as well as morphological characteristics of other rivers. These rivers are characteristically very dynamic with frequent river erosion and accretion. From 1973 to 2014, around 90,000 ha of land have been eroded along the Jamuna River, but only 16,000 ha of land have been accreted. Along the Padma River, the erosion and accretion are approximately 33,000 ha and 12,000 ha, respectively. See Fig. 9.3.

So far, the river management of these three major rivers has not been done comprehensively; instead, sporadic interventions have been made over the decades to protect the towns, growth centers, and embankments from river erosion. These are site-specific interventions in the form of hard points, revetments, groins, spurs, dredging etc. A long-term strategy for these rivers has therefore national importance. Some of the strategies that Bangladesh Delta Plan adopted for major river management include the following.

9.2.2.1 Provide Adequate Room for the River and Infrastructure to Reduce Flood Risk:

BDP 2100 emphasized on providing adequate room for the rivers to reduce flood risks. It proposed construction, repair, upgrading, and maintenance of embankments and conduct strategic dredging. For implementing this strategy, GoB has initiated

---

[11] Ibid.

[12] Arif, 2022.

[13] Ibid.

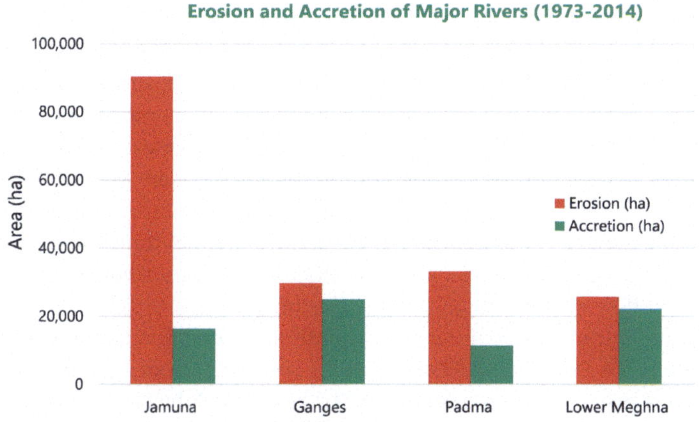

Fig. 9.3 Erosion and accretion along the major rivers. *Source* BDP 2100, Baseline Study 1, Page 74

a dredging master plan to develop the navigability of 10,000 km of river routes by excavating 178 rivers across the country.

9.2.2.2 Improvement of the Conveyance Capacity as Well as River Stabilization

The riverbank erosion in Bangladesh is expected to become a significant issue with increasing economic development. Currently, the stabilization and channelization of rivers is applied locally at the most critical spots. However, in the future, the government has to decide whether to prioritize river bank protection in economically valuable areas, which can be expensive, or limit urban and economic development to areas where river stabilization works can be done at a reasonable cost. The preferred approach is to align urban and economic development with river management, requiring an integrated cross-sectoral approach.

In the Jamuna and Padma rivers, there is an opportunity for land reclamation on a lot of locations, utilizing the high sedimentation rates. The Bangladesh Delta Plan proposed a project on the Integrated Jamuna-Padma Rivers Stabilization and Land Reclamation issue. It has been estimated that an area of 0.157 million ha of land could be reclaimed as a result of the stabilization of the Jamuna-Padma River systems. The reclaimed land could be used for agricultural, industrial, settlement and recreational purposes, but for that it is essential to conduct extensive studies. Therefore, the Delta Plan proposed a study on the long-term effects of river channelization and land reclamation on river morphology and discharge capacity. Once the land is reclaimed, it will always require protection against the highly dynamic river. In this regard, it was emphasized to start with small initiatives, focus on economically viable spots, and carefully balance land reclamation with the sediment balance.

9.2.2.3 Provide Fresh Water of Sufficient Quantity and Quality

The Bangladesh Delta Plan aims to ensure fresh water of sufficient quantity and quality. This can be achieved through transboundary cooperation and maintaining flows in major rivers and floodplains. Especially, Bangladesh needs to focus on transboundary negotiation and cooperation in order to ensure sufficient freshwater inflow in the future. For transboundary water management, it is essential to first understand the current status and future impact of potential transboundary projects. Therefore, BDP 2100 proposed a study on the impact of transboundary projects on the dry season flow of the Ganges and the Brahmaputra.

Besides, the optimization of water distribution by retention and flow diversion is very important for Bangladesh. This involves construction of barriers and dams to retain or divert water. In this regard, BDP 2100 proposed the construction of a head-work's structure with proper dredging and training at the Gorai offtake. However, for constructing such infrastructure, it is essential to conduct necessary studies. Therefore, a study has been indicated on the impact and feasibility of large-scale water retention.

9.2.2.4 Maintain Ecological Balance and Values of the Rivers

Bangladesh Delta Plan aims to maintain the ecological balance of rivers through preservation of ecosystem and restoring ecosystems by ensuring a wise use of resources. For ensuring this, BDP 2100 focused on prioritization of certain initiatives as the preservation of ecosystems which is considered a key priority and has therefore been prioritized over developing sustainable ecotourism. For instance, BDP proposed the development of Riverine Ecosystem Habitat Preservation Program. This program will ensure preservation for plants, wildlife, fish, birds and assess endangered species.

9.2.2.5 Allow Safe and Reliable Waterway Transport in the River System

Bangladesh Delta Plan emphasized on improving navigability for ensuring safe and reliable waterway transport. The plan proposes different water transport investments ranging from improving local cargo facilities to creating high-standard marine logistics hubs in high-income areas. Dredging at strategic locations is recommended in the BDP to maintain navigability. In this regard, key decisions arise when waterway transport intensity increases: should rivers be fully stabilized and channelized or should river dynamics be tolerated with advanced monitoring systems? Advanced monitoring and marine communication systems can offer the possibility to cope with riverbed dynamics—within limits—while maintaining navigability. Vessels must then be equipped with GPS systems and with updated electronic bathymetric information so that the position of the main channel and its actual depth is known to each vessel at any position.

BDP 2100 also highlighted that capital dredging may not be effective for the large and high energy rivers like the Jamuna, Ganges, Padma, Teesta, and the Dharla. Instead, annual recurrent dredging would be more effective to maintain the river navigability. In a very wide river, the dredged-coarse sediment may be dumped into the riverbed in a strategic suitable location, so that it cannot return back into the dredged channel before the onset of the monsoon. To implement this BDP strategy, the Executive Committee of the National Economic Council (ECNEC) has approved a Taka 560.07 crore project to protect Singrabari, Patagram, Baoikhola areas under Kazipur upazila of Sirajganj district from the erosion of the Jamuna River. Upon successful implementation, the project will protect assets and properties from the erosion of Jamuna River on its right embankment at Sirajganj Sadar and Kazipur upazilas. In addition, the Bangladesh Water Development Board (BWDB) has devised an ambitious plan to reclaim approximately 1600 km^2 of land over the next 25 years by strategically narrowing down the width of the Jamuna and Padma rivers. This vast undertaking is captured in a comprehensive project titled 'Comprehensive Plan for Stabilization of the Jamuna-Padma River and Pilot Intervention for Land Reclamation'. The project aims to achieve an average width of 7 km for both rivers, significantly reducing their current width of 16 km, ultimately leading to the reclamation of the targeted land area. The envisioned plan is expected to be executed in multiple phases, with a projected cost of Tk 400.00 billion. Notably, this project is intended to complement the ongoing efforts of the Flood and Riverbank Erosion Risk Mitigation Investment Program (FRERMIP), which has already successfully reclaimed 16 km^2 of land from the Jamuna in Sirajganj. Another reclamation process is under preparation at the Harirampur area of Manikganj, showcasing the board's dedication to address land reclamation and river management challenges in the region.

9.2.3 Coastal Zone

The extended coastal zone is very heterogeneous and has a tremendous potential to create opportunities of national importance. The examples are intensification of agriculture, aquaculture and marine fishery, exploration of gas and oil resources, development of the shipbuilding industry, ecotourism, renewable energy, and deep sea port development. Unfortunately, the coastal zone of Bangladesh is also a disaster-prone area. Cyclones, storm surges, droughts, floods, waterlogging, and salinity intrusion greatly impact people and their livelihood. Figure 9.4 shows potential modeled flood hazards for the coastal region, providing indicative maximum water depths after overtopping in extreme events (100 year. probability). It is important to note that riverine floods show a potentially larger flood extent than storm surges, but that storm surges may lead to larger water depths, particularly in the sea-facing areas. A rapid analysis, based on the modeled information from IWM, shows that a substantial part of current total coastal flood risk (combination of hazard and vulnerability) can be found outside the current polder areas, making these areas as such particularly hazardous for extreme flood events.

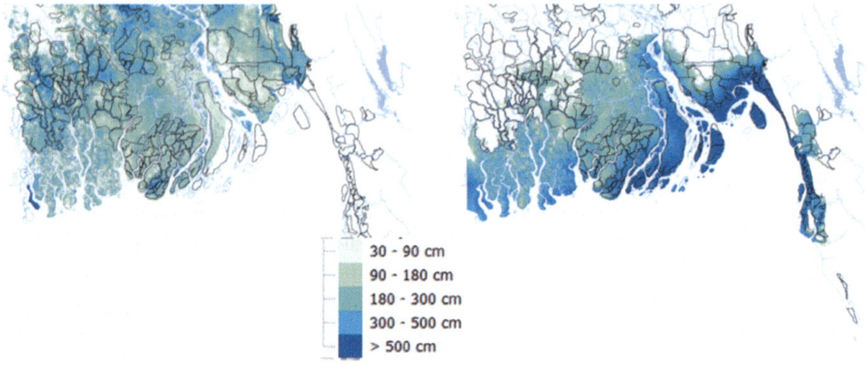

Fig. 9.4 Indication of potential flood hazards at maximum water depth for 100 year water-level return periods. *Source* IWM

Along with natural disasters, poor communication, lack of education and health care facilities, prolonged absence of safe drinking water, and insufficient cyclone shelters contribute and multiply the dimension of vulnerability in the coastal region. Furthermore, increasing population pressure increases the competition for limited resources. For addressing these issues, Bangladesh Delta Plan emphasized that existing and new interventions in the coastal area need to be developed in an integrated and inclusive way. Subsequently, BDP 2100 proposed six strategies for coastal zone for ensuring integrated and inclusive development; see Fig. 9.5. Further elaboration is incorporated in the Extended Coastal Zone Implementation Program (De Heer & Choudhury, 2023).

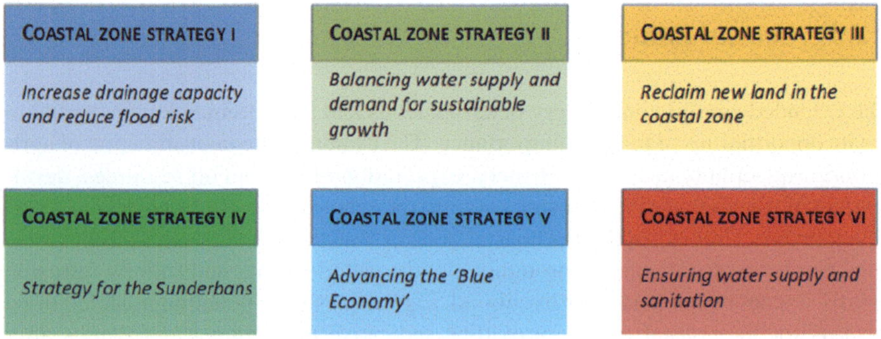

Fig. 9.5 Coastal zone hotspot strategies. *Source* SIBDP (2023)

9.2.3.1 Increase Drainage Capacity and Reduce Flood Risk at Coastal Zone

The coastal zone of Bangladesh will remain hazardous for coastal floods in the foreseeable future. Cyclones and accompanying storm surges will continue to develop in the Bay of Bengal, as well as high river discharges and monsoon precipitation will continuously put an enormous pressure on the drainage capacity of the coastal delta. Besides, possible future changes in discharge regimes or sea-level rise, the coastal zone will be under increasing pressure of socio-economic change, with economic development and demographic changes as main drivers. Sub-strategies identify measures to diminish drainage congestion, flood risk protection, flood risk prevention and flood risk preparedness. 13 projects mentioned in the BDP 2100 Investment Plan are related to this strategy; for example, Study on Tidal River Management (CZ 1.5), Rationalization of Polders (CZ 1.40/1.44/1.47) and constructing multi-purpose disaster shelters and a Management Information System (MIS) (CC 1.4).[14]

9.2.3.2 Balancing Water Supply and Demand for Sustainable Growth

The Bangladesh Delta Plan focused on improving living conditions, sustain agricultural and aqua-cultural production as well maintain environmental sustainability. For the short term, the preferred strategy includes revitalizing surface water supply sources from the Ganges as much as possible. For the coast, the construction of the Ganges barrage to restore the freshwater flow through the Gorai and other rivers would be advantageous, certainly with respect to drinking water and salinization in the Sundarbans. But additional study is needed to ensure a long-term sustainable situation at the regional level.

Current FCD(I) infrastructure needs to be modernized, as insufficient maintenance and changing external conditions have led to a substantial degree of malfunctioning. A main aspect to be solved is management (especially of operations and maintenance), which needs innovative forms of institutions.

Year-round availability of freshwater for different users is currently specifically a problem in the southwestern region, as freshwater flow and precipitation excess in the Lower Meghna-dependent area is still sufficient. Overall, irrigated crops in the dry seasons are the largest consumers of freshwater throughout the coastal zone. Measures to relieve pressure on these demands are recommended. Drought and salinity early warning systems with localized impact and water pricing are considered no-regret. Competing main users of freshwater in the coastal zone are natural ecosystems (e.g. wetlands, beels, fisheries, and the Sundarbans) as well as navigation for regional waterway transport. For ecosystems, it is recommended to improve the connectivity between wetlands and floodplains and restore the natural capacities of beels and natural waterbodies where possible. The unique largest Mangrove forest, the Sundarbans, need to be protected by (inter-)national laws and programs

[14] The codes relate to Project codes in the BDP 2100 Investment Plan (2018).

and should be managed to maintain its unique characteristics. Some polders on the northern border of the Sundarbans should be investigated to restore the area to historical proportions, which will improve environmental sustainability, as well as the natural wave attenuation capacities of Mangroves in times of storm surges. In this sense, it is also fundamental to monitor and understand the impact of land accretion for the functioning of the Sundarbans, also in terms of flood hazard attenuation. From environmental point of view, Mongla port is poorly located and its activities could be limited or guaranteed to be not destructive for the vulnerable ecosystem of the Sundarbans. Finally, industrial facilities and farmers upstream of the coastal zone, need to follow strict regulations in their treatment of effluents and fertilizers to the surface water. Last, but not least, the tidal rivers are part of an extensive network for transport of goods and people over water. Several measures are to be taken; on the one hand to improve navigation possibilities by excavation/dredging and adapted cargo facilities, on the other hand measures to optimize alternative ways of transport; specifically multi-modal opportunities can be explored. This will decrease pressure on the yet exhaustive use of the water system as a whole.

9.2.3.3 Reclaiming New Land in the Coastal Zone

The Ganges–Brahmaputra–Meghna (GBM) system is carrying huge volumes of sediment to the Bay of Bengal though several estuaries.[15] Especially in the Meghna estuary, there is considerable potential of reclaiming new land. In the Meghna estuary, the areas of accretion and erosion were 1810 km^2 and 1205 km^2, respectively, during 1973–2015. As net accretion area is about 605 km^2 during last 42 years, land development may be possible in the Meghna estuaries and in the coastal zone[16]; see Fig. 9.6

Already through the Char Development Settlement Project (CDSP) project, government has taken land reclamation initiatives. Building further on the successes of these land reclamation projects in the past, Bangladesh Delta Plan emphasized on exploring natural sedimentation processes in the Meghna estuary at strategic locations. Where natural process may take 60–80 years to build the new land, the human interventions may accelerate the process to 35 years.

Through this strategy, an enormous multi-sectoral potential can be realized on the new land. In addition, these new lands can be used for tidal and wave energy production, wind energy parks, new deep sea port development, exclusive housing as well as tourist developments. But for unlocking this potential, extensive research needs to conducted to gather knowledge about the influence of land reclamation and possibility of artificial interventions for acceleration process. Accelerating the natural accretion processes with cross dams and other infrastructure can provide

[15] Estuaries are dynamic (in terms of sedimentation, erosion, currents) delta areas where rivers meet the sea. They are home to unique plant and animal communities that have adapted to brackish water, often a sandy mixture of fresh water draining from the land and salty seawater.

[16] Bangladesh Delta Plan 2100 (2018).

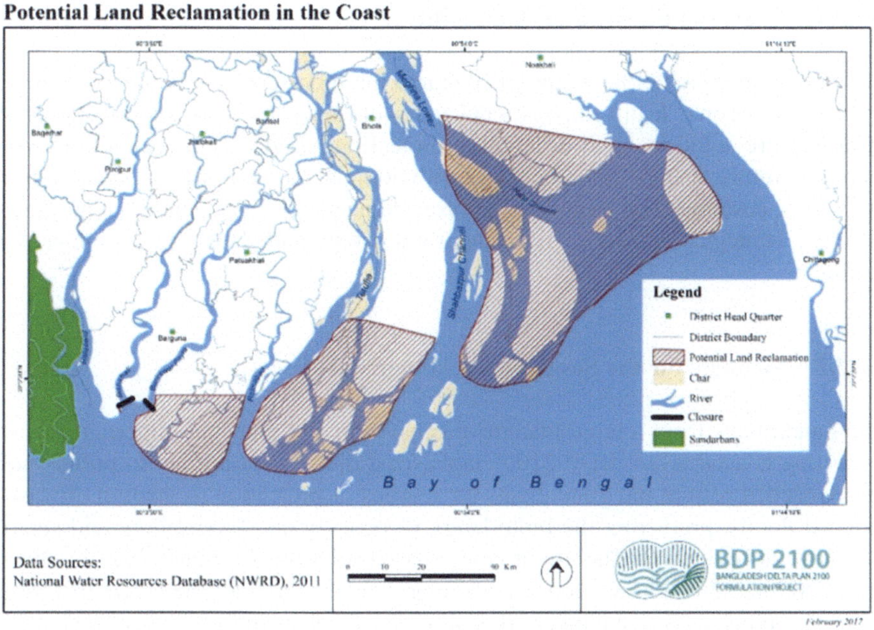

Fig. 9.6 Potential land reclamation in the extended coastal zone. *Source* BDP 2100 (2018)

Bangladesh additional land, free to be developed in the preferred way. Bangladesh Delta Plan stressed that accreted and reclaimed land should be put to value added roles, particularly for urban and industrial needs. However, for using these lands for these needs, the flood risk needs to be reduced, which will require significant fund mobilization. The flood risk challenge may be dealt through a combination of land filling up to a sufficient level above sea level for the most critical infrastructure by using the sediment from the Meghna estuary and through building dykes. Besides, it is also essential to introduce different nature-based solutions like mangrove belts to reduce wave run up.

In relation to this, the government has formulated a comprehensive plan to establish a mangrove forest covering an extensive area of 17,000 hectares. This has also a significant meaning for safeguarding and conserving the rich biodiversity of forestry and wildlife. The Planning Commission approved furthermore the construction of the cross dam between the area south of Noakhali and Urir Char. Advantages of the cross dam on the chosen location is its role in accelerating the natural sedimentation process, accretion of new land, protection and further char development and settlement.

9.2.3.4 Ensuring Freshwater Availability

The Bangladesh Delta Plan highlighted that in the short term, the preferred strategy will be to focus on risk-informed rationalization of current projects, low-maintenance infrastructure solutions, and differential protection based on cost-benefit analysis including future change. Inland polders with dominant agricultural or aqua-cultural use will be sustained at a basic level of protection. Bangladesh Delta Plan proposed different strategies including restoring rural rivers and canals and options for retention.

9.2.3.5 Strategy for the Sundarbans Mangroves

The Sundarbans form a large mangrove ecosystem and plays an important role in protecting coastal areas. BDP 2100 emphasized that it is important to protect and where needed restore the Sundarbans for reasons of protection (reduction of wave-induced floods, protecting the hinterland), of ecology and biodiversity and socio-economic development. The use of nature-based solutions for Sundarbans has been proposed, including simultaneously planting native trees in all the layers of forest. Also, the importance of maintaining perennial tidal flow in the Sundarbans region has been highlighted.

9.2.3.6 Advancing the 'Blue Economy'

BDP 2100 indicates the importance of developing the so called 'Blue Economy'. Different high priority sectors and approaches are identified to create and maintain a prosperous and sustainable Blue Economy based on unlocking potentials of shipping, marine fisheries, tourism, land reclamation, renewable energy, etc. Further research, capacity building and elaboration of governance. Institutional aspects here are that development of the 'Blue Economy' is in an early stage and need to be coordinated by a government body for several involved sectors. Besides, increasing near coast and seaward activities make dedicated regulations and, e.g., marine spatial planning necessary.

9.2.4 Haor and Flash Flood Areas

The haor area is an insecure region of Bangladesh covering about 1.99 million ha of area. This area is highly vulnerable to flash floods from the hills of India. In 2022, this area experienced several flash flood incidents because of low conveyance capacity of flashy rivers and intense rainfall. As haor and flash flood areas are flooded for almost six months of May to October, access to safe water and sanitary provisions become very difficult. Moreover, during this period, the settlements are exposed to wave attack.

The extended period of annual flooding means only boro crops can be grown (see Fig. 9.7), thus increasing food security vulnerabilities. With the effect of climate change, farmers are experiencing early rainfall and flash floods, making it difficult to protect even their single annual crop. The submersible embankments and dykes generally obstruct flood in this region, but they are after damage during the monsoon or otherwise often not properly repaired. The rising population and over-exploitation of swamp forests and fish have disturbed the ecosystem of this region. In recent years, many FCD/I projects have been constructed in this region, but the disturbance to pre-monsoon spawning migration due to FCD/I projects has also impacted the availability of fish in this region. With the effect of climate change, minimum opportunity of crop cultivation, and reduction of fisheries, this region is one of the most vulnerable regions of Bangladesh.

Navigation is not possible during dry season due to the reduction of water in the rivers. On the other hand, during wet season, road networks cannot be used because of being inundated by flood water. The Bangladesh Delta Plan 2100 has developed the haor and flash flood area strategies through optimal utilization of natural and human resources following the principles of the IWRM. The main strategies are as follows.

Fig. 9.7 Boro rice harvesting in haor area. *Photo* courtesy of Amirul Hossain

9.2.4.1 Protect Agriculture and Vulnerable Communities from Flood

BDP 2100 proposed that the existing FCD/FCDI projects should be rationalized in response to future climatic conditions. It is also essential to increase the conveyance capacity of the flashy rivers through strategic and maintenance dredging and re-excavation. Wave erosion would be minimized through building wave breakers to protect settlements from strong wave action. In this regard, different nature-based solutions including plantation of hizol trees can be used. Moreover, elevated village platforms and other sustainable interventions have been proposed to cope with flood water. For protecting the post-harvest crop from rotting in flash flood areas, private and public initiatives need to be encouraged. Traditional farming dominates in this area; see Fig. 9.7. In this regard, mechanization in the haor area has been emphasized for quick land preparation, planting, weeding, harvesting, processing, seed drying, storage, and logistics.

9.2.4.2 Achieving Water Security

Although the haor region remains under water for approximately six months, the region experiences water shortage during the dry season. In most of the cases, the farmers use pumps for extracting groundwater, but it is not an affordable solution for marginal farmers. In this regard, BDP 2100 stressed the importance of re-excavation of canals to serve the twin purposes of moisture retention and water distribution for irrigation purposes. The haor region experiences drinking water shortage, both in the dry and monsoon season. While in the monsoon, there is abundant water in the surrounding areas, there is a shortage of drinkable water. For addressing this issue, Bangladesh Delta Plan proposed the introduction of piped water supply systems in districts, upazila and growth centers. For enhancing water quality, wastewater treatment and sanitation solutions could be introduced in cities. Moreover, sustainable and flood-proof hygienic sanitation provisions have been emphasized.

9.2.4.3 Improving River Management

Presently, most of the rivers in the haor region have siltation issues and these rivers have very limited conveyance capacity. Estimated sediment yields for the streams of Tripura hills as observed in the Khowai River are 1.7 million ton/year. The Manu River has even higher sediment yields with 4.9 million ton/year. The sediment yields from the Meghalaya streams ranged between 139 and 229 ton/km^2 and the highest amount was about 360 tons/year on the Dhansiri River. The highest sediment concentration has been found in case of Dhalai River followed by Manu, Surma, Khowai, Someswari, Kushiyara, and Upper Meghna Rivers. This causes higher sedimentation rate in case of flashy rivers in haor and flash flood areas. Bangladesh Delta Plan 2100, thus, emphasized on improving the conveyance capacity and maintains the navigability of the Upper Meghna, Surma, Kushiyara, Kangsha, Bhogai, Jadukata,

Pagla, Buri, Jamuna, Titas, Mogra, Monu, Piyain, Sari Gowain, Ghora Utra, Khowai, and Sutang river systems through dredging with vegetation cutting. In addition, Bangladesh Delta Plan also suggested necessary projects for increasing the navigability of the Kushiyara River upstream of Markuli by building weirs and navigation locks.

Both the strategic and maintenance dredging of medium and minor rivers should be maintained for improving the navigability of rivers all year round. Priority is given to the main transboundary rivers entering the region. While river dredging and re-excavating is urgently needed, only limited progress is possible due to capacity constraints. In addition, it is often difficult to store dredged silts as land owners oppose such storage on their land. A market could be stimulated by utilizing the sand and sediment material to elevate low-lying areas for settlements and homesteads.

9.2.4.4 Sustainable Haor Ecosystem and Biodiversity Management

The rich wetland biodiversity of the haor and flash flood area makes it a unique wetland ecosystem. The natural resources need to be protected and Bangladesh Delta Plan proposed different strategies for conserving the ecological balance of this region. It proposed different awareness building programs for making people conscious about biodiversity management. It also emphasized the strict compliance of existing environmental laws to protect the rivers, haors, and wetlands from agricultural pollutants and dumping of wastes both from urban and rural areas. For protecting environment from industrial pollutions, establishing Effluent Treatment Programs (ETP) and Waste Treatment Programs (WTP) for every industry in this Hotspot area have been proposed.

9.2.4.5 Integrated Water and Land Resources Management

While the haor region is currently experiencing urbanization and tourism development, BDP 2100 emphasized integrated water and land resources management for the region. Especially for building roads and railways, it is required to consider the rich ecosystems and to keep adequate space for flood passage in order to avoid drainage congestion. In addition, boat passes or navigation locks could be built in existing FCD/I projects. Furthermore, it is important to modify and reconstruct bridges and culverts, keeping adequate free board for plying river crafts and ensuring water drainage.

9.2.5 Urban Areas

Most of Bangladesh's urban areas are in an early stage of development and experiencing rapid uncontrolled growth and industrialization. Uncontrolled urban growth and haphazard development is creating different challenges like flood risk management. Sea-level rise, salinization of agricultural land, desertification and growing water scarcity are forcing many people to migrate to the urban areas of Bangladesh. Especially from the southwest coast, many poor male household members temporarily migrate to the urban areas, while the women left behind face the full range of challenges caused by climate change impacts and stresses. Although it is clear that climate is influencing patterns of migration, directly attributing migration to climate change is difficult as people move for a wide variety of reasons and currently national and international databases related to migration do not contain consistent climate and environment-related attribution data. For addressing the urban issues, Bangladesh Delta Plan emphasized on the following strategies.

9.2.5.1 Increase Drainage Capacity and Reduce Flood Risk at Urban Area

For reducing flood risk in the urban areas, Bangladesh Delta Plan proposed the restoration of existing urban drainage systems and re-excavation of natural khals. In this regard, it is argued that the improvements in urban drainage are inextricably linked with solid waste management initiatives and the importance of regularly removing waste and silt from urban waterbodies has been emphasized. Subsequently, solid waste management projects in Khulna and other selected urban towns have been included in the Investment Plan. Bangladesh Delta Plan also highlighted that existing wetlands, khals and retention areas need to be formalized as water storage areas and protected from encroachment through land use regulations. In this regard, along with focusing on new projects, BDP 2100 indicated securing of the maintenance budgets for ensuring proper O&M of these infrastructures. The Delta Plan also proposed the redevelopment of public space to enlarge the retention capacity of the drainage system.

BDP 2100 proposed the application of land use regulations and risk sensitive land use management to prevent unwanted development outside the protected urban centers. For securing space for future reinforcements, it proposed that embankments in high density urban areas can have spatial reservations and building restrictions. Urban growth needs to be considered with respect to flood risks, space for water resources (see Fig. 9.8) and drainage capacity.

9.2.5.2 Enhance Water Security and Efficiency of Water Usage

The Delta Plan emphasized on limiting leakage and pursue full metering to reduce the percentage of non-revenue water. Development of piped water systems to allow more conformable water consumption and central water treatment (including arsenic removal) is important. Besides, development is needed of surface water as additional drinking water source, including water treatment plants, specifically for Dhaka, Chattogram, and other major urban centers.

Fig. 9.8 Urbanization near Dhaka. *Photo* J. M. de Heer

The BDP 2100 focused on promoting fecal sludge management, entailing sludge collection, disposal, and treatment. Suitable cost recovery principles including disposal premiums and municipal taxes are elaborated in the cross-cutting water supply and sanitation strategy. Reallocation of small and medium heavy polluting industries from residential and commercial areas to dedicated economic zones could be done through land use regulation and enforcement. Improvement of Dhaka's and Chattogram's sewerage system has been proposed, linked with improvements in urban drainage to prevent sewerage overflows. Education programs on water pollution, environmental degradation, and raising of public awareness are needed (see, e.g., Fig. 9.9). Research is proposed on the potential of ground water recharge and (managed) aquifer recharge in urban areas, as well as to participate in research and experiments towards new, future-proof water quality and sanitation concepts.

9.2.5.3 Ensure Sustainable and Integrated River Systems and Estuaries Management

River stabilization and control is aligned with the distribution of economic activities and population centers on national and regional scale. For ensuring this alignment, the Bangladesh Delta Plan focused on the preservation of the unique character of

Fig. 9.9 Fishing next to discharge of polluted urban water. *Photo* courtesy of Naymuzzaman Prince

Bangladesh's river landscape, through integrated urban and landscape design for land reclamation.

9.2.5.4 Conserve and Preserve Wetlands and Ecosystems and Promote Their Wise Use

The Bangladesh Delta Plan emphasized establishing the required water retention and discharge capacity in (peri-) urban wetlands, related to the urban drainage strategy and the river/flood risk strategy. It indicates also:

- Reach broad consensus on water retention and discharge targets.
- Define ecological targets for the wetlands; establish the formal function and conditions (e.g. water levels).
- Make use of placeholder functions and integral wetlands in qualitative public space (e.g. parks) to protect wetlands in the short term.
- Take care of integral development in public–private partnerships for wetlands that experience high levels of spatial pressure due to high land prices and uncontrolled urbanization.

The integrated planned area development allows for both economic growth and housing as well as maintaining the wetlands' ecological and hydrological values.

The Delta Plan also focused on promoting urban green and blue spaces. It suggests a regional spatial plan in which green and blue ecological and recreational spaces and networks are defined related to an urban strategy for green, blue, and recreational networks of different scales (urban parkways, city parks, neighborhood parks, and green routes.

9.2.5.5 Develop Effective Institutions and Equitable Governance for In-Country and Transboundary Water Resources Management

The Bangladesh Delta Plan focused on strengthening of the institutional capacity of city corporations, Pourashavas, and other urban areas. Research is useful into the options for decentralization, focused on subsidiarity in infrastructure development and spatial planning. For ensured implementation of urban plans the BDP 2100 emphasized on looking after possibilities for local urban government to increase local revenue streams for providing green and blue urban services and sustainable urban development.

9.2.5.6 Achieve Optimal and Integrated Use of Land and Water Resources

The BDP 2100 proposed the introduction of integral spatial planning in spatial policy making on national and regional scale. Formulation of generic principles for sustainable urban development as guidelines for urban planning provide a sustainable, qualitative, and integral perspective on key planning aspects as mobility, essential urban spaces, protection of waterbodies, risk management, and others. Strategic spatial reservations are important to prevent encroachment of critical open areas (water bodies and green space). For this, it is needed to develop a strategic vision on spatial reservations in urban areas. This includes the early acquisition of yet undeveloped land for future use as recreational space or for infrastructure purposes. To guard these areas from land grabbing, 'soft functions' (e.g. sports fields, green recreational space) can be assigned to temporarily occupy the area as a 'placeholder'. Review of land use regulations and the use of incentives may help to avoid unwanted effects. A review is needed on how current land use regulation steers urban development, and what the intentional and unintentional effects are on the built environment.

The Delta Plan highlighted the importance of making advancements in local urban planning and land use regulations. In this regard, the establishment of a national spatial strategy and plan (including links to the FYP's and Annual Development Programming) is important. This can include a vision and guidelines for sustainable urban development. Town and city planning could address improved assessment of socio-economic factors, disaster vulnerability factors and environmental factors, creation of maps that show the relation between the natural landscape, infrastructure and occupations.

9.2.6 Chattogram Hill Tracts

In the Hotspot ‘Chattogram Hill Tracts’, both the hills and related coastal plains are considered. Problems experienced in the hills, like landslides, are related to coastal plain problems like sedimentation and recurrent dredging. While for the larger rivers in Bangladesh, much of the catchment areas are outside the country, in the eastern hill area, most of the catchment areas of the important rivers are within the country’s borders. This offers the required conditions for integrated river basin management and opportunities to address problems holistically. BDP 2100 strategies include the following.

9.2.6.1 Protect Economic Zones and Towns from Flood and Storm Surge

The BDP 2100 focused on river flood risk management in the Chattogram Hill Tracts, especially for rivers like Chengi, Karnafuli, Sangu, Matamhuri, Bakkhali rivers, and hilly streams. It also emphasized rationalization of existing sea and river facing polders and extends soft coastal defense and shelters against extreme floods, including flood resilient housing (together with Hotspot Coastal Zone).

9.2.6.2 Ensure Water Security and Sustainable Sanitation

The Delta Plan also suggested establishing flow control and water reservation structures for water availability in the dry season, including small-scale (cross)dams and reservoirs for irrigation, livestock and aquaculture (e.g. by rubber dams) for the CHT region. Besides it also proposed the development of sustainable and safe water supply and sanitation systems. This includes installation of location sensitive water supply mechanisms to ensure safe drinking water (including gravity flow systems, water harvesting, and tube wells where applicable). The Bangladesh Delta Plan’s Investment Plan includes sustainable water supply and sewage system in large cities and financial sustainability and cost recovery (linked to elaboration of city master plans).

9.2.6.3 Ensure Integrated River Management

The BDP 2100 emphasized the integrated sediment and erosion management. For achieving this, it proposed creating a green belt, ecofriendly structures, soil and water conservation, and reforestation. Another proposed strategy has been the integrated management of rivers and hilly streams through the development of catchment and sub-catchment management plans.

9.2.6.4 Maintain Ecological Balance and Values (Assets) at CHT

The Delta Plan focused on protecting and restoring water-related ecosystems, including forests and hill sides. It also advocated the promotion of sustainable cultivation practices and regeneration of degraded areas. The BDP 2100 Investment Program includes creating awareness on the role of forests in water management, replace mono-plantation and promote agroforestry with native (fruit) trees in support of livelihoods and water availability.

9.2.6.5 Increase Institutional Capacity for Integrated Water Resources Management

The Bangladesh Delta Plan highlighted the importance of strengthening institutions and cooperation for integrated holistic watershed management. Therefore, it emphasized developing river basin management plans in cooperation with involved departments and communities. For ensuring disaster preparedness, the BDP 2100 proposed revision of flood warning systems and flooding-related procedures and also establish disaster management groups and community-based forecasting and early warning systems.

9.2.6.6 Develop Multi-purpose Resources Management System for Sustainable Growth

This focuses on development of safe, reliable and optimal use of land, water ways and energy. To this belongs the Kaptai Lake rehabilitation project and lake-level optimization (rehabilitation of reservoir and surroundings). This project includes pilots for Lake Compartment management and optimized lake-level management for hydropower, agriculture, fishing, and water communication.

Besides, strengthen the resilience of livelihoods and of sustainable food production is proposed. It is suggested to create markets and work on value chain development for income diversification, e.g., by growing high value crops, food processing, fisheries, agroforestry value chain, and local storage and transportation facilities.

9.3 Thematic Strategies

The Bangladesh Delta Plan 2100 also developed thematic strategies to address the delta vision and goals in specific priority areas and challenges within the context of the country's deltaic region. For instance, ensuring food security and nutrition is a key objective of the Government and for addressing this priority issue the BDP 2100 developed a specific thematic strategy for sustainable agriculture, food security, nutrition, and livelihood. In a similar manner, BDP 2100 developed thematic strategies for other priority issues like urban waste management, advancing Blue Economy and managing transboundary water issues. Ultimately, the development of thematic strategies ensured a focused, comprehensive, and tailored approach to address the challenges and opportunities within specific priority areas.

9.3.1 Urban Sanitation and Wastewater Management

In many urban areas of Bangladesh, there is no appropriate systematic sewer disposal and treatment system. For instance, the capital Dhaka has a sewer system that has a limited capacity to serve the city. However, this existing situation is supposed to worsen in the upcoming days considering the rapid urbanization of Bangladesh. For ensuring required sanitation and wastewater management in the urban areas, BDP 2100 proposed the following strategies.

9.3.1.1 Increase the Application of Fecal Sludge Management

To increase the application of fecal sludge management, GoB has already taken several sectoral initiatives. These include the collaboration of the public sector with different international NGOs through public–private partnership (PPP) arrangements. An example is the Citywide Inclusive Sanitation Engagement (CWISE) project, a three-year urban sanitation project with Khulna City Corporation, Jhenaidah, and Kushtia Pourashavas. However, to make these efforts more fruitful, BDP 2100 has proposed some sub-strategies, among which most notable is targeting the financial sustainability of the FSM sector and optimization of working processes. In general, one of the critical cost components of the FSM business is the O&M cost, which varies significantly with the frequency of desludging and the quantity of fecal sludge transported to the Fecal Sludge Treatment Plant. If not appropriately managed, this O&M cost can put the sustainability of the PPP-business at risk.

9.3.1.2 Ensure Freshwater Supply and Wastewater Treatment to Enhance Industrial Growth

Many industries in Bangladesh are infamous for ill-regulated water abstraction and illegal or uncontrolled disposal of waste and wastewater. In light of this, BDP 2100 has proposed the provision of zoning and the establishment of restrictions on illicit wastewater disposal to ensure freshwater supply and improved environmental health. At the same time, BDP 2100 suggested imposing restrictions on water abstraction to help recover groundwater levels. In contrast with the proposed restrictions, it also advocated incentive programs for good industrial practices.

9.3.1.3 Strengthen the Governance Framework

BDP has proposed strengthening the governance framework to make institutional responsibilities in the sanitation sector firmer. At present, the institutional responsibilities, especially the costs of pollution abatement, have not yet sufficiently materialized in major cities, towns, villages, and rural areas. To address this situation, BDP 2100 has recommended the adoption of polluter pay principles so that the polluters pay the cost of pollution abatement. Besides that, it mentioned involving the private sector via public–private partnerships (PPPs) to optimize initial investment costs and future maintenance and operation costs.

9.3.1.4 Institutionalize Cost Recovery Principles

In the sanitation sector, cost recovery has yet to be suitably established. At present, the house owners are mainly responsible for paying for emptying their septic tanks. In major cities (i.e. Dhaka), there is an arrangement where house owners pay for the sewer connection or the vicinity of available sewerage. However, full-service re-payment, either by direct billing or municipal tax levies, requires further development. Therefore, to address this issue, BDP 2100 has advised measures such as municipal tax development to ensure cost recovery.

9.3.2 *Solid Waste Management*

Solid waste management is one of the major environmental challenges in the urban areas of Bangladesh. Dhaka has to deal with around 6,500 tons of waste daily, which is projected to increase to 8,500 tons by 2032. In addition, hazardous medical and e-wastes further burden the ineffective waste management system. However, the overall waste management system is not sustainable at all, except for some pilot projects. This leads to piling up waste and creating pressure on landfills in the long term. BDP 2100 has developed solid waste management sub-strategies to address these issues, which includes the following.

9.3.2.1 Integrated Wastewater and Solid Waste Management

Conventional SWM is not sustainable and is often too costly. Besides, it brings undesirable health impacts, environmental problems, and adverse social consequences (such as informal communities or waste pickers working in unsafe conditions). To address this issue, BDP 2100 has proposed Integrated Solid Waste Management (ISWM) to achieve sustainable solid waste management. ISWM is a comprehensive approach that minimizes the amount of waste generated, promotes the reuse and recycling of materials, and ensures that any residual waste is disposed of safely and effectively. In addition to ISWM, BDP has proposed to strengthen institutional capacity and increase community participation to increase awareness as well as enhance a sense of responsibility.

9.3.2.2 Ensure Proper Radioactive Waste Management

The disposal procedure of radioactive and nonradioactive waste is different and therefore requires distinctive management options. Considering this, BDP 2100 has proposed a strategy emphasizing the need for legislation and regulation of waste management activities in the national waste management policy. According to BDP, the policy may state a preference for one management option over another. For instance, storage for decay and then disposal as nonradioactive waste versus immediate disposal as radioactive waste. Correspondingly, the policy may define the funding and management responsibilities for the programs as well as the legal/jurisdictional roles of various government departments.

9.3.3 Transboundary Water Management

Being within the downstream position of the Ganges—the Brahmaputra—and the Meghna basins and the recipient of a substantial amount of trans-transboundary flows from several neighboring countries, cooperation regarding water resources management with India, Nepal, Bhutan, Myanmar, and China is essential for Bangladesh to foster further development. Closer cooperation can be ensured with neighboring countries in view of the general understanding that joint cooperation outweighs individual development. Strategies in this field concern the following.

9.3.3.1 Progressing on Ganges Treaty

In 1996, the Governments of Bangladesh and India signed the Ganges Treaty on water sharing for a period of 30 years. To date this is the only transboundary river treaty signed by Bangladesh. The two countries have been sharing the Ganges waters since 1997. Discussions are being held with India for augmentation of the dry season flows of the Ganges at Farakka in accordance with the provision of the Treaty. The issue of augmentation of the Ganges flow at Farakka by constructing a reservoir in Nepal was also discussed in the 37th meeting of Joint River Commission in March 2010.

One of the main observations on the Ganges Treaty is that it only addresses the quantity of water shared and does not consider the quality of water discharged to Bangladesh. It also does not have any flood management provisions. The absence of a guarantee clause is also essential as it does not limit India from developing water-intensive agriculture or constructing barrages upstream. Moreover, the agreement has been signed between both national governments while Indian water interventions are developed at a relatively autonomous sub-national (state) level, complicating the treaty arrangements' conduct. Furthermore, no single project has been implemented to take advantage of the Ganges Treaty in Bangladesh and for augmentation jointly as enunciated in the Treaty.

9.3.3.2 Framework Agreement on Cooperation for Development

The Governments of Bangladesh and India signed a 'Framework Agreement on Cooperation for Development' in September 2011. The aim is proper utilization and management of the river basins through equitably sharing knowledge and resources and planning holistic and integrated approaches at the river basin scale. Both governments agreed on enhancing cooperation in sharing of water resources of common rivers and integrated river basin management in light of this framework. This has opened the opportunity for the establishment of joint River Basin Organizations (RBOs). The RBOs aim to address the issue of common river basin management by implementing Integrated Water Resources Management at the river basin level and comprising representatives from the respective member riparian countries.

9.3.3.3 Water Diplomacy Approaches

Water diplomacy 'tries to harmonize water policies and standards (of co-riparians) and brings into focus issues at different levels and across sectors, encompassing not only technical but also social and economic implications' (Pangare & Nishat, 2014). Water diplomacy ranges from 'conflict prevention, to treaty-making, to ensuring regional integration' (Clingendael Institute, 2011). Regardless of the amount of new interventions in the river basins, a cooperative attitude results in many more benefits. By working together, cooperation's social, economic, and ecological benefits will be greater than the sum of the current individual national benefits.[17]

Bangladesh is downstream and therefore extremely affected by co-riparian actions upstream. Generally speaking, most of the future developments are not in Bangladesh's direct sphere of influence. Policies of key regional players like India and China will undoubtedly have huge implications for Bangladesh's situation in 2100. In order to control the rivers' conditions, Bangladesh can, utilizing a well prepared water diplomacy approach, influence Indian, Chinese, and other co-riparian water policies and river development plans.

[17] See also: Pangare & Nishat (2014).

9.3.3.4 Other Strategies

- Continuing efforts for signing of a treaty regarding the sharing of water for the Teesta and all other transboundary rivers.
- Demand-based common river basin management schemes can be initiated.
- Third party involvement (multi-lateral or bilateral development partner or country) to resolve transboundary water-related issues.
- Improved basin-wide flood forecasting.

9.3.4 Sustainable Land Use and Spatial Planning

Sustainable land use and spatial planning are crucial for overall economic development as they help to protect ecosystems, promote sustainable use of resources, reduce greenhouse-gas emissions, encourage social equity, and reduce risk from natural disasters (Alam et al., 2018b). Considering its significance, BDP 2100 has developed sustainable and commonly agreed upon strategies, along with all relevant stakeholders, for optimum land resources management, water safety, food security, and sustained economic growth in Bangladesh. Strategies include the following.

9.3.4.1 Preserve/Conserve Agricultural Land from Floods or Erosion to Sustain Food Grain Production

BDP 2100 proposed the strategy of preserving/ conserving agricultural land from a food security point of view. This strategy has several sub-strategies that involve important priorities and strategic interventions. For instance, BDP suggests that hotspot areas (e.g. the coastal zone, the Sylhet region, the char areas, and drought-prone areas) must receive priority in crop sector development plans for crop intensification. Besides that, it emphasizes the need for measures to protect agricultural land from inundation, river erosion, and other negative impacts. Additionally, BDP 2100 recommends utilizing the unutilized Haor land in the northeast part of Bangladesh to meet the growing demand for additional food for the increasing number of people in the country.

9.3.4.2 Prevention of Salinity Intrusion and Desertification

BDP 2100 has recommended provisions to increase freshwater flow from upstream in the coastal zone to prevent salinity intrusion. Regarding the reduction in desertification, it has emphasized the need for further vegetation in the Barind area. Along with vegetation, BDP has also developed a sub-strategy focusing on reducing stress on the water resources system.

9.3.4.3 Management of Newly Accreted Land in the Meghna Estuary.

To ensure proper management of newly accreted land in the Meghna estuary, towards the sea and coastal zone, BDP has proposed increasing land availability through accretion or land reclamation. On another note, to overcome the land loss due to sea-level rise and riverbank erosion, BDP developed sub-strategies such as new land formation towards the sea through land reclamation, construction of closure, cross dams, etc. According to the BDP-suggestions, these measures can be accelerated with building with nature techniques.

9.3.4.4 Development of Land Information System for Land Administration and Management

BDP 2100 has suggested the development of a Land Information System (LIS) as a strategy for sustainable land use and spatial planning. Though mainly used for land administration and management, LIS has multi-fold purposes. Such purposes include providing access to land and property rights to all citizens, preparing digital maps and using modern equipment, creating authentic land records, etc. In addition to LIS, BDP recommends proper zoning laws with institutional enforcement in place.

9.3.4.5 Development of a Digital Land Resource Management System

To ensure sustainable land use, BDP 2100 has proposed a strategy that involves the computerization of the land resource management processes in Bangladesh. The target should be the accessibility of information to the citizen, reducing unnecessary steps. Nonetheless, BDP also stresses the need for clear linkage between the different agencies to obtain desired results.

9.3.4.6 Other Strategies

- Reviewing and updating/enactment of laws/regulations relating to alluvion and diluvion to improve efficiency of land administration of accreting and reclaimed land.
- Formulation of necessary laws and acts to form land zoning.
- Increase climate change adaptation capacity for land management.
- Spatial land use planning for urbanization; optimization of land use.
- Formulation of Spatial Planning and Land Resource Management Act.
- Enhance afforestation and plantation in the coastal zone for stabilizing land.
- Restoration and protection of soil health, erosion, and land loss.
- Integrated management of coastal water infrastructures to protect land.

9.3.5 Sustainable Agriculture, Food Security, Nutrition, and Livelihood

During the recent decade, the overall GDP of Bangladesh has considerably shown a positive trend; however, the growth in agriculture GDP slightly declined. This is remarkable, considering the fact that more than three-quarters of the total population in rural areas of Bangladesh derive their livelihood from the agricultural sector. At the same time, about 48 percent of the labor force is employed here. Therefore, agriculture is seen in Bangladesh as being the engine of growth of the economy, and that it is important to develop this sector to alleviate poverty.

Keeping this in mind, BDP 2100 has proposed several strategies to ensure sustainable agriculture, food security, nutrition, and livelihood; see Fig. 9.10. Strategies are indicated below.

9.3.5.1 Addressing the Impact of Climate Change in Agriculture

Climate change can reduce crop yields and livestock productivity, thereby threatening overall food security. BDP 2100 proposed a strategy for developing awareness and responses to address climate change in agriculture. This strategy can help farmers and other stakeholders to adapt to climate change and maintain economic stability. In this regard, adaptations to climate change can be thought of as incremental changes to

Fig. 9.10 Rice harvesting in Sonargaon area. *Photo* J. M. de Heer

existing systems or more systemic changes that bring new components to (or remove old parts from) systems.

9.3.5.2 Increasing Resilience of Agricultural Production Systems

To achieve sustainable Agriculture, BDP 2100 has proposed to increase the resilience of agricultural production systems. In other words, ensure increased yields and resilience of crops while reducing the need for often costly and inaccessible synthetic fertilizers. For this purpose, it has mentioned several measures such as appropriate soil and nutrient management through composting manure and crop residues, more precise matching of nutrients with plant needs, controlled-release and deep-placement technologies, using legumes for natural nitrogen fixation, etc.

9.3.5.3 Diversification in Agricultural Output and Livelihoods

BDP 2100 has proposed agricultural output and livelihood diversification to achieve sustainable food security. Agricultural output diversification helps producers to become less vulnerable to the impacts of a single crop or livestock failure. Similarly, livelihood diversification has various benefits, including risk reduction, income stability, poverty reduction, increased resilience, skill development, and gender equity.

9.3.5.4 Devise a Longer-Term Adaptation Mechanism Against the Impacts of Climate Change and Particularly of Sea-Level Rise in Food Security

Considering the upcoming challenges of climate change, BDP 2100 has emphasized the need for a different direction for sustaining food security over a longer time frame as a strategy. In this strategy, it advocates that strategic needs of future-developed Bangladesh must gradually replace traditional food security approaches. For that purpose, BDP mentions several measures, such as implementing climate-smart agriculture, expanding cropped areas with saline-tolerant varieties for the coastal zone, investigating the enhancement of productivity of traditional floating agriculture.

9.3.5.5 Improved Resources Management for Ensuring Irrigation Facilities

BDP 2100 has developed the strategy of improving delta management in conjunctive use of surface and groundwater. This strategy involves the augmentation of surface water for irrigation by developing water reservoirs, recharging groundwater especially during monsoon time, and reducing groundwater use. As part of this strategy,

BDP has also proposed several activities, including reducing water losses in existing schemes through improved water management (capacity building of water management cooperatives), development of water-saving techniques such as drip irrigation, buried pipe irrigation, 'fertigation'; or rehabilitation of existing schemes. Vertical farming would be promising as it is a practice of growing crops in vertically and horizontally stacked layers. Water efficiency and productivity are ensured due to a controlled-environment, which aims to optimize plant growth and new techniques as hydroponics, aquaponics and aeroponics.

The following issues have been covered as well:

- Increasing resilience of agricultural production systems.
- Diversification in agricultural output and livelihoods.
- Lower emissions (GHGs) from agricultural land.
- Encourage establishing of commercial farms
- Introduction of aquaponics farming system to culture fish and plants together.
- Using nanotechnology in agriculture.
- Introduce precision agriculture model.
- Encouraging solar power in irrigation.
- Improved farm practices and technologies on impacts of climate change.
- Preservation of ecosystems for plant, wild animals, fishes, birds, etc. and encourage fruit tree plantation.
- Improve wetland management in Haor areas for development of fisheries.
- Maintaining biodiversity to ensure long-term fish availability.
- Sustainable increase of marine fisheries (Blue Economy aspect) and marine resources management.
- Production of climate resilient livestock.

9.3.6 Dynamizing the Inland Water Transport System

The inland water transport (IWT) sector of Bangladesh is of crucial importance in the whole transportation system. However, it is crippled with several challenges, such as incremental dredging demand to maintain navigability, lack of infrastructure, non-compliance of guidelines for the construction of bridges over the rivers, encroachment in the rivers, lack of safe vessels and skilled workforce, lack of policy guidelines, poor governance, inadequate budget allocation, lack of intermodal coordination, and a poor level of international cooperation. BDP 2100 has proposed several strategies for dynamizing IWT and port development. Some of the strategies on IWT and port development are as follows.

Fig. 9.11 Always busy Dhaka river port. *Photo* J. M. de Heer

9.3.6.1 Developing Reliable Water Systems Conditions for Long-Term Sustainable Economic Development

A reliable water transport system is critical for economic development. Many industries, such as agriculture, manufacturing, and tourism, rely on water for their operations. Considering the economic significance, BDP 2100 has stressed the need for a flood and climate change resilient operational environment for shipping, ports, and related facilities. Along with that, for dynamizing the Inland Water Transport (IWT) System, BDP has highlighted the need to ensure the performance of key societal and economic functions of IWT, especially with respect to economic centers and large cities; see, e.g., Fig. 9.11.

9.3.6.2 Cooperate and Coordinate with BWDB to Provide Optimal Levels of Surface Water for Navigation

Since Bangladesh Water Development Board (BWDB) is primarily responsible for large-scale water infrastructure projects, cooperation and coordination of the IWT-sector with BWDB are necessary to provide optimal surface water levels for navigation. For this purpose, BDP has highlighted several areas where coordination can occur. These include water-level coordination, river classification and standards renovation, advanced water supply and sanitation, and wastewater treatment in all ports and ships.

9.3.6.3 Increase Water Use Efficiency Across the Whole IWT-Sector and Ensure Sustainable Water Use

For dynamizing the Inland Water Transport (IWT) System, BDP 2100 has proposed to increase water use efficiency across the whole IWT-sector and ensure sustainable water use. And as a part of this strategy, it has included measures such as the promotion of recycling and safe reuse of water with the latest technology, small-scale water harvesting techniques at large ports including recycling and safe reuse of water, full coverage small-scale water harvesting in all ports and landing places, advanced water supply and sanitation and wastewater treatment in ports and ships.

9.3.6.4 Develop the Navigation Network According to the Societal and Economic Demands

BDP 2100 has highlighted the need for developing a navigation network according to societal and economic demands for meeting transportation needs, promoting economic growth, reducing transportation costs, and enhancing environmental sustainability. To achieve this, BDP has proposed several measures as: conducting traffic surveys to establish passenger and cargo requirements on the main rivers. Doing so will help create a more efficient and sustainable future for designated communities and relevant businesses that rely on inland waterway transportation.

9.3.6.5 Contribution to integrated management of rivers

Another strategy for dynamizing the Inland Water Transport (IWT) System is the integrated management of rivers. As suggested by BDP, drawing up programs of dredging requirements and priorities for efficient maintenance of existing navigable waterways can help implement this strategy effectively. In addition, the development of new channels and canals for navigation can also help improve transportation efficiency by providing new or improved routes for vessels to travel. Moreover, this can help reduce transportation time and costs, as well as reduce congestion on existing waterways.

9.3.6.6 Items Covered by the (Sub)strategies

- Regular dredging for maintaining water flow and transport in the rivers.
- Develop reliable water system conditions for long-term sustainable IWT through the capital and maintenance dredging of the rivers Padma, Meghna, Jamuna, Brahmaputra, Dharla, Arial khan, Kushiyara, Gorai, and Manu.
- Regular dredging should also be considered for Ghashiakhali and other channels in the Sundarbans.

Fig. 9.12 Shipping in flooded areas. *Photo* courtesy of Amirul Hossain

- Ensure efficient and equitable use of sand through the regular shifting of the 'Balu-mahal' (sand quarry). The local administration should take necessary steps accordingly. Specific guidelines should be developed for the management of soil or sediment resultant from dredging.
- BIWTA to cooperate and coordinate with BWDB to provide optimal levels of surface water for navigation and, e.g., to include navigation aspects in preparation of land reclamation or accretion projects.
- Develop the navigation network according to the societal and economic demands with attention for shipping during flooding; see Fig. 9.12.
- Develop, maintain and operate inland river ports, landing ferry ghats, and terminal facilities in ports or ghats.
- Contribute to dealing with transboundary water aspects by developing mutual understanding and cooperation.
- Development of riverine and maritime ports.
- Initiatives for activating transboundary waterways.

9.3.7 Renewable Energy

Despite the significant potential of renewable energy (RE) in Bangladesh, the development of grid-connected RE technologies has been slow to materialize. There are several regulatory, financial, and technical barriers that, if addressed, could accelerate renewable energy investment in the country. Improved regulations, such as establishing a standard feed-in tariff and provisions for compensating mini-grid investors after transmission expansion, would reduce risk and send strong signals to investors.

The following strategies could contribute to the development and dissemination of successful renewable energy programs in the country:

- Development of a long-term renewable energy policy as well as strategies and formulation of a master plan to harness the potential of renewable energy resources in the country involving public and private sector investments.
- Promotion of research on the development of technology in the field of renewable energy in universities and research institutions as well as building capacity for its application.
- Enhancing Green Growth through research and development of renewable technologies including Clean Development Mechanism (CDM).
- Devising innovative financing packages for grant funding and low interest financing to address affordability for both grid and off-grid renewable energy projects.
- Targeting on significant energy production from renewable sources by 2041 in the context of a growing economy.

9.3.8 Dealing with Earthquakes

Considering the high risk of major earthquakes in Bangladesh, BDP 2100 has proposed strengthening the existing earthquake management system and enhancing the capacity of relevant stakeholders. The pre-disaster management plan includes designing earthquake-proof (infra)structures in conformity with the Bangladesh National Building codes or any other approved standards. Aside from that, BDP emphasizes the need to formulate a proper land use plan for building constructions in municipal areas.

9.3.9 Advancing the Blue Economy

Bangladesh has a 720 km coastline, and this vast coastline offers immense tourism, renewable energy, fishery, agriculture, biodiversity, and ecosystem services potentials for the country (Rahman & Bhattacharya, 2014). In 2014, Bangladesh won more than 1,18,813 km^2 of waters comprising territorial sea, exclusive economic zone extending out to 200 NM across sizable area (Islam et al., 2018). However, the country has not been able to effectively utilize its marine resources.

The advancement of the Blue Economy in Bangladesh is mainly hindered by numerous reasons such as the inadequacy of knowledge and assessment of ocean resources, the lack of skilled and trained workforce, the absence of a dedicated Blue Economy governance, institution and marine spatial planning, inadequate coastal ship fleets, lack of investment opportunities, absence of environmental friendly ship recycling, etc. Besides that, the fragmentation of initiatives and of institutional actions and responsibilities is a concern. Considering the existing challenges and to create and maintain prosperous and sustainable Blue Economy bases, BDP 2100 has proposed several practical strategies as follows.

9.3.9.1 Keeping the Coasts and Sea Ports Pollution Free

Environmental ship recycling generally refers to the process of dismantling and disposing of end-of-life ships in an environmentally and socially responsible manner. The concept of environmental ship recycling is essential for Bangladesh, considering the fact that Bangladesh derives 80–90% of its steel from end-of-life ships (Rabbi & Rahman, 2017). Besides, the whole shipbreaking industry generates a number of employment opportunities for Bangladesh too, where workers' safety, child labor, and environmental issues are often brought to light. Considering these, BDP 2100 has proposed the strategy of enhancing environmental ship recycling with an aim to improve the existing working conditions by ensuring workplace safety. Likewise, BDP aims to improve environmental conditions through this strategy by preventing the release of hazardous materials into the environment.

Among the proposed strategies, one is the enhancement of the coastal ship fleet and capabilities, including modern port facilities. If this strategy is implemented properly, it will lead to increased knowledge and skills and capacity for transportation. Besides, it will provide improved access to remote areas that may not have other means of transportation. Some of the other strategies concern:

- Quick completion of multi-dimensional survey of marine resources.
- Increase the number of sea going vessels and modernization and capacity building of the sea ports.
- Increase both shallow and deep sea fishing.
- Introduction of ecotourism and private sector initiatives in sea cruise.

References

Alam, S., De Heer, J. M., & Choudhury, G. A. (Eds.) (2018a). *Bangladesh Delta Plan 2100: Baseline Studies on Water Resources Management. Volume 1*. General Economics Division, Planning Commission, Government of Bangladesh. https://plancomm.gov.bd/

Alam, S., De Heer, J. M., & Choudhury, G.A. (Eds.) (2018b). *Bangladesh Delta Plan 2100: Baseline Studies on Land Use and Infrastructure Development. Volume 3*. General Economics Division, Planning Commission, Government of Bangladesh. https://plancomm.gov.bd/

Arif, H. (2022). Government to launch Tk 249cr project to facilitate irrigation in Rajshahi-Rangpur division. *The Business Post*.

BanDuDeltAS. (2017). Draft BDP 2100 Report with annexes.

Clingendael Institute. (2011). Van Genderen, R., & Rood. J. *Water Diplomacy Report*.

De Heer, J. M., & Choudhury, G. A. et al. (2023). *Extended Coastal Zone Implementation Program, 2023*.

IPCC. (1990). *Strategies for Adaptation to Sea Level Rise*. Report of the Coastal Zone Management. Subgroup, IPCC Response Strategies Working Group.

IPCC. (2022). Climate Change 2022: Impacts, Adaptation and Vulnerability. In Pörtner, H.-O., Roberts, D. C., Tignor, M., Poloczanska, E. S., Mintenbeck, K., Alegría, A., Craig, M., Langsdorf, S., Löschke, S., Möller, V., Okem, A., Rama, B. (eds.) *Contribution of Working Group II to the Sixth Assessment Report of the Intergovernmental Panel on Climate Change* (3056 pp.). Cambridge University Press. https://doi.org/10.1017/9781009325844

Islam, M. K., Rahaman, M., & Ahmed, Z. (2018). *Blue Economy of Bangladesh: Opportunities and challenges for sustainable development.*

Haasnoot, M., Diermanse, F., Kwadijk, J., De Winter, R., & Winter, G. (2019). *Strategieën voor adaptatie aan hoge en versnelde zeespiegelstijging. Een verkenning.* Deltares report 11203724-004.

Ministry of Environment, Forest and Climate Change. (2022). *National Adaptation Plan of Bangladesh (2023-2050).*

Ministry of Planning. (2028). *Bangladesh Delta Plan 2100.* 2018.

Netherlands Institute of International Relations' Clingendael' (2011). *Clingendael annual report 2011.*

Pangare, G., & Nishat, B. (2014). Perspectives on Hydro diplomacy In: *Hydro-Diplomacy: Sharing Water Across Borders.* Academic Foundation.

Rabbi, H. R., & Rahman, A. (2017). Ship breaking and recycling industry of Bangladesh; issues and challenges. *Procedia Engineering, 194*, 254–259.

Rahman, M. M., & Bhattacharya, A. K. (2014). Saline water Intrusion in Coastal Aquifer: A case study from Bangladesh. *Journal of Engineering, 4*(1), 7–13.

Rashid, M. H. A., Islam, A. B. M. J., Shirazy, B. J., & Shahidullah, S. M. (2017). Cropping systems and land use pattern in Rajshahi region. *Bangladesh Rice Journal, 21*(2), 237–254.

SIBDP. (2023). Implementation Program for the Extended Coastal Zone.

Part IV
Realization of Aspirations

Institutional Transformation and Provisions

Chapter 10
Investment Plan and Financing of Measures

Abstract Characteristic for the Bangladesh Delta Plan is that it is not only a plan but also takes a step further into elaboration of the proposed measures into investments, to find out the financial requirements to implement it. This chapter describes and explores financial and funding aspects. The formed investment plan (IP) is based on a large number of in line with the formulated strategies proposed measures, which have been elaborated into Project Concept Notes. A selection of Project Concept Notes is included in the IP. The financing strategy consists of several contributions from GoB and different partners. Here, recent developments and funding constraints are highlighted. An important background for the financing strategy is the government commitment in BDP 2100 to spend 2.5% of the GDP in Delta Plan-related investments. Of this, approximately 2% of GDP would be for new investments, and 0.5% of GDP is for operations and maintenance (O&M). 0.5% of GDP allocation of O&M is important as historically, most of the water projects of Bangladesh obtained very limited O&M funding. Due to a lack of O&M funding, and inadequate investment in upgrading existing infrastructure, much of the investments made in water resource management in the form of embankments, polders, and to improve inland waterways navigation do not remain in good shape to develop them with respect to climate change-related investments. Besides GoB contributions, additional financial resources are explored. To make use of the Green Climate Fund, six proposals have been submitted. Also funding is sought in the form of private sector investments, by innovative contracting and public–private partnership.

Keywords Investment plan · Funding · Financing strategy · Public–private partnership · GDP · Green climate fund · Loss and damage fund · Blues bonds · Taxation · Beneficiary pay principle · Service charges · Financial constraint · Funding constraint · Development partners · Financial institutions · Bank · NGO · Foundation · Water user institutions · Water management organizations

J. M. de Heer and G. A. Choudhury, *Strategies for Adaptation to Climate Change in a Transformative Approach*, Springer Climate,
https://doi.org/10.1007/978-3-031-88884-7_10

10.1 Investment Plan and Funding

10.1.1 Investment Plan Project Proposals

During the strategy process of BDP 2100, each strategy has been elaborated in a number of measures. Further detailing resulted in a large number of about 400 Project Concept Notes of which initially 130 and finally 80 were selected. The formed Investment Plan (IP) consists of a total of these 80 projects: 65 are physical projects, and 15 are institutional and knowledge development projects for the first implementation phase up to 2030. Its total capital investment cost is in 2018 BDT 2,978 billion (US $37 billion) (Ministry of Planning, 2018). Selected projects can according to the IP be started within the next eight years; though given the scale and programmatic nature of some investments, implementation may extend over longer periods. The investments plan indicates that approved projects or programs will be guaranteed funding for the full life of the project or program and a proper balance between new projects and completion of ongoing projects will be ensured. Operations and maintenance fund will initially come from the GoB/ Delta Fund but over time it may be phased out by transferring responsibilities to the cities, municipalities, and local or regional water management institutions. Similarly, over time, the Delta Fund will manage only a small number of national high priority projects, innovation, and research. The financing of smaller projects could transfer to cities, municipalities, and local or regional water management authorities with quality assurance on standards and inspection functions performed at the national level.

10.1.2 Funding Aspects of BDP 2100

Bangladesh's long-term water safety and security, food security, and socio-economic performance would critically depend on how it manages the effects of climate change and other environmental challenges. Already the country is experiencing impacts of climate change, especially sea-level rise, salinity intrusion, and erratic and heavy rainfall. According to the latest IPCC report, nearly 26 million people are currently exposed to very high (> 1500 $\mu S\ cm^{-1}$) salinity situations in shallow groundwater in coastal Bangladesh.[1] The proper implementation of the Bangladesh Delta Plan can be crucial in addressing climate change's impact, but to realize that, the country needs to mobilize significant financial resources. According to the Bangladesh Delta Plan, the government's annual spending in delta-related program/projects in agriculture, water resources management, environment, and climate change is approximately 0.8% of GDP ($ 321 bln in 2018). This amount is inadequate to address the impact of climate change in these sectors. Realizing this issue, BDP 2100 indicated a target, block provision, of securing at least 2.5% of GDP per year by 2025 to implement the

[1] IPCC (2022).

BDP 2100.[2] Of this, 2% of GDP would be for new investments, 1.5% funding from government resources and 0.5% investments from private sector involvement.[3] In addition, 0.5% of GDP will be allocated for operations and maintenance (O&M). This 0.5% of GDP allocation of O&M is important and historic, since most of the water projects of Bangladesh obtained limited O&M funding. Due to constrained O&M and inadequate investment in upgrading existing infrastructure, much of the investments made in water resource management in the form of embankments, polders, and to improve inland waterways navigation do not remain in good shape. It will be a challenge for GoB given the financial demands from many societal sectors to fulfill the annual funding target of 2.5% of GDP, and the country will need strong support from the private sector. A solution to guarantee or safeguard the 2.5% block provision for Delta Plan investments is the formation of a Delta Fund, as indicated in BDP 2100. Such Delta Fund currently takes the form of prioritizing and earmarking Delta Plan investments in the regular annual budget. A step further would be to form a separate, dedicated multi-annual Delta Fund and Fund management to ensure that BDP 2100 goals and vision can be achieved over time.

The macroeconomic and sectoral costs of not implementing or under-implementing the Delta Plan can be substantial (Alam et al. 2018). At the macroeconomic level, under moderate climate change conditions, this can annually reduce GDP growth by 1.3% through the loss of physical assets and reduced economic activities. In addition, Bangladesh will also face other costs in terms of health risks and loss of biodiversity. The costs would be commensurately higher under a worst-case climate change environment. Considering this reality, GoB indicated the target, block provision, with public and private sector contributions. The GoB is also trying to access different climate funds to address the impact of climate change. In the COP 28 held in Dubai in 2023, over $50 billion in new funding was announced, including a $3 billion commitment by the USA to the Green Climate Fund. COP 27 and COP 28 led to the development of the Loss and Damage (L&D) fund, and COP 28 facilitated the creation of a $30 bln ALTÉRRA climate fund. Bangladesh wants to tap into existing and new climate financing channels.

The implementation of the Delta Plan in the year 2025 will need to involve total spending on delta-related interventions through new projects and operation and maintenance of new and old projects, in an amount of about 2.5% of GDP per annum. However, the existing spending is lower, with an annual spending of a mere 1.1% of GDP. It may be difficult for GoB to mobilize large-scale additional financing and it will require mobilizing funds from other potential domestic and external sources. The in 2018 expected overall delta investment percentages are highlighted in the following Table 10.1. Global or national economic and political circumstances will of course influence the investment possibilities.

A number of BDP 2100 Investment Plan projects or related parts are already in preparation or execution, as shown in Table 10.2.

[2] Bangladesh Delta Plan 2100. Volume 1: Strategy; Volume 2: Investment Plan.

[3] Ibid.

Table 10.1 BDP 2100 Investment Requirements (% of GDP). *Source* BDP 2100 Projections, GED, Ministry of Planning, 2018

Investment components	FY2016	FY2020	FY2025	FY2031
Total delta investment (water related)	0.8	1.8	2.5	2.5
Of which:				
- Additional public investment	–	0.5	0.7	0.7
- Public investment (baseline)	0.8	0.8	0.8	0.8
- Total public investment	0.8	1.3	1.5	1.5
- Maintenance cost (recurrent budget)	0	0.3	0.5	0.5
- Private investment	0	0.2	0.5	0.5

Table 10.2 BDP 2100 Investment Plan projects under implementation. *Source* GED, Ministry of Planning, February 2, 2023

No.	Name of the project	Executing agency
1	Drainage Improvement of Dhaka-Narayanganj-Demra (DND) project phase 2	BWDB
2	Revitalization of Khals all over the country	BWDB and LGED
3	Development of Water management infrastructure in Bhola island	BWDB
4	Char development and settlement project-V (CDSP)	BWDB, M/o Land, Forest Dept
5	West Gopalganj integrated water management project	BWDB
6	Improvement of drainage congestion and flood control for Chittagong City	CDA
7	Improvement of sanitation systems in city corporation areas of Bangladesh	DWASA, CWASA, Khulna WASA
8	Kurigram Irrigation project	BMDA
9	Southern Agriculture Improvement project	DAE and BWDB
10	Village protection against wave action in Haor area and improved water management in Haor basins	BWDB
11	Water supply and sustainable sanitation in Pourashavas of CHT	CHT Board and DPHE
12	Improvement of drainage network, flood control and solid waste management for Khulna City	Khulna City Corporation
13	Greater Dhaka integrated water and sewage improvement project	Dhaka WASA
14	Water supply project for city corporation areas in Bangladesh (Phase I and II)	Chattogram WASA, Dhaka WASA, Rajshahi WASA
15	Piped water supply project in 100 Pourashavas	DPHE
16	Village piped water supply system project	DPHE
17	Water supply project in the urban areas of Bangladesh (secondary towns)	DPHE

Table 10.3 Delta Plan-related PPP projects in the ADP 2023–2024. *Source* ADP, 2023–2024

Implementing agency	Project title	Current situation
Rajdhani Unnayan Katripakkha	Development of Water Distribution and Supply Facilities at Purbachal New Town	Operational
Gazipur City Corporation	Wastewater treatment in Gazipur City Corporation	Feasibility study
Chattogram WASA	Sewerage Treatment Plan at Patenga, Chattogram WASA	Initial stage after approval in principle
Bangladesh Economic Zone Authority	Central Effluent Treatment Plant (CETP)	Feasibility study

For addressing the required BDP 2100 financing, Bangladesh is exploring potential collaboration with the private sector through public–private partnership (PPP). In the Annual Development Program of 2023–2024, the government has listed four PPP projects related to different delta sectors; see Table 10.3. The implementation of these projects through the PPP framework can play a crucial role in attracting private investments in BDP 2100 realization.

10.2 Financing Strategy

10.2.1 Modalities of Public Funding

Traditionally, Bangladesh relied on public sector funding for delta investments. The strategy for public funding typically includes the combination of tax financing, application of some cost recovery, based on beneficiary pays principle and obtaining donor funding. Mainly from these three sources of financing, the GoB is providing the existing allocation of 1.1% of GDP for delta investment. For implementing BDP 2100, GoB needs to scale up funding from all these three sources of funding and explore other possible funding sources.

10.2.1.1 Increase of Taxation

The Government of Bangladesh focuses on domestic resource mobilization through more effective tax collection. Bangladesh has long struggled to generate adequate financial resources from domestic sources largely because of weaknesses in the tax system. Some key weaknesses include the narrow tax base, low tax elasticity, weak tax administration, a low tax-GDP ratio, and a low direct-to-total tax ratio

(Mujeri & Mujeri, 2022). While the country needs to increase its overall tax collection, increasing water-related taxes can also provide a sustainable and reliable funding source to implement the Bangladesh Delta Plan 2100.

By increasing water-related taxes, the government can generate additional revenues specifically dedicated to water management, infrastructure development, ecosystem conservation, and other aspects of the plan. Water-related taxes can capture different externalities and environmental costs of water use. For example, industries that discharge pollutants into water bodies can be subjected to higher taxes, reflecting the environmental impact of their activities.

Higher water-related taxes can act as an economic incentive for individuals, industries, and agricultural practices to adopt more sustainable and efficient water use practices. By increasing the cost of water usage, the government encourages users to conserve water, adopt water-saving technologies, and promote responsible water management practices. This helps address water scarcity, promote water use efficiency, and reduce the overall demand for water resources. However, the implementation of water-related taxes should be accompanied by transparency, accountability, and proper governance mechanisms to ensure that the generated funds are effectively utilized for the intended purposes and benefit the delta region as a whole.

10.2.1.2 Beneficiary Pay Principal

Over the years, GoB has developed many infrastructures across the country. These infrastructures need regular monitoring, adequate operations and maintenance to keep them functional. The earthen embankments are frequently eroded by wind, rainfall, human intervention, and animal grazing. Drainage channels are recurrently silted up during monsoons and floods, requiring yearly maintenance excavation. The sluice gates are often corroded due to salinity and other reasons. A substantial amount of money is required every year for proper operation and maintenance.

In order to keep the existing infrastructures in good shape, substantial fund is required for O&M. However, in most cases, limited O&M fund is allocated. For instance, if we review the experience of BWDB, it highlights that while the organization indicated significant formal demands for O&M funds, it receives very limited allocation. The proportion of BWDB's fund demand and availability is shown in Fig. 10.1.

Good practice international experience with project funding shows the strong role of the application of the beneficiary pays principle in financing water investments. In the Dutch Delta Program, much of the funding of flood control, irrigation, water supply, sanitation, and waste management investments is financed through the application of the beneficiary pays principle. The O&M is fully funded through cost recovery. In the Netherlands, water management is heavily decentralized. In fact, 80% of water spending is done by lower levels of government. In addition, 100% O&M is managed by local bodies. The bulk of this spending is done by the local water bodies that are managed and run as autonomous organizations, with application of the beneficiary pays principle-based decentral water tax.

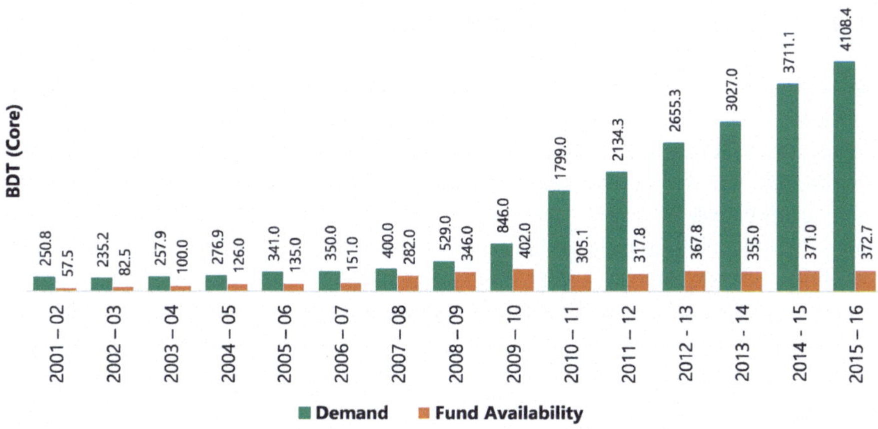

Fig. 10.1 O&M demand and fund availability of BWDB. *Source* BDP 2100, Baseline Study

The scope for application of the beneficiary pays principle in Bangladesh Delta management is huge. O&M funding of urban water and sanitation is prevalent in significant ways. But even here, full cost recovery of O&M is not prevalent. Moreover, the cost recovery of O&M of waste management is negligible. By default, there is no cost recovery of the capital cost of urban water supply, sanitation, and waste management. Regarding flood control and irrigation, there is hardly any cost recovery of either capital cost or O&M in Bangladesh.

This is mainly because of the absence of local water bodies. Once these local water bodies are created, e.g., under the proposed Delta Act, the application of the beneficiary pays principle will become a reality. Regarding urban water and sanitation, the institutions and regulations are in place. They need to be applied more rigorously. A time-bound policy could be adopted whereby charges for public urban water and sanitation services covers the O&M cost. Over time, consideration may be given to recovering capital costs, starting with the relatively well-off service areas of the four WASAs. Regarding solid waste, cost recovery can happen through an annual service charge linked with the setting up of a modern property tax system.

In the future, water management cost is expected to increase in Bangladesh. For instance, climate change impacts, such as sea-level rise, more intensive rainfall events and increased risks of water shortage in the summer, may increase future water resource management costs. Especially, future costs for flood defenses could substantially increase future water management costs. More significant costs are related to the improvement of sewage systems in order to reduce the impact of flooding, especially in the coastal zone.

10.2.1.3 Funding by Development Partners

Donor funding include loans, grants, and technical assistance from international development partners, financial institutions, and donor countries. These supports can play a crucial role in BDP implementation by supplementing domestic resources, providing technical expertise, and ensuring technology transfer. Since its independence, Bangladesh received support from World Bank, the ADB, and bilateral donors like the Netherlands and Japan for financing large water-related projects. The middle income status of the country has impact on the development support. However, Bangladesh can continue collaborating with these development partners with other arrangements to realize large-scale flood control, irrigation, urban water, and sanitation projects.

10.2.2 Public–Private Partnership

While Bangladesh needs to spend 2.5% of its GDP annually on delta projects, some 0.5% of GDP could be funded through public–private partnership (PPP). Therefore, the BDP emphasized using public–private partnership (PPP) for implementing delta projects. With the increasing needs of delta investment and growing gaps in public expenditure, it is essential to promote the role of the private sector in delta projects. The GoB has been gradually moving towards public–private partnership (PPP) water infrastructure projects in recent years.

Along with mobilizing funding for project construction, the private sector can play a crucial role in project operation and management. After the construction phase, water projects need to be efficiently managed. With limited human resources, it can be difficult for government agencies to manage and maintain water projects properly. Over the years, the private sector has already played an important role in Bangladesh's water sector by assignments and mostly traditional contracting.

Bangladesh has implemented several water treatment PPP projects. Private pump operators have played a crucial role in the irrigation sector in Bangladesh by operating fee-based shallow and deep tube well pumps. However, private entrepreneurs often charge excessive water tariffs from the farmers and utilize the limited supply of irrigation water to maximize their profit.

There is a growing realization that given their respective strengths and weaknesses, neither the public nor private sector alone can deliver the water services to the people in their best interest. The water sector related to water safety shows the importance of embankments, sea defenses, etc. which are costly and generally seen as public goods to be delivered by government agencies. In water supply and sanitation and water quality areas, the private sector can play already an important role. In some water sectors, PPP can offer a middle-path solution—where the public and private sectors collaborate and jointly work to achieve a sustainable solution. Together they work for achieving a common goal and, in doing so, jointly share the risks and benefits. Thinking about separate roles can clarify the government responsibilities and risks and benefits while it is also clear that the private sector works on the basis of business cases and profits on investments.

Through public–private partnership (PPP), the government and private sector collaborate to ensure sustainable water infrastructure projects. Recently, the GoB has signed a landmark PPP project to develop 'Installation of Water Supply, Sewerage, Drainage System and Solid Waste Management System in Purbachal New Town'. The project is currently in the construction stage. Through such water sector PPP projects, GoB can introduce private sector financing in the water sector. The government can also efficiently control the service delivery standard and monitor the private sector's performance.

Since water sector PPP projects are a new concept in Bangladesh, just like in other countries implementing initial projects can be challenging. At the initial stage, most projects may face difficulties obtaining financing. There are limited corporate debt instruments in Bangladesh, and most of the corporate sector borrowings are from the banking sector. Without an active corporate bond market, Bangladesh's private investors do not have any other option but to rely extensively on banks for providing debt financing.

Bangladesh's banking sector had considerable non-performing loans (NPLs), accounting for 11.60 percent of the total volume of classified loans (Akter & Roy, 2017). Despite the progressive improvement of the sector, most private commercial banks have limited capacity to allocate resources to provide long-term credit. Most infrastructure projects require significant construction time and therefore require long-term financing. However, in most cases, local banks do not provide long-term financing, which is a mandatory condition for implementing large-scale infrastructure projects. Predominantly, most local banks are interested in lending short-term trade finance debt. In trade finance, banks often keep the land document from the borrowers as the mortgage. For instance, if a private investor wants to construct a factory on one-acre land and asks for a bank loan, possibly the lender will keep the factory's land document as a mortgage. Bangladesh has a significant demand for trade finance loans, and most banks are willing to lend in this market. However, it is difficult to show the land document as the mortgage in a PPP project. For instance, in an irrigation PPP project, the project land belongs to the government, and it is not feasible for the private partner to claim that land as private property and use it as a mortgage. In fact, during the PPP project contract signing, the preferred bidder has to create a new Special Purpose Vehicle (SPV) company. The SPV is a distinct company with its assets, liabilities, and legal status. The balance sheet of the SPV is separate from its larger parent entity.

In a PPP project, lenders extensively review the PPP contract agreement between the private and government partners. If the project risks are appropriately shared between the private partner and the government, lenders consider providing debt financing. One possible strategy for addressing this concern and winning lenders confidence in water PPP projects is to start with small-scale pilot projects which do not require extensive debt financing. If Bangladesh can successfully implement and manage these small-scale water PPP projects, then it will enhance the private sector's confidence, and in the future, investors and banks will be willing to participate in large-scale PPP projects. Development of trust, transparent supporting investment agencies and an attractive project pipeline will enhance the private sector's confidence and establish a stronger investment culture, beneficial to all involved partners. It is therefore important to address three key priority issues to attract private investments in the water sector. First, it is essential to develop sector-specific PPP guidelines. An important factor behind water sector PPP projects in Bangladesh is the need of a clear sectoral framework for promoting these projects. For instance, looking at the Bangladesh's energy generation sector, where the private sector plays a leading role, the GoB clearly defined the sectoral framework for implementing Independent Power Producer projects in the 1990s. To encourage the private sector in the country's power sector, the Government of Bangladesh adopted the Private Sector Power Generation Policy of Bangladesh, 1996 (revised 2004) and Policy Guideline for Enhancement of Private Participation in the Power Sector, 2008. The government also took a pragmatic step to introduce the Electricity Act 2016, where adequate provisions have been kept facilitating private companies in developing the country's power sector. Bangladesh has no sectoral policy guidelines for irrigation, sanitation, and dredging PPP projects. Without such guidelines, it is still possible to develop water PPP projects, as evident from the Purbachal water supply PPP project. However, it takes considerable time to develop projects without sectoral policy support. Commercial banks consider these projects risky and usually require support from development partners. In addition, without sectoral PPP guidelines, attracting many qualified bidders during the bidding process can be challenging. If minimum investors bid during the bidding stage, the project may not achieve the value-for-money objective. Before making any PPP investment decision, the private sector assesses whether the relevant sector has a conducive regulatory framework. The multi-disciplinary nature of the water sector demands coordination between various government departments. Implementing an infrastructure project under a PPP framework would require addressing some key issues and structural considerations, including the broad scope of engagement, roles, and responsibilities of various stakeholders, the robustness of the revenue model and operating framework, and the market appetite for these kinds of projects. Private investment requires a set of established rules and processes to allow reasonable confidence about the project's commercial viability. A well-defined legal and sectoral framework, allied with a PPP policy, allows contracts to be determined with certainty and allows the parties to understand their interaction boundaries. The consequences of not having this certainty have been demonstrated to result in greater project risk and limited response from potential investors. With no sectoral policy guideline, often private investors are uncertain regarding the Government's future plan, project development process, and risk-sharing mechanism. Along with private investors, a policy guideline is also important for lenders. Most large-scale infrastructure projects are capital-intensive and cannot be logically implemented without loan financing. While providing loan support, lenders check the bankability of certain infrastructure projects and the future business prospect of that sector. Without a clear sectoral guideline, it is difficult for lenders to assess the viability of a loan proposal from a potential PPP investor.

Furthermore, it is crucial to properly coordinate projects. Water PPP projects can be multi-sectoral and -regional, involving multiple government agencies. The granting Government Ministry often does not have the power to act over authorities in other sectors and bring all stakeholders to one table. For instance, the Bangladesh Water Development Board (BWDB) under the Ministry of Water Resources (MoWR) is leading Bangladesh's medium and large irrigation, flood control, and drainage projects. At the same time, the Ministry of Agriculture (MoA), through its Bangladesh Agriculture Development Board (BADC) and Barind Multi-purpose Development Authority (BMDA), plays a vital role in several irrigation projects. Depending on the PPP project context, these two ministries may need to work together in an integrated way. Besides, DAE also needs to play an important role through the extension services. Primary and secondary responsibilities may vary or shift during different project stages. For instance, BWDB may be responsible for building the water reservoir, and BMDA may be responsible for managing the water reservoir and supervising a management contract with a PPP operator. If these two agencies do not coordinate, the irrigation PPP operator may face delays, leading to a business loss for the private partner. In order to minimize the project risk for private partners, it's essential to ensure that the PPP project agreement clearly spells out the responsibilities of different government agencies and the coordination mechanisms among them. In the Guerdane Irrigation project of Morocco, the contract commitment of the Government to provide a minimum allocation of water was even supported by an additional intra-government contract between the contracting authority, Regional Office for Agricultural Development, and the regional water authority (Varma et al., 2012). Such an intra-government contract arrangement assured the potential private partners regarding the project's commercial viability. Similar approaches may be required for large-scale irrigation PPP projects in Bangladesh.

Arranging capacity-building training and field visits for government officials is essential. Understanding and managing PPP contracts would involve a significant amount of capacity on the part of the implementation agency. Although these agencies are already closely working with private contractors, the private sector's role and scope would change in a water PPP project. For instance, the private contractor gets paid for implementing dredging work in the public procurement process. However, the private sector operates as a partner with its own earning or business model in a dredging PPP project and assumes much higher risks and responsibilities. The monitoring process of PPP projects is also different from public sector projects. Therefore, arranging capacity-building events for implementing agency officials and field visits to demonstrate the implementation process of PPP projects is essential.

10.2.3 Blue Bonds

A blue bond is a new form of a sustainability bond issued to support investments in sustainable blue economy projects. This debt instrument targets investors who aim to fulfill their corporate social responsibilities and create certain blue benefits.

Concerning the issuance, any government, bank, or corporation can issue these bonds according to the Asian Development Bank on the mechanism of Blue Bonds.[4] Though a relatively new instrument, different models of blue bonds have started to emerge in the global markets. The first blue bond was issued in October 2018 after the World Bank facilitated a bond agreement to offload a small portion of Seychelles' debt in exchange for marine protection.[5] It served the dual purpose of stabilizing the country's credit rating and investing in its economy. The second Blue Bond—the Nordic Sea Blue Bond—was issued around three months after the Nordic Investment Bank (NIB) launched the first Blue Bond. NIB used a model known as a 'use of proceeds' bond, where blue bonds fund projects generate financial returns.[6] Later, the Bank of China also issued blue bonds using the proceeds model.[7]

Blue Bonds can play an important role in implementing BDP 2100.[8] However, before issuing the bond, Bangladesh needs to fulfill several prerequisites. The four prerequisites[9] for blue bonds suggested by IUCN are as follows:

1. A pipeline of acceptable large projects: The fundamental requirement of issuing blue bonds is the availability of a pipeline of acceptable projects large enough in size.
2. Growing awareness of market actors: An important constraint identified for blue bond issuance is the lack of knowledge of the potential market. Therefore, before issuing blue bonds, there has to be adequate awareness of participants in this space—both issuers and investors.
3. Adequate monitoring and verification procedures: Proper monitoring and verification procedures are a fundamental pre-condition for blue bond issuance and subsequent growth. According to IUCN, both issuers and investors may face reputational risks and potential accusations of 'greenwashing' if the proceeds are not used for their intended purposes.
4. Adequate impact management procedures: There has to be adequate attention to the impact management process. This may relate to the type of reporting (ex-ante/ex-post), the frequency, and the choice of the reporting agent (in-house or independent), among others.

Along with the prerequisites mentioned above, Bangladesh can address several key challenges. First, the bond market of Bangladesh has limited liquidity and is currently in the initial phase of development. The private bond market is largely limited to commercial banks issuing subordinated debt to meet capital adequacy requirements. Second, the newly introduced sectors, such as coastal tourism, gas exploration, salt production, and renewable energy production, are not

[4] Asian Development Bank (2021).

[5] Ministry of Planning, General Economics Division (2021).

[6] Asian Development Bank (2021). f.

[7] Ibid.

[8] Ministry of Planning, General Economics Division (2021).

[9] Ibid.

well-researched or managed holistically. In many cases, scientific information, a fundamental prerequisite of sustainable management, is insufficient in Bangladesh.[10]

For launching blue bonds, three distinct approaches can be used[11]: (i) private placement to dedicated impact investors; (ii) enlistment in the sustainable stock exchanges; and (iii) a mixed approach combining (i) and (iii). According to GED, a safer approach for Bangladesh would be to follow the two tranches approach adopted by Fiji. At the end of 2017, Fiji issued a sovereign green bond, equivalent to USD 50 million, which was the first ever green bond issued by a developing country. This green bond was aimed at both climate mitigation and adaptation, with some use of proceeds having a direct and indirect positive impact on the blue natural capital of Fiji. The first tranche was privately placed, while the second dual-tranche green bond transaction was listed on the London Stock Exchange. Bangladesh could explore this approach and when positive seek possibilities to introduce it.

10.3 Climate Change-Related Financing

10.3.1 Green Climate Fund and Other Funds

The Green Climate Fund has been established to mobilize climate finance to support scaled-up mitigation and adaptation action in developing countries under the auspices of the United Nations Framework Convention on Climate Change (UNFCCC) involving 194 countries. So far, GCF has approved about nine projects for Bangladesh; among these nine projects, two involve multiple countries. The total value for these projects is US$ 441.2 million.[12] This is a small sum, especially when broken down by yearly commitments. Actual yearly disbursements are even smaller. Nevertheless, GoB wants to continue to mobilize grants from GCF and other climate-related international trust funds. Overall, Bangladesh's success rate in accessing GCF funding has been moderate. Being a vulnerable country to climate change, Bangladesh considers GCF a potential source for getting funds for climate-resilient development. Particularly for implementing BDP 2100, GCF can play an important role. Given the scale and urgency of the climate change, the country has received funding for very limited projects. Therefore, Bangladesh needs to scale up its GCF funding initiatives through Multi-lateral Implementing Entities (MIE) and National Implementing Entities (NIE).

Obtaining GCF funding would enable the Bangladesh government to leverage additional funding from other sources, such as the World Bank and the Asian Development Bank. This could help accelerate the delta plan's implementation and increase its overall impact. GCF funding would also support the Bangladesh government in its

[10] Ibid.

[11] Ibid.

[12] See SIBDP Consortium report on Green Climate Fund submitted to GED.

National Adaptation Plan and promote the country's efforts to achieve its Nationally Determined Contributions under the Paris Agreement.

To obtain funding from the Green Climate Fund (GCF), it is essential to prepare Project Concept Notes and submit them to the GCF Secretariat. Through the National Designated Authority, exploring potential opportunities of receiving Project Preparation Facility support from GCF is essential.[13] If Bangladesh can receive Project Preparation Facility support, it will ensure the funding for completing the feasibility study. Completing the detailed feasibility study is one of the key requirements for submitting a GCF funding proposal. While it has taken a long time to set up NIEs and build NIEs capacity to follow all the application procedures, Bangladesh is going through the knowledge curve for GCF. The potential government and private NIE agencies could to focus on their capacity-building initiatives for enhancing their understanding regarding GCF funding. For developing Bangladesh's capacity to access more climate funding, sharing knowledge between existing NIEs and potential NIE applicants is essential. Bangladesh's access to potentially larger amounts will depend on effective knowledge-sharing mechanisms among the different stakeholders.

10.3.1.1 Global Environment Facility (GEF)

The Global Environment Facility (GEF) is a multi-lateral fund that helps confront various environmental issues, including biodiversity, climate change (mitigation and adaptation), international waters, waste management, and land degradation. In most of the projects, GEF mainly provides grant financing. Besides, GEF also provides full and partial credit guarantees. The GEF guarantees the entire commercial loan or a pre-defined portion depending on the chosen instrument.[14] This is done by covering general credit risk during a particular project phase.[15] In the case of a partial credit guarantee, the rest of the risk of potential losses is shared with the lender and other investors.[16]

10.3.1.2 Adaptation Fund

The Adaptation Fund (AF) provides adaptation finance to developing countries.[17] Though AF started on a small scale, the fund size has grown with time. As of November 2020, the fund has received approximately USD 1.05 billion[18] in financial contributions. AF-funded projects are implemented through various entities, such as accredited Multi-lateral Implementing Entities, Regional Implementing Entities,

[13] See SIBDP Consortium Proposals submitted to GED.

[14] IISD, Global Environment Facility (GEF) (2020).

[15] Ibid.

[16] Ibid.

[17] https://www.thegef.org/who-we-are/funding.

[18] GEF, Adaptation Fund, 2019.

and National Implementing Entities.[19] In addition to providing finance, the fund has a growing Readiness Program that arranges capacity-building workshops and provides small technical assistance grants.[20] Bangladesh received project funding from the Adaptation Fund in 2021. The Adaptation Fund approved the 'Adaptation Initiative for Climate Vulnerable Offshore Small Islands and Riverine Charland in Bangladesh' project and provided grant support of approximately 9.9 million USD. While Bangladesh has limited success in accessing the Adaptation Fund, ultimately, this fund is unique as part of its total funding is financed by an international revenue source. Climate change mitigation projects registered under the clean development mechanism (CDM) allocate 2% of the 'certified emission reductions (CERs)' into the account of the Adaptation Fund.

10.3.1.3 Mitigation Action Facility

The Mitigation Action Facility is a grant-based multi-donor fund for providing technical support and climate finance for mitigation projects. The facility primarily focuses on three priority sectors—energy, transport, and industry, but remains open to cross-sectoral projects linked to one of the priority sectors. So far, Bangladesh did not access funding from the Mitigation Action Facility. However, other South Asian countries, including Nepal, Pakistan, and India, have successfully accessed funding from the Mitigation Action Facility.

The Mitigation Action Facility each year calls for partner countries or organizations to apply for funding. The most ambitious and feasible climate change mitigation projects are selected for support through a comprehensive assessment process.

10.3.1.4 *REDD+*

Reducing Emissions from Deforestation and Forest Degradation (REDD+) is a UNFCCC-developed framework to provide financial incentives to reduce greenhouse-gas emissions. As REDD+ is focused on reducing deforestation and forest degradation emissions, REDD+ can contribute to implementing the Bangladesh Delta Plan. The public sector has pledged most of the REDD+ and forest-related funding through bilateral and multi-lateral channels. In addition, developed countries and the private sector are channeling finance through dedicated multi-lateral funds targeting REDD+ and sustainable forest management. Some of these funds include among others the Forest Carbon Partnership Facility (FCPF) Readiness Fund, Carbon Fund, the Forest Investment Program (FIP), the Amazon Fund, the Congo Basin Forest Fund (CBFF), and the Biocarbon Fund Initiative for Sustainable Forest Landscapes (ISFL). REDD+ offers results-based payments to communities that successfully reduce deforestation and forest degradation emissions. These

[19] GEF, Funding (2023).

[20] https://www.thegef.org/who-we-are/funding.

payments are based on the amount of carbon emissions that are reduced or avoided. Common carbon-financed REDD+ activities include:

- Farmer training on sustainable land management and biodiversity conservation.
- Non-timber forest product cooperatives.
- Clean water and sanitation.
- Technology access.

The carbon credits generated from REDD+ projects can improve the lives of local people in Bangladesh. REDD+ can provide economic incentives for vulnerable communities to adopt sustainable forest management practices. These practices can increase community members' income while protecting forests. By protecting and restoring the forests, REDD+ can help conserve biodiversity and maintain ecosystem services in Bangladesh. REDD+ projects could help ensure clean water availability for vulnerable communities by protecting and restoring the forests.

10.3.1.5 Land Degradation Neutrality Fund

Bangladesh is highly vulnerable to the impacts of climate change, including temperature changes, rainfall patterns, and increased frequency of extreme weather events. These factors are exacerbating land degradation issues. Land degradation leads to soil erosion, nutrient depletion, and reduced soil fertility. This directly affects agricultural productivity, a critical component of Bangladesh's economy.

Worldwide, different funds are operating to promote private finance in Land Degradation Neutrality (LDN) projects. The Land Degradation Neutrality (LDN) Fund is an investment fund promoted by the UN Convention to Combat Desertification (UNCCD) and Mirova. The UNCCD is focused on land, promoting practices that avoid, reduce and reverse land degradation and they are the driving force behind Sustainable Development Goal 15 and Land Degradation Neutrality. Mirova is an affiliate of Natixis Global Asset Management dedicated to responsible investing. Officially launched at UNCCD COP13, the LDN fund leverages public money to raise private capital for sustainable land projects. The fund mainly targets private sector firms and contributes to addressing LDN through sustainable land management and land restoration projects. By leveraging long-term non-grant financing, the LDN fund invests in financially viable private sector initiative projects on sustainable land management (SLM) and land rehabilitation.[21] Its focus also includes sustainable agriculture, sustainable livestock management, sustainable forestry, and agroforestry.[22] Bangladesh did not access any funding from LDN fund. However, Bhutan has successfully obtained funding from LDN fund in South Asia. In Bhutan, the LDN fund is creating employment opportunities for vulnerable communities. LDN fund could also create job opportunities for the vulnerable people in different areas

---

[21] See: https://www.unccd.int/land-and-life/land-degradation-neutrality/impact-investment- fund-land-degradation-neutrality.

[22] Ibid.

of Bangladesh, thereby lifting them out of poverty. Besides, the fund promotes social inclusion and gender equality from project design to implementation.[23] Such inclusion will increase the household incomes of vulnerable households. Along with direct investment in sustainable businesses, the LDN fund works with financial intermediaries, especially for channeling capital to businesses through local financial institutions.[24] The LDN fund could use a similar mechanism in Bangladesh to enhance the capacity of local financial intermediaries. In the long run, these intermediaries can support local businesses in green financing. With collaboration experience with the LDN fund, the local banks' green investment assessment capacity would increase.

10.3.2 NGOs and Foundations

NGOs and Foundations can play a crucial role in supporting the implementation of the Bangladesh Delta Plan by leveraging their water sector expertise, and community connections. In Bangladesh, NGOs like BRAC, RDRS, ASA, Proshika have a long track record of working in areas of water supply and sanitation, livelihood and employment generation, nature and environment protection. NGOs can attract philanthropic funds and contribute to small-scale delta projects, such as implementing water management solutions or undertaking conservation efforts. Besides, NGOs also have the experience and technical expertise of obtaining international grants from different development partners. These grants can support Bangladesh's climate resilience, environmental conservation, and sustainable development.

Foundations, on the other hand, can also access international funds by utilizing their own internal funds. In Bangladesh, foundations like the Sajida Foundation have a track record of working with different government and international agencies. Sajida Foundation currently holds 51% shareholding of Renata Pharmaceuticals Ltd., and the yearly dividend from Renata Pharmaceutical has facilitated Sajida Foundation's root-level operation.

As pilot projects, NGOs and Foundations can introduce innovative technologies and best practices for climate resilience, water management, and sustainable agriculture. These agencies can assist in developing and implementing disaster preparedness and response plans, which are crucial for the Bangladesh delta region, prone to natural disasters and the impacts of climate change. NGOs and Foundations can facilitate coordination and collaboration among various stakeholders, including government agencies, private sector entities, academia, and other civil society organizations. They have the ability to mobilize community people and implement locally led climate adaptation. Mobilizing funds at the community level can empower local initiatives and enhance the resilience of vulnerable communities. This may involve

[23] https://catalogue.unccd.int/1076_LDN%20Fund_Brochure_Q1_2019.

[24] https://www.unccd.int/land-and-life/land-degradation-neutrality/impact-investment-fund-land-degradation-neutrality.

supporting community-based adaptation projects, sustainable agriculture practices, and renewable energy initiatives.

10.4 Funding Constraints

After the approval of the Bangladesh Delta Plan 2100 in 2018, Bangladesh initiated the delta plan implementation process. The overall progress of the BDP 2100 Investment Plan shows an expected pattern of gradual start-up, increase of projects and proposals, followed however by a rather slow continuation. The latter caused by institutional aspects and mainly because of funding constraints. These constraints developed during COVID-19 when exports of products reduced (especially from the garment industry) and income from remittances from Bangladeshi abroad declined. Besides, because of the increased global tensions, interest rates and inflation increased. Uncertain geopolitical factors are also affecting the availability and price of food on the world market. Like many emerging countries, Bangladesh is facing economic challenges that are affecting its ability to allocate sufficient funds for major delta investment projects. Limited financial resources and competing priorities in sectors like health and education is limiting the funds available with GoB for the delta plan. The country is also relying on external sources of funding for delta investment projects. However, external donors have their own agendas and criteria, which can be stringent, leading to challenges in securing funding. Moreover, bureaucratic hurdles, complex approval processes, and legal challenges are slowing down the funding process. Lengthy bureaucratic procedures often delay the disbursement of funds, leading to project implementation delays.

To overcome these funding constraints, Bangladesh focuses on several measures, such as engaging in diplomatic efforts to secure external funding, promoting public–private partnerships with attractive incentives for investors, and addressing legal and bureaucratic obstacles to facilitate efficient fund allocation. Collaboration with international organizations and forging partnerships with climate finance initiatives can also play a vital role in overcoming funding challenges for the Delta Plan projects. An option is also to issue green and blue bonds to raise funds for environmentally sustainable projects, such as climate adaptation and renewable energy initiatives.

Along with constructing new delta infrastructures, it is essential to operate and maintain these and existing infrastructures. However, operations and maintenance (O&M) are confronted with low budgets as well. GoB expects local people to contribute for the benefits they have from embankments, sluices, etc. Water user institutions with adequate regulations and linked with local Water Management Organizations can form a basis for peoples participation and in O&M or project implementation. Further introduction of service charges for water services will take time. In many cases, local people do not actively take part in routine maintenance and get involved in the emergency maintenance process after considerable structural damage has been done to the embankment. Local people often use community funding mechanisms ('collections') to make last-minute attempts to save the embankments. Instead of last-minute efforts, it is essential to include community members in the project's implementation and O&M process from the planning stage on.

References

Akter, R., & Roy, J. K. (2017). The impacts of non-performing loan on profitability: An empirical study on banking sector of Dhaka stock exchange. *International Journal of Economics and Finance, 9*(3), 126–132.

Alam, S., De Heer, J. M., & Choudhury, G. A. (Eds.). (2018). *Bangladesh delta plan 2100: Baseline studies on governance and institutional development* (Vol. 6). General Economics Division, Planning Commission, Government of Bangladesh. https://plancomm.gov.bd/

Asian Development Bank. (2021). *Sovereign blue bonds*. Publication 731026.

International Institute for Sustainable Development. Global Environment Facility. (2020). *Credit-enhancement-instruments*.

International Institute for Sustainable Development. Global Environment Facility, GEF, Funding. (2023). From: https://www.thegef.org/who-we-are/funding.

International Institute for Sustainable Development. Global Environment Facility, GEF, Adaptation Fund. (2019). Retrieved from https://climatefundsupdate.org/the-funds/adaptation-fund.

IPCC. (2022). *Climate Change 2022: Impacts, adaptation and vulnerability. Contribution of working group II to the sixth assessment report of the intergovernmental panel on climate change*.

Ministry of Planning. (2018). *General Economics Division, Planning Commission, Bangladesh Delta Plan, Volume 1: Strategy; Volume 2: Investment Plan, Dhaka*.

Ministry of Planning, General Economics Division. (2021). *Promoting sustainable blue economy in Bangladesh through sustainable blue bond. Assessing the feasibility of instituting blue bond in Bangladesh. Blue economy in Bangladesh through sustainable blue bond*. Retrieved from https://gedkp.gov.bd/wp-content/uploads/2022/06/Assessing-the-Feasibility-of-Instituting-Blue-Bond-in-Bangladesh.pdf?post_id=1988

Mujeri, N., & Mujeri, M. K. (2022). South-South cooperation: Addressing development finance in Bangladesh. In: *The financial express*.

UNCCD. www.unccd.int/land-and-lifeetc

Varma, H.K., A.S. Dhingra, & Raghu Rama Swamy, D. T. V. (2012). *Exploring public–private partnership in the irrigation and drainage sector in India*. Asian Development Bank.

Chapter 11
Administrative and Institutional Aspects

Abstract How to organize oneself for adaptation to climate change and natural hazards? This question indicates that besides technical and financial matters, adaptation is closely related with governance and organization challenges. Effective policy making and realization require an enabling government and societal setting with good governance. Especially multi-sectoral plans that involve many ministries, agencies, and stakeholders give these needs even more importance. Delta areas have been the subject of global experiences, also in the preparation of a plan for protection and development of these areas. Bangladesh applied these experiences to its own situation in order to avoid the pitfalls and apply the beneficial lessons learned. Bangladesh has an interesting background of addressing the challenges and issues associated with its delta through water-related and other institutions. The institutional framework for adaptive management of the Delta Plan has been built on both, international experiences and the own rich background, plans and rules. Government organizations are mostly stable, show sometimes even inertia but are also gradually changing due to modernization and influx of new staff or are influenced by changing political situations. Governmental and governance changes will further evolve while climate change and adaptation are ongoing. In other words, institutional reform happens and can be utilized in the transformative approach of adaptation. The aim is to develop an institutional framework which stimulates and facilitates adaptive realization of the BDP 2100. This chapter provides insight in the institutional backgrounds, in analysis and framework needed for execution of the Delta Plan and realization of the long-term vision. Clarity of role, interdependence of actions, and a coordinated approach are essential requirements of the institutional set up for adaptive realization of the Delta Plan.

Keywords Governance · Government · Administrative culture · Institutional · Institutional framework · Institutional reform · Institutional strengthening · Legal framework · Mandate · Responsibility · Execution · Transparency · Accountability · Reform · Transformation · Delta governance council · Project/Program selection committee · Delta wing · Focal point · Water user institution · Executing agencies · Information and knowledge portal · Implementation

J. M. de Heer and G. A. Choudhury, *Strategies for Adaptation to Climate Change in a Transformative Approach*, Springer Climate,
https://doi.org/10.1007/978-3-031-88884-7_11

11.1 Existing Institutional Framework

The existing institutional arrangements and governance with respect to the water sector and delta governance aspects belong to the larger political and administrative constellation of Bangladesh.[1] As indicated before, BDP 2100 covers a broad institutional domain belonging to Delta governance as formulated in the Delta Vision, the three National level and six Delta Goals. Part of that domain belongs to water resources management, other parts of the Delta Plan institutional domain belong to closely related sectors of local government and rural development, land use, inland water transport, water supply and sanitation, agriculture, fisheries, environment and more as planning and finance. These last two concern 5-Year Planning, programming and budget allocation as well as the approval process by the Planning Commission with respect to water resources projects and other projects.

The existing institutional setting on water resources has been developed in the recent past as discussed in Chapter 3. The physical characteristics and frequent disaster focused attention to protection, flood management and utilization of water for agriculture. Because of this, climate change, navigation, urban water and drainage and water quality issues got less attention and investments. Important aspects of water resources management have been covered by the Water Act 2013. The main function of this Act is to "make provisions for integrated development, management, abstraction, distribution, use, protection and conservation of water resources".[2] Although the Act is from 2013, it is not completely implemented with respect to its regulatory policies and institutions. For BDP 2100, which is multi-sectoral thus broader than water resources, the Act provides a basis for moving forward in close connection with other sectoral institutional arrangements. The mandates and roles which are important for BDP 2100 implementation are, besides the Ministry of Planning/Planning Commission, that of the Ministry of Water Resources, the Water Resources Planning Organization (WARPO) and especially the Bangladesh Water Development Board (BWDB) which prepares and executes major surface water development projects and takes care of operations and maintenance, the Local Government Engineering Department (LGED) focused on smaller local projects, the Inland Water Transport (IWT) Division which is responsible for river transport including river dredging to maintain river navigation.

11.1.1 Progress with Water Policies and Programs

Bangladesh has as more than sixty years of experience in managing water resources that is reflected in a growing number of water policies and water programs. A chronological list of the major initiatives are provided in Annex 1. This is an impressive list of policies and programs covering most aspects of water management. Evaluation

[1] See for broader background e.g. Muhith (2007) and Barkat et al., (2015).

[2] The Water Act 2013, Government of Bangladesh.

studies of implementation of these policies and programs suggest some common issues and challenges. First, in the area of policies, the most comprehensive policy is the 1999 Water Policy. It is a well thought out policy paper that provided guidance on regulations as well as institutions such as emphasizing the need for local level representation in water management. The main problem area is in implementation. Many of the recommendations have not been implemented, including participatory water management. Second, the most ambitious project was the adoption of the 1990–1995 Flood Action Plan that was prepared with the help of the World Bank. The FAP involved many studies and pilot projects to be supported by a large number, about 17, development partners. The project was the first attempt to have a coordinated approach to flood management involving the government and the donor community. However, the project came under strong scrutiny and criticism on a number of weak links including weak institutional capacity and coordination, lack of involvement of stake holders, inadequate impact assessment with possible negative effects on fisheries and the natural environment, and inadequate attention to operations and maintenance arrangements including financing. Consequently, the project was only partially implemented with the main focus on the studies. Third, coordination of water investments have been attempted through a number of rounds of national water plans (NWP 1986, NWP 1991 and NWMP 2001 (adopted in 2004)). A common feature of these plans has been the development of ambitious investments plans that is not based on a realistic assessment of available resources and implementation arrangements. Additionally, these plans were sectoral plans, all prepared by the Ministry of Water Resources and did not have the buy-in of all line ministries dealing with water resources. In the absence of a coordinating mechanism, the programs have remained on the shelves and did not have much impact on actual investments by line ministries done through the annual development plan (ADP) process. Coordination problems between water policies and programs prepared by the Ministry of Water Resources and actual decisions in other water-related institutions has been a perennial concern that led to the establishment of the NWRC in 1985, ECNWRC in 1997 and the Water Act in 2013. Fifth, many specialized area studies done so far, such as the regional area studies, the individual city protection plans, the Master plans for Southern Area and the Haor Area, the District Water Development Plans, etc. remained more or less on the shelf.

This shows some difficult governance problems, where many sectoral studies are done and plans are formulated with limited central government ownership and consequently limited funding. Besides there were few realistic assessments of how the plans would be implemented. Finally, policies, studies and plans have tended to be done piece meal with single line ministry involvement, not taking a holistic approach to water, food and development as cross-sectoral issues. Also the demand side or careful assessment of socio-economic impacts are only partly taken into account. The institutional aspects of integrated water management tend to be side-tracked or weakly covered. Consideration of readiness and implementation capacity constraints are not well understood and managed. Most importantly, the implications

Table 11.1 Legal framework for environmental management. *Source* Ministry of Environment, Forest and Climate Change

Sl.	Law, Year
1	The Bangladesh Environment Conservation Act, 1995
2	The Bangladesh Environment Conservation Rules, 1997
3	The Brick Burning (Control) (Amendment) Acts, 2001
4	The Clean Dhaka Master Plan, 2005
5	Noise Pollution (Control) Rules, 2006
6	The Environment Court (Amendment) Act, 2010
7	Hazardous Waste and Ship Breaking Waste Management Rules, 2011

for resources and how they will be funded in the context of a budget constraints and conflicting priorities is seldom reviewed or discussed.[3]

11.1.2 Legal Framework and Issues of Environmental Protection

As in the case of water management, a number of legislation and rules have been established to protect the natural environment. These are summarized in Table 11.1. The main issues are limited implementation capacity and inadequate application of these laws to integrated delta management.

11.1.3 Policies and Programs in Delta-Related Ministries

As in the case of water management, there is a range of policies and programs in related delta management areas of climate change, environment, land, agriculture, fisheries and forestry. These are summarized in Table 11.2. A broad array of topics of relevance to adaptive management is covered here. As in the case of water management, the policies and programs have emerged in individual ministries without integration with related parts or being mainstreamed at the central level. The possible impacts of these sectoral policies on the delta as a whole has not been analyzed. Also here implementation capacities are limited.[4]

Given the multiple dimensions of water supply and demand, transboundary water issues and multiple sources of water-related risks and hazards as well as water and food and socio-economic development, many government agencies are involved in tasks of delta management. However, several key ministries dominate in the actual implementation of water and delta programs. So far as water management is concerned, the MoWR is the largest and most important player. In recognition

[3] Alam et al. (2018c).

[4] Ibid.

Table 11.2 List of important policies and programs in delta-related ministries. *Source* SIBDP, Compiled from various government sources

Sl	Policies and programs	Responsible ministry
A	*Agriculture, Fisheries and Livestock*	
1	Master plan for agriculture for the southern region	Ministry of Agriculture
2	National Agriculture Policy Draft (2009)	Ministry of Agriculture
3	New Agricultural Extension Policy (1996)	Ministry of Agriculture
4	National Seed Policy (1993); Seed Rules (1998)	Ministry of Agriculture
5	National Integrated Pest Management Policy (2002)	Ministry of Agriculture
6	National Fisheries Policy (1998)	Ministry of Fisheries and Livestock
7	National Livestock Development Policy (2007)	Ministry of Fisheries and Livestock
B	*Environment and Forestry*	
1	National Environment Policy and Implementation Plan (1992)	Ministry of Environment, Forest and Climate Change
2	National Forestry Policy (1994)	Ministry of Environment, Forest and Climate Change
3	National Wetland Policy (draft)	Ministry of Environment, Forest and Climate Change
C	*Land Management*	
1	National Land Use Policy (2001)	Ministry of Land
2	Jalmohal Management Policy (2009)	Ministry of Land

of this, the main implementation arm is the Bangladesh Water Development Board (BWDB) established in 1959. The BWDB is the most significant water management organization in Bangladesh with offices all over the country. It handles a substantial amount of budget in the water sector. BWDB is responsible for preparing and implementing development projects related to flood control, drainage and irrigation, riverbank erosion, coastal development, and land reclamation. It is also responsible for the collection and dissemination of hydrologic and hydraulic data, provides flood forecasting and warning information through the management of the Flood Forecasting and Warning Centre (FFWC). While the BWDB is the implementing arm of MoWR, the Water Resources Planning Organization (WARPO) is the macro-planning arm. WARPO formulates policies, plans, strategies, guidelines, instructions, acts, and rules relating to the development and management of water resources. It has been involved in formulating the National Water Policy (1999), a Coastal Zone Policy (2005), the National Water Management Plan (2001/4), and the more recently the Bangladesh Water Act (2013).

Additionally there are a number of specialized institutions in MoWR, dealing with specific areas of surface water management including coordination and research. The specialized institutions fall under three broad categories: First, concerns institutions responsible for coordination of specific areas of water management, e.g., the Haor and Wetland Development Department. The second relates to the coordination of policy dialogue and information on transboundary water issues. The concerned institution is the Joint River Commission (JRC) serving as secretariat of the Ministry for dealing with cross-boundary water issues. The third relate to institutions that support policy through research: the River Research Institute (RRI) which deals with physical and mathematical water modeling; the Institute of Water Modeling (IWM) which is a government trust organization concerned with mathematical water modeling and water-related research and development; and another trust organization, Centre for Environmental and Geographic Information Services (CEGIS) which provides integrated environmental analysis involving GIS, remote sensing, IT and data management as well as broader research and development. Both trusts have become important knowledge institutions and contributed to BDP 2100 formulation and follow-up.

The responsibility for water supply and sanitation lies with the Ministry of Local Government, Rural Development and Cooperatives (LGRD). Much of the effort is focused on urban water supply and sanitation. There are few institutionalized water and sanitation services in the rural areas. Piped water and sanitary toilet facilities are generally not available yet in rural areas. The main effort is to address rural drinking water safety concerns, emerging pollution and soil-based arsenic poisoning carried through tube wells. Urban water and sanitation is provided through a range of institutions as: Water and Sewerage Authorities (WASAs), city corporations, municipalities and the Department of Public Health Engineering (DPHE). The four WASAs are: Dhaka WASA, Chattogram WASA; Khulna WASA and the Rajshahi WASA. Typically, WASAs have received substantial international assistance, are well staffed and provide relatively high levels of service compared to the city corporations and municipalities. Many of the medium sized cities are experiencing very high rates of population and industrial growth. Yet, urban water and sanitation management institutions are mostly understaffed and receive only modest support from central organizations. Their capacity for planning and enforcement of the few regulations that exist and those that are planned is limited.

Efforts to decentralize irrigation and drainage management have led the government to distinguish between large and small schemes based on area of coverage of the schemes. Large schemes are those that encompass services for 1000 ha or more; small schemes are less than 1000 ha. The large schemes are managed by BWDB; the small schemes are managed by LGRD through the local government engineering department (LGED). The aim is to eventually transfer the operation and maintenance (O&M) of these smaller schemes to the community organizations.

Inland water transport is the accountability of the Ministry of Shipping and Inland Water Transport through the Inland Water Transport (IWT) Division. The IWT has two major institutions: the Inland Water Transport Authority (IWTA) and the Inland Water Transport Corporation. The IWTA is responsible for dredging, navigation routes maintenance, river port facilities and river transport regulations including safety standards. The IWTC is responsible for running the river transport facilities for cargo and passengers. The Ministry of Agriculture has responsibility for managing small scale mechanized irrigation facilities (deep tube wells) through the Bangladesh Agricultural Development Corporation (BADC) and the Barind Multipurpose Development Authority (BMDA). The BADC role has diminished as private initiatives emerged and largely took over the BADC role in irrigation. The connection between agriculture and water resources is however becoming more important as climate change and trans-boundary related challenges increase.

All the involved institutional domains and respective policies, plans and actions impact delta management or governance. Because of this, effective coordination, sound planning discipline and joint implementation with overall monitoring and evaluation are required.

11.2 Observations on Aspects of Governance

11.2.1 Governance Lessons Learned from Water Resources Management

The BDP 2100 Study 'Sixty Years of Water Resources Development in Bangladesh: Lessons Learnt' clearly indicates an urgent need for multi-sectoral and multi-level interaction and coordination between ministries and agencies during the national planning, fund and resources allocation as well as preparation of implementation. Practical examples have been given below, showing a lack of coordination and collaboration, leading to unnecessary but tough problems and more demands from the government (Ministry of Planning, 2016).

The policy intention is that sources of water must be protected at any cost. The scarcity of water is apprehended to increase day by day. Water in the lakes, ponds and other water bodies should be preserved through the drier months and rainwater should be harvested in every household, public organization or private company. The citizens and parties in the different sectors may be given information and instruction on that. It was observed that the large irrigation projects required huge funds and long time for completion, the groundwater seemed to be a cheaper solution. So exploitation of groundwater through shallow and deep tube wells was started by the public sector and later by the private sector. As a result, the irrigation-dependent dry season crop has become the major crop. However, the exploitation of the groundwater has many limitations such as over-exploitation and drawdown of the groundwater table. As the groundwater is progressively going down, more sophisticated pumping sets are used. This decreases the availability of groundwater by hand pumps used for domestic purposes and the bio-diversity is also affected. Moreover, in many places, the arsenic, iron and other industrial chemical contamination of groundwater has limited its uses.

Another observation in this area is that here has been a general lack in multi-sectoral planning of Integrated Water Resources Management (IWRM). Surface water and groundwater are being developed independently and not integrated. It is always recommended to apply conjunctive use of both surface and groundwater, but hardly practiced. It is according to documents and plans evident that the focus of IWRM must go beyond flood control, drainage and irrigation and that environmental consideration must be integrated into water resources management but it seems to be hard to realize this in practice. FCD and FCDI projects have played vital role in achieving food security and protecting assets and infrastructure from damages by floods. Flood embankments are being used as the roads, but the construction of FCD and FCDI projects generally have negative impacts on capture fisheries. However, the culture fisheries and shrimp cultivation have increased tremendously within the flood protected areas. The remedial measures could be building fish passes as well as practicing controlled flooding of the protected lands. In this way, fish fries can enter during pre-monsoon and sedimentation of the floodplains within the protected areas could be achieved partially.

The rivers should be maintained and conserved properly to sustain its functions of providing drainage and navigation. However, in the past, river management was practiced on piece-meal basis, without considering and/or gaining real understanding about the morphological processes happening in the delta. The upstream withdrawal of water during monsoon and post-monsoon periods has decreased the capacity of sediment transport of the rivers to the Bay of Bengal. As such, the sedimentation becomes another major problem on river beds, resulting in loss of conveyance capacity and navigability. Moreover, the agricultural floodplains are not drained due to sedimentation of rivers, resulting in water-logging, which increases the soil salinity, and damage crops and alter the whole social structure of that particular area. Any intervention in the upstream location in the river impacts another location at the downstream. As such, it is essential that an integrated river management plan should be prepared after thorough investigation and research. The availability of time-series satellite images has enabled morphologists to conduct in-depth studies to gain better insight into the morphological processes of rivers in the delta. Other tools available are physical and mathematical modeling. It is better to carry out a careful research and plan rather than the sporadic interventions in the rivers on assumptions, which may not work. There is a need here of coordination among BWDB and other agencies for the integrated river management. In response to the 1998 flood, Local government institutions constructed in 2000 many roads east–west in the greater Jessore area. It appeared, that some of these roads obstruct the natural flow of the rivers and channels which flow north to south and that bridges were built too low for navigation. Later it was concluded that national institutions as BWDB, Road and Highways or other relevant organizations were not consulted.

The Baseline Studies and the '60 Years Water Resources Management in Bangladesh: Lessons Learnt' report[5] show that problematic consistency over time and implementation problems are strongly related to major governance issues. The

[5] BDP 2100 Formulation Project, 2016.

challenge lies in creating realistic and integrated long-term strategies and plans including decision making with already during the plan formulation process real involvement and commitment of the government and stakeholders. Better preparation of an adequate institutional set up and of timely financial arrangements are most important. The need for coordination to this has been identified at the national planning and implementation level, the so called horizontal coordination between sectors. Complicated is however also the vertical coordination within national institutions and also from national level to decentral levels of the government and on the ground during project execution. This latter form of coordination is a form of multi-level governance aimed at involvement of different government bodies, Water Management Organizations (WMOs) and agencies in adequate execution of projects.

Given the governments ambition to attain high middle-income country status by 2030, to achieve the SDG's by 2030, to become developed country by 2041 and to address the expected impacts of transboundary developments and climate change on the longer term, there is a strong need for adequate governance and institutional development to achieve these strategic ambitions by effective and efficient implementation of the strategies and selected projects or measures. In addition, the fragmented sector- and agency-wide composite governance setting of BDP 2100 formulation and implementation requires consistent coordination and systematic focus on addressing of Bangladesh's future challenges.

11.2.2 Understanding Governance Issues

The expectation was, that the Water Act 2013, which included also the specific items of the National Water Policy 1999, would provide an adequate governance and institutional basis for integrated water management in Bangladesh. The issue is that the Act has not been implemented well within the water sector and between sectors. Because of this, the relationship with national plans as Five-Year Plans and BDP 2100 is weak and too limited for coordination of BDP 2100 goals in relation to climate change, environment, bio-diversity, agriculture, fisheries, forestry, inland water transport and land management as well as stimulating socio-economic development. Another challenge concerning Water Act 2013 is developing the main organizations covered in the Act to full swing adequate entities.

11.2.2.1 Need of Coordination and Collaboration

Institutions responsible for water management in Bangladesh are fragmented and characterized by multiple inefficiencies and capacity constraints. Although the responsibility for water management is vested among three ministries, the Ministry of Water Resources is a dedicated water ministry; the other two, especially the Ministry of Local Government, Rural Development and Cooperatives and Ministry of Shipping have some water tasks but also other areas of accountability. The differentiation of water- and delta-related tasks is broad and require coordination and collaboration, especially with respect to joint planning, generating funding and execution of programs. Attention for new investments is high while operations and maintenance tasks are seen as less interesting and important. The implication is that each agency works on its own tasks instead of delivering an integrated service. For example, water institutions tend to follow response-based actions instead of working on a long-term, sustainable solution to flooding and erosion. This is encouraged by budget constraints and limited human resources, a limited number of development projects, insufficient maintenance budget, failures, e.g., due to river erosion, limited planning capacity at agency level and traditional performance and culture. Management practices are also inflexible in dealing with the dynamic nature of the river system and adapting to climate change. This has increased the vulnerability to flood disasters and undermined opportunities to safeguard economic development. Much of the water research is focused on engineering and physical aspects of water flows and environmental impact. The economics of water management and related climate change in terms of effects on economic output, employment and poverty could get more attention. The same counts for issues of investment requirements and financing options (Alam et al., 2018a).

11.2.2.2 Capacity Constraints

As said before, the Water Act 2013 and even the NWP 1999 are not completely implemented in the water sector and beyond. This means that capacity constraints also exist in other delta-related agencies in sectors as Agriculture, Environment and Forests, Shipping, Land, Fisheries and Livestock. Moreover, capacities in the general ministries of Planning, Finance, Establishment and Local Government are also limited. Capacity constraints in all Delta Plan related policy making and implementing institutions are importance to the realization of BDP 2100.

11.2.2.3 Genuine Stakeholder Participation

Global delta and adaptation experiences show the importance of community participation. For a long time, a top down engineering approach was prevalent to water management. Many projects could be realized all over the country but only with limited stakeholder involvement. In cases where participatory approaches were organized, they can be seen as add-ons with few sustainable results. Although this is known for a long period and the NWP 1999 already provided directions for community involvement, real stakeholder participation was limited. The MoWR established a guideline known as Guidelines for Participatory Water Management which was the basis for the Rule on Participatory Water Management published in 2015 (Ministry of Water Resources, 2015). Community participation improved to some extend but is dependent of projects instead of giving involved stakeholders a clear role and voice in decision preparation and making.

11.2.2.4 Effective Research and Knowledge

Research and knowledge form valuable input for adaptive BDP 2100 which is basically science and knowledge driven. Research on delta-related issues is primarily carried out by research institutions of the MoWR and MoA, which focus on water-related and agriculture research including soil technology. The research conducted by research agencies is determined by their own interests. For instance, there is a scarcity of collaborative or interactive research that is essential for comprehending the effects of climate change on comprehensive aspects of water and, e.g., on agricultural productivity and vulnerability. Consequently, the global and local research conducted in the private sector, as well as in universities, on a variety of delta-related issues is not widely recognized or disseminated for policy purposes. As a result, there is a knowledge deficit regarding delta vulnerabilities, challenges, institutions, outlook, and solutions that require attention and action.

11.2.2.5 Result-Based Monitoring and Evaluation (M&E)

Result or output based monitoring and evaluation (M&E) of water and other delta-related programs shows another deficit. Because of uncertainties climate change, river course movements and behavior of sea and occurrence of natural disasters, it is important to not only know financial figures about the progress of projects but to know the results and understand outcomes of interventions in relation to goal achievement. Such reflective monitoring is rarely seen yet. The development of appropriate responses in the context of adaptation to climate change and natural hazards is fundamentally dependent on the successful management of water and delta factors, as well as a correct understanding of the outcomes of interventions, the progress, risks, and vulnerabilities. Effective adaptation will be challenging in the absence of a well-developed results and outcomes-based M&E system and a systematic approach to data collection, updating, and research.

11.2.2.6 Coordinated Approach to Adaptive Planning and Investments

In the context of the agreed delta plan, it will be necessary to mobilize appropriate resources and ensure their optimal utilization, as resources in Bangladesh will continue to be restricted. Prioritization of investments is of great importance and requires coordinated action. The point is that normally line ministries have been responsible for advocating for investment priorities and ensuring that they are subject to the formal evaluation process, which involves the Planning Commission and the Executive Committee of the National Economic Council. The Ministry of Finance annually allocates non-development budget (non-investment funding) in accordance with discussions with relevant department ministries. The budget funding is however an annual feature with no carryovers, the distinction between non-development and development budgets continues, and the allocation system for these two categories of spending remains largely disjointed. The allocation of funding is skewed in favor of new initiatives on the wish list of every ministry. Balance between investment and O&M-funding is not a policy factor in funding allocations. While climate change is ongoing and measures are needed with limited financial resources, it is clear that prioritization and decision making of investments have to be organized between the in BDP 2100 involved ministries in a coordinated way.

11.2.2.7 Seeking Collaboration and Public–Private Partnership

In addition to Chapter 10, public–private partnership in water investments forms a point of attention. Private participation in water resources management as such is already taking place for a considerable time. This is to a large extent based on tendering and contracts with companies working in the water sector. Besides, other forms of private participation have been developed in irrigation, water supply and sanitation. Here, private investments and pay back mechanisms have been explored and applied on a relatively small scale. There are still many opportunities in these areas and in other BDP 2100 related applications as well. Although PPP in large water infra structure projects is not really common in Bangladesh, there is still an unused potential. Public–Private Partnership in the field of climate change adaptation, e.g., concerning investments in protection, flood control, and drainage as well as for example new well protected land development or climate smart agriculture could be further opened up and practiced with respect to private investments and fair business cases. This could offer interesting opportunities in addition to the government budget constraints.

11.3 Requirements of BDP 2100 Institutional Framework

The purpose of the BDP 2100 Institutional Framework is to address the lessons learnt and governance issues and to improve the governance structure in order to bring the existing institutional framework and government agencies in a better position of consistent and coordinated governance and to play an active and required role in the BDP 2100 process and needs of its implementation. The BDP 2100 Institutional Framework encompasses necessary and agreed transformations of the current institutional framework and parts of the governmental organizations involved. These organizations need to be involved in such a way as to be prepared to participate in the BDP 2100 realization and updating for renewed adaptation.

A summary of characteristics of integrated governance which should be supported by the Institutional Framework and transformations is as follows.

- Establishing a long-term integrated adaptive planning and funding approach, and addressing major challenges as climate change, water safety, access to fresh water, food security and socio-economic development. Since this is an approach of timely

adaptation, and frequent updating of the Delta Plan is therefore required, result based monitoring and reflective evaluation of progress are important for decision making with respect to the whole process of Delta Vision achievement.
- Preparing institutional arrangements for adaptive planning and funding over a long-term, multi-sectoral and multi-level coordination, long-term funding and progress, e.g., by arranging a Delta Act with respect to dedicated delta governance, an adequate funding approach and decision making on adaptive programming and design. Frequent updating should be incorporated in the Act.
- Giving high priority to realize Delta Plan goals and execute measures by implementing agencies. The Delta Fund should provide resources for building the required capacities for which additional and innovative ways need to be explored.
- Stronger involvement and utilization of regional and local planning and implementation capacities. Starting point could be delegating or decentralizing implementation to City Corporations and, e.g., to the District level who already developed capacities. On local scale Water Management Organizations and communities can be involved which need to be better aligned for that with the local and regional government. Organizing participation of stakeholders and beneficiaries is required to raise awareness of relevant items, to balance the different interests and to utilize local knowledge which all will lead to more support for planning and implementation. This will lead to multi-level water governance responsibilities, regional and local financial contributions and related coordination mechanisms with decentral alignment to national strategies, delivering contributions in planning, implementation and funding also partly based on service charges and or water tax.
- Giving high priority to water safety and security by differentiated application of safety standards based on value—risk—protection level approach and based on secured fresh water supply.
- Delta Plans should always be science and knowledge driven to better understand the trends and developments, the up to date drivers behind it but also possible innovative ways to respond with new technologies or approaches. An Information and Knowledge strategy is needed to address the knowledge gaps and needs, to gain and accumulate knowledge in research programs and education, to set knowledge available and to apply knowledge.

Institutional development and good governance are probably the most important factors for achieving the mentioned ambitions for safety and security of life, livelihood improvement and economy growth in the Bangladesh delta. So there is an urgent governance challenge which can be characterized as 'collaborative planning and execution'. This is an issue on national level requiring so called horizontal coordination but also in terms of multi-level governance, requiring so called vertical coordination between national and local government. The recent and future anthropogenic changes in the hydrological cycle due to, e.g., climate change, construction of dams and barrages in the upstream countries in combination with increasing water demand are expected to make future water governance and management of collaborative implementation more challenging. The development culture could be transformed more profoundly from a short to mid-term sector and project approach to a longer term multi-sectoral program approach with a strong emphasis on coordination of planning and resource allocation as well as on adequate collaborative implementation.

In Bangladesh, large-scale programs/projects are usually implemented with funding assistance from development partners. The Government of Bangladesh finances a number of large-scale program from its own resources. An early indication and decision of funding volume of BDP 2100 has been considered as important signal. In the 7th Five-Year Plan (2016–2020), the GoB has committed 2% of the GDP annually for implementation of the BDP 2100. It has been observed that if funding is not secured in time, the implementation of the plan will suffer.

In addition, based on experiences and the analysis, the following challenges are considered as vital for the governance of BDP 2100.

- Reducing fragmentation and balancing the predominantly sectoral approach of planning and implementation with a long-term holistic governance approach including the concepts belonging to that (e.g. delta vision, scenarios, strategy pathways, delta institutional framework).
- Long-term strategic planning, including a long-term vision which needs to be back casted into medium and short term strategies and measures.
- Existing weak governance practices can be transformed to deal with a longer-term holistic planning and implementation effectively. Transparency and integrity require attention.
- Development of existing institutional abilities and capacities for effective coordination on BDP 2100, decision making on measures, investments and budget allocation in an integrated manner.
- Localization with decentral resource allocation i.e. from Central Budgeting System including ADP allocation to the District level budget allocation (resource utilization accounting).
- Involving development partners and International Financing Institutions in the BDP 2100 project preparations and execution.
- Involving private sector parties in proposals on interventions, improving required resource capacity and skills.
- Increasing regional cooperation between countries, e.g., with attention for strengthening water diplomacy.
- Effective and efficient (pre)feasibility and execution of measures, setting priorities, managing for results.

Governance Elements which are part of an adequate institutional framework are summarized in Fig. 11.1 per element including the issues and challenges to be addressed.[6]

[6] De Heer et al. (2016).

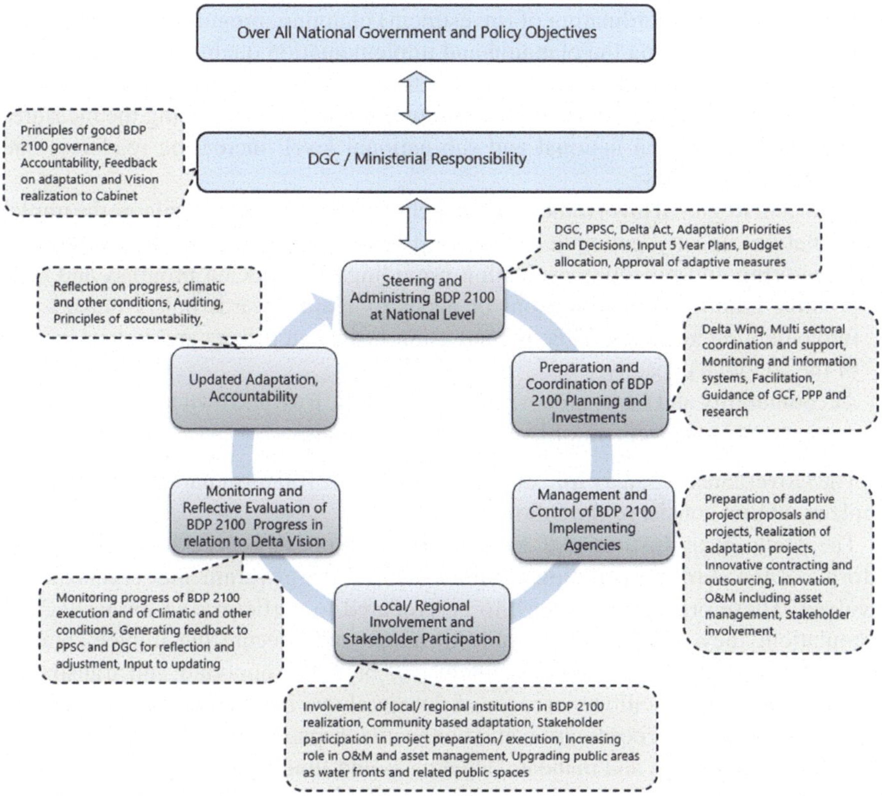

Fig. 11.1 BDP 2100 governance framework with issues and challenges per element. *Source* Study on Governance, BanDuDeltAS, 2016

11.4 Transformation of Governance and Institutions

11.4.1 Governance Strategy for Administering BDP 2100

Given the issues and challenges, the development culture regarding delta governance could be transformed more profoundly from short term and sector wise to a medium–long-term, multi-sector and multi-level approach with a consistent programming, funding and implementation process and with strong emphasis on coordination of planning and resource allocation as well as on adequate collaborative implementation. In this respect, the BDP 2100 governance approach addresses the following six basic elements of the governance cycle:

- Steering and Administering BDP 2100 at national level: increasing the connectedness and adaptive policy making, coordination and funding and spending capacity.

- Preparation and coordinating of (investment) planning, programming and implementation: increasing the planning and implementation quality and coordination and connective capacity.
- Management and control of implementing agencies, implementing the measures and interventions on national and sub-national level: increasing readiness and execution capacities.
- Local and Regional involvement and stakeholder engagement to enlarge the implementation capacity, get input and develop ownership on adaptive interventions.
- Monitoring and reflective evaluation providing feedback on progress and on a possible additional modifications or needs to adapt towards policy makers and implementing agencies. This by increasing insight in drivers and indicators and of the adaptive capacity.
- Accountability and transparency, increasing public trust and ownership in adaptive capacity and achievement of Delta Vision and goals.

The Governance Strategy for Administering BDP 2100 formulation, updating and implementation consist of the following items.

The Delta Plan Institutional Framework encompasses necessary and agreed reforms of the current institutional framework and the governmental organizations involved. These organizations need to be prepared to participate in the BDP 2100 formulation; they also should be able to adopt and execute the implementation program in an integrated, targeted, inclusive, broadly supported and transparent way. To enable and facilitate increase of the six above mentioned capacities, especially with respect to coordination and connective capacity, a central national body with adequate position and mandate, legal basis and fund is proposed to coordinate, facilitate and monitor the sectoral and decentral implementing agencies. A Delta Governance Council, Delta Fund and Delta Act are therefore proposed.

In addition, also the following items are important.

a. Embedding BDP 2100 in core governance of Bangladesh is of crucial importance because:
   - BDP 2100 aims to make Bangladesh climate change resilient by 2100 and requires continuous consistency, priority and emphasis to achieve that.
   - Steering, coordinating, prioritizing and deciding strategically about investments and resource allocation and implementation. Therefore, a multi-sectoral coordination mechanism is proposed and a financial mechanism has been recommended. Both coordinating and financial mechanisms require to be formalized in a Delta Act of which a draft is proposed.
   - Formal alignment is needed with existing, acts, policies, strategies and institutions, utilizing, strengthening and adding to these as much as possible.
b. Alignment of BDP 2100 with planning and coordinated action at the national level concerning Perspective Plan 2021 and 2041, Five-Year Planning Cycle, also in relation to responding to trends and developments, signaling tipping points with respect to main drivers and elaborating Adaptive Pathways when needed.

c. Preparing implementation in an integrated, multi-sectoral way. Implementation by Agencies at National Level urgently require strengthening and innovation. Implementation at Sub-National Level (e.g. City Corporations and Water Management Organizations) require also strengthening and innovation while Water Management Organizations need to be formed all over the country in line with the NWPo, 1999 and the Rule on Participatory Water Management.
d. Addressing Transboundary Issues via regional cooperation and water diplomacy while making contingency plans to cope with possible Transboundary problems.
e. Monitoring and reflective evaluation of progress and outcomes in relation to expected results, time frame and possible tipping points.
f. Conditions of accountability and transparency for optimal utilization of budgets and quality of work.

Establishing effective governance and institutions for adaptive management by BDP 2100 is obviously a huge challenge. The adoption of the Water Act 2013 is however positive in providing a legal framework for establishing the required institutional arrangements. It is a good starting point that can be modified appropriately to serve the needs of the Delta Plan. In this regard, Bangladesh can also learn from the experiences of other delta countries. With respect to this, strategic considerations are as follows.

11.4.1.1 Assigning Topmost Priority to Safety and Security from Water Disasters

This is the main organizing principle also in relation to food security and socio-economic development. Given the growth of population and economy, the type of disasters that happened are not acceptable anymore for GoB and preventive as well as adaptive measures will have to be taken to safeguard the life and property of citizens and society as a whole. This resolve will further grow over time as the per capita income and wealth grows and along with this the potential cost of large-scale flooding. Sensitivity to climate changes has similarly grown in view of the added risk they pose to flooding and loss of coastal land from sea level rise. This needs to be institutionalized in water planning by doing systematic cost benefit analysis of flooding and by imposing a large value per life lost to floods (Alam et al., 2018b).

11.4.1.2 Need for Adaptive Planning and Management

The rationale for adaptive planning is the major uncertainty in climate change and related natural forces. While there is considerable scientific progress in measuring temperature rise, the rising sea level and other adverse climate change factors, the specific magnitudes of occurrence in each countries and the likely adverse effects on climate variables like rainfall, flooding, hurricanes/cyclones and other natural disasters are not fully known while the possible impacts happen over a long period of time (Alam et al. 2018c). Given these uncertainties, delta planning inevitably involves a long-term vision with periodic updates of underlying climate parameters and investment responses.

To support adaptive management, investment of a considerable amounts of resources is required to gather data and conduct research on a range of delta-related subjects including climate change aspects. The knowledge should accumulate, be stored and shared with policy makers, knowledge institutions and the public. Innovation can be supported through research and pilot projects. The knowledge agenda is important in relation to knowledge gaps and can be managed in relation to national and international research projects. This proactive knowledge management is a critical factor for finding innovative approaches for adaptive management. Bangladesh has a 50 year plus delta management experience, although not well organized or integrated. As noted before in Chapter 3, there are many studies on different aspects of delta management, often prepared with support of development partners or financial institutions. Although they lack coherence and consistency with each other as they were each done in isolation and mostly independent of each other rather than as a part of a coordinated incremental approach, there is nevertheless a wealth of knowledge available. Critical research on impact assessments is however missing. Most importantly, the data infrastructure and knowledge management can be improved. This is an important area of institutional reform moving forward. In this situation, it was possible for CEGIS and IWM, both being Government Trusts, to come forward as knowledge driven supportive and modeling organizations. Besides, universities can incorporate curricula on adaptation to climate change and related aspects.

11.4.1.3 Importance of Inclusive and Participatory Delta Planning and Management

A feature of the quality of adaptive management is the extent to which strategizing and planning take place through an inclusive and consultative process. Local or regional as well as academic knowledge and experiences but also interests can be taken into account while lessons learnt from earlier planning processes can be incorporated in improving the subsequent planning process. The delta planning can in this way be informed by a rich body of practical experiences, scientific research, water modeling, climate change analysis, impact assessment and quantitative detailed cost–benefit analysis of different approaches and initiatives. Moreover, differences in opinion lead to debates and possible useful modifications or solutions.

11.4.1.4 Innovation as the Basis for Adaptive Planning and Management

There are two broad considerations with respect to this item for the Bangladesh delta. First, flood control and major irrigation investments comprise capital intensive interventions. The social and economic benefits can be maximized by making these investments multi-purpose. Investments for coastal flood management might be combined with initiatives to develop fishery facilities and coastal tourism. Similarly, management of fresh water wetlands can be combined into areas of eco-tourism. Another example is that lake cleaning and restoration in urban areas can be converted into recreational park areas. These multi-purpose investments can help cost recovery of capital investments. Secondly, the natural course of rivers can be maintained through an adaptive design approach that involves flexible and expandable solutions wherever practicable. The scope for this will obviously depend on specific geographical and social context of Bangladesh, the limits of which are set by the heavy population pressure and shortage of land. The important point is that in designing adaptive interventions, natural processes and building with nature could be followed as much as possible.

11.4.1.5 Considering Regional and Local Participation

An interesting point for consideration is that delta management, certainly routine operation and maintenance tasks, can be taken care of by local and regional institutions. Water User Institutions have been explored and proposed in BDP 2100.

Participatory water management is envisaged as an instrument to harness water resources for their optimal use. The country faces the problem of flood and drainage in the monsoon months and acute shortage of water in many areas in the post-monsoon period. Increasing salinization and arsenic contamination of water have worsened the situation. This problem is bound to be multiplied with the pace of climate change, urbanization, industrialization and intensive agriculture. BDP 2100 is an attempt in this respect to provide strategies and measures as well as institutional transformation aimed at water user involvement. The quest for a workable and enduring system of participatory water management is an ongoing process which has been addressed in BDP 2100 and elaborated in the SIBDP project, which proposes the following steps for considerate attention. First, Water User Institutions representing all stakeholders, be it the community organizations or the implementing agencies, are lacking and could be embedded in the administrative structure of the State starting from the lowest stratum of the Local Government to the apex level. Second, WMOs can be justified in the pretext of their engagement in operations and maintenance (O&M) and their role as interface between the community and the implementing institutions. To carry out such role effectively, they need institutional support, as well as financial resources. Sustainability of WMOs can be ensured if there are adequate resources for investment and O&M. WMOs are not an end, but a vehicle for successful O&M. Third, as resources are scarce, their prudent utilization is necessary to reap the maximum benefit. This requires good planning by and coordination among key implementing agencies at the central, as well as at the local level. Fourth, the need of the hour is to get out of the small piecemeal and ad hoc decision-making practices and to absolve the culture of longer-term planning, also for the local and regional levels. Fifth, GoB could focus on the possible coordination mechanism in the water sector, especially coordinating the work of different leading agencies including BWDB, LGED, and DAE. For appropriate O&M, it is indispensable to ensure coordination among different project stakeholders. In the last two decades, Bangladesh has established many Water Management Organizations (WMO, at small local level), which form a collective in Water Management Associations (WMA) who are represented on the regional level in a Water Management Federations (WMF) together forming a participatory water management system. Integrating (representatives of) these organizations, WMO-WMA-WMF of a particular region with other stakeholders and creating a regional-level stakeholder's platform is essential for multi-sectoral planning as in BDP 2100 and integrated interventions. All water users would be involved in required interventions and water-related services to be delivered. This would also be a strong basis for considering water value for the stakeholders. Bangladesh does not have a regional water users platform nor any such national-level platform where all water user stakeholders can jointly meet and collaborate and, e.g., generate advice to the government on basin wise interventions. Along with creating a regional stakeholder's forum, it is equally important to coordinate the activities of the regional stakeholder's platform with a national stakeholder's platform. While the regional platform can actively focus on urgent regional-level water management issues, the national stakeholder platform could take a more macro approach.

11.4.1.6 Innovative Water Financing Mechanism Based on Beneficiary Pays Principle

Cost recovery from users can be considered with respect to the benefits and in relation to per capita income. This two-step innovative financing principle (tax based plus beneficiary pays principle) can be formed as additional way of funding of a safe delta with well-maintained infrastructure. Quality of work of water bodies and proper maintenance of the water infrastructure is than assured through approvals of design standards and inspections by the national or district authorities. By linking water financing directly to water and flood control services and benefits, the government would be able to mobilize substantial resources for water management that would not have been possible from direct state tax resources alone.

Delta Plan investments, maintenance, institution building and research needs are huge and there may be some constraints to meet all these requirements. The ability to establish local Water Management Organizations and to develop them into more professional water user bodies, assigning them a role in the local planning, implementation, and maintenance of these local investments can be an option for developing sustainable financing for adaptive management under the Delta Plan.

11.5 Key Elements of Reforms Under the Bangladesh Delta Plan

Drawing from the above noted considerations, the past Bangladesh experience, and the socio-economic-political realities of Bangladesh, the proposed approach to delta governance and institutional transformation is as follows (Ministry of Planning, 2018).

11.5.1 The Legal Institutional Framework for Delta Plan Management

The NEC in its meeting on approval of BDP 2100 in September 2018 gave the direction for (a) formation of a Delta Governance Council (DGC); (b) formation of a Projects/Programs Selection Committee (PPSC); (c) establishment of a Bangladesh Delta Fund; (d) establishment of Focal Points and at subsequent stages; (e) establishment of regional/local Water bodies. A new Delta Act is considered be enacted to provide a sound legal coverage for the long-term nature of BDP 2100 including establishment of the Delta Fund, establishment of the Delta Governance Council and the Plan itself. This Delta Act is expected to be not conflicting with the National Water Act 2013, because it only provides legal coverage to the mentioned institutional transformations.

11.5.1.1 Bangladesh Delta Governance Council

An important way the government may signal its top priority to adaptive delta management including issues and challenges related to climate change, coastal zone and river management, water supply and sanitation, environment, bio-diversity, agriculture, fisheries, Blue Economy, navigation and land management is to establish a Delta Governance Council to coordinate, facilitate and monitor and evaluate planning and implementation of the Delta Plan in an integrated and holistic manner. The Bangladesh Delta Governance Council would be an independent entity of Government of Bangladesh (GoB) which would be linked, for administrative purposes, with the Ministry of Planning. A small higher level forum, the 'Delta Governance Council (DGC)' is proposed, which would be chaired by the Prime Minister. The DGC would function as a formal linkage for achieving political commitments regarding BDP 2100, provide directions and makes decisions. It would provide strategic advice and policy guidelines. The Delta Governance Council would steer continuous planning and coordinate and facilitate the implementation of the Bangladesh Delta Plan 2100 (BDP 2100) by related ministries and implementing agencies, including its investment plan, to ensure that Bangladesh is able to pursue climate-resilient sustainable development. Key functions of the Delta Governance Council would be national level coordination, planning, programming, resource mobilization, facilitating program and project preparation and facilitating implementation, monitoring and evaluation, conducting research and innovation studies, preparing policy and guidelines, knowledge management and coordination. Involved Ministries and Agencies shall implement the programs and projects included in the Delta Plan 2100.

11.5.1.2 Projects and Programs Selection Committee

In compliance to the above mentioned decision, a Project/Program Selection Committee (PPSC) has been constituted on 25th October 2022 with the Member, GED as chairperson as well as representatives of relevant ministries/divisions and departments for selection and evaluation of investment projects or programs of BDP 2100. The ministries and divisions would process project proposals related to BDP 2100 to the relevant sector/ divisions through the PPSC. The Project/Program Selection Committee (PPSC) is proposed as a coordinating committee to facilitate specific project/program selection for inclusion in the Annual Delta Investment Program. The BDP 2100 investment projects or programs will be considered for project or program selection and prioritization by the PPSC following dedicated guidelines or Delta Appraisal Framework. In general, the project approval process will normally remain the same as is being practiced before. The PPSC will facilitate critical scrutiny of only delta-related projects/programs with assistance from and in consultation with the Planning Commission and relevant ministries/divisions and agencies.

11.5.1.3 Delta Wing as Coordinating, Facilitating and Monitoring Body

The Delta Wing is seen as a significant institutional reform as it mainstreams adaptive delta planning, management and implementation in the national government structure and ensures long-term commitment of the Government of Bangladesh. Main responsibilities assigned to GED are: coordination, facilitation as well as M&E of BDP 2100 implementation. These responsibilities form the basis for the structuring of the Delta Wing. Key functions of the Delta Wing in GED are among others: (1) ‘Programming’ with activities related to approved BDP 2100 Investment Plan implementation. Besides, it prepares program coordination with ministries and agencies; prepares working papers to support meetings of the Delta Governance Council and the Projects/Programmes Selection Committee. (2) ‘Policy making, Planning and Coordination’ with actions concerning policy option briefs on issues for BDP 2100 adaptive management and, e.g., creating platforms for debates on critical issues for adaptive management. Interacts proactively with sectoral agencies to develop harmonized guideline documents and proposals. (3) ‘Financing’ focused on develops strategies for public and private sector financing options; funding availability through internal and external resources including loans and project/program aid for implementation of the BDP 2100; projects/programs for Public Private Partnership (PPP) and for financing under the Green Climate Fund and other global funds. (4) ‘Capacity development', aimed at capacity building of relevant stakeholders for the new adaptive management approach and systems with extensive training at national institutes; facilitates capacity development/strengthening of core Delta Institutions of the Bangladesh Delta Plan 2100; organizes strengthening of capacity of Focal Points and the 'Delta Wing or Cell' established in different implementing agencies; develops training modules on adaptive BDP 2100 related aspects.

11.5.1.4 Participation of Local Water Institutions

In Bangladesh both the Local Government Institutions (LGIs) and Water Management Organizations (WMO) are institutionally weak. WMOs are still seen as temporary set ups and linked to donor-assisted projects in BWDB or in LGED. Furthermore, many other agencies like DAE, DPHE also establish their own user groups. For example, Tube well user groups are established (DPHE) to operate and maintain communal water supply, markets and other community facilities. Farmers forum is established (DAE) as focal point for the agricultural development activities of DAE. More over Social Forestry Groups have also been formed (FD) to carry out tree plantation, look after the planted trees and also to safeguard the infrastructure and climate as well.

Involvement of Water Management Organizations (WMO) in connection with the Local Government Institutions (LGIs) in implementation of the BDP 2100 is foreseen in future but these institutions need to develop to be able to address challenges on their level and scale. Establishment of WMOs all over the country with necessary arrangements is needed which could be taken up as part of the Rationalization Programs proposed in the BDP strategies and Investment Program. There is an on-going discussion on stronger linkages between Union Parishads and WMOs. It is possible that over time, this linkage can emerge as 'decentral water bodies'. The Water Management Organizations represent local stakeholder's interests. Also community based adaptation could be an important aspect, together with local government. The full governance principles including membership, selection process, functions, financing and accountabilities must be considered in totality in relation to the relevant delta issues, similar to the concept of hotspots that share broadly similar hydrological concerns.

11.5.1.5 Bangladesh Delta Fund

Bangladesh spends about 1% of GDP for water resources, mostly for new investments. O&M funding is very limited which is inadequate compared to the needs of the BDP 2100 on maintenance requirements. A minimum financing need of about 2.5% of GDP has been indicated by GoB, of which 2.0% would be new investments and 0.5% for annual O&M. Given the competing needs, the increase from 1% of GDP to 2.5% of GDP is considerable and also requires alternative sources of financing. Given the limited resources and the need for a coordinated and integrated approach to Delta

Plan spending, a Delta Fund is considered in the following way. The Bangladesh Delta Fund would be an earmarked fund for financing the implementation of the Bangladesh Delta Plan 2100. Sources of this fund may be GOB, DPs, Environment and Climate Change related funds, PPP, etc. The Government has the intention to allocate about 2.5% of GDP annually to the Delta Fund gradually—as indicated in the 7th and 8th Five-Year Plan. The Delta Fund would consist of capital expenditure for investment projects and recurrent budget funds for operation and maintenance to ensure sustainability of delta plan related projects. It was decided to follow as first step clear indication of the BDP 2100 investments, while further following the usual way. The step towards a separate dedicated Delta Fund could come when needed and after establishment of the legal basis for such fund.

11.5.1.6 Developing the Delta Act

GoB considers that the institutional framework requires a legal basis in a Delta Act. The Bangladesh Delta Governance Council, PPSC and Delta Wing and the Delta Fund would be legally established and mandated through enactment of the 'Bangladesh Delta Act'. A draft Act has been prepared and legislative process is required for completing this.

11.5.2 Strengthening Core Delta Governance Institutions

Success of any institutional arrangement depends upon the quality of the involved institutions. These in turn depend upon clarity of rules of engagements, clearly identified functions, measurable outputs and deliverables, clear accountability rules, quality of staffing and adequacy of budgetary resources. Bangladesh has as indicated considerable capacity constraints in its public administration. For the Delta Plan to succeed and to realize the institutional transformations, it is necessary to strengthen the core delta institutions in order to increase their implementation capacity considerably. BDP 2100 indicates that some organizations require attention, like BWDB for the largest part of BDP 2100 execution, but also WARPO, DoE, DAE and others in the field of, e.g., disaster management, shipping and inland water, local government, environment and forestry and fishing.

Especially BWDB needs considerable strengthening in terms of man power and project execution skills but also in new technology, innovation, integrated planning, research, economic management and consultative processes. These gaps can be addressed with proper staffing of experts and necessary tools and provisions in the missing areas. Attention is needed for adaptive management, adaptive programming and design and no-regrets solutions. Besides the O&M programming and execution of Rationalization programs is important take up urgently.

The demand for water safety and security, including drinking water, sanitation, irrigation and navigation services in Bangladesh far exceeds the capacities of the water institutions and local government. While parts of the gap can be met through private participation, the capacity can be substantially strengthened by establishing cost recovery, based on beneficiary and polluter pays principles.

11.5.3 Strengthening Transboundary Dialogue and Related Institutions

As the lower riparian for all international rivers, Bangladesh faces some real risks from adverse developments upstream. The transboundary issues were discussed in Chapter 4. The main concerned upper riparian country is India. It is understood that politics takes the upper hand in transboundary water diplomacy. Many of the past dialogues can be explained by adverse political relations. More recently, a major positive development that holds considerable promise of better cross-boundary water outcomes is the sharply improved political relations of Bangladesh with India. In India, water issues and disputes are jointly managed by the national government and the involved state governments.

The water dialogue between Bangladesh and India concerns in practice Bangladesh, India and the related bordering states (West Bengal, Assam, Tripura, Mizoram and Meghalaya). Bangladesh could further strengthen its relations with these Indian states as much as with the federal government in Delhi. Lessons of international experience suggest that a joint river basin approach has the best prospects of success. A further consideration would be to think of river agreements in terms of providing multi-purpose benefits. When water sharing is viewed as a zero-sum game where more for one riparian means less for other, disputes become dominant and negotiations tend to fail. When water sharing involves multiple benefits (flood control, water storage, irrigation and hydro-power), the zero-sum game feature can be converted to a win–win situation for both parties and the dialogue takes the shape of negotiating the best deal in terms of sharing of costs and benefits.

To improve the negotiation position of Bangladesh, a combination of water diplomacy approaches could be applied during negotiations on transboundary water with India, Nepal, Bhutan and China. (1) The basin approach is centered on the concept of benefit sharing to make optimal use of the river and its services by constructing infrastructure on the most optimal hydrological and ecological location in the river basin. (2) Negotiating multiple treaties simultaneously is the second approach Bangladesh can implement in order to increase beneficial options during the negotiation process. (3) The mutual gains approach entails the incorporation of non-water issues into the negotiation process in order to augment the basket of benefits to consequently other tradable benefits next to water can be taken into account as well. (4) As part of the Multi-layer water dialogue approach civil society and academics are invited to the negotiation table as well. Multi-layer dialogues enable participating countries to start the negotiations from a new entry point and connect water negotiations with other river interests. (5) The fifth approach involves the active participation of a third party, either an international organization or a country. In line with this, water diplomacy approaches, a total of nine non-water issues are identified that could function as leverage during negotiations on shared water bodies, including data and knowledge sharing, climate change, migration, Sundarban Forest, inland navigation and transportation, trade, electricity/hydropower, port development and tourism. The ability to think innovatively requires both diplomacy and technical skills. The latter are required to come up with well-thought multi-purpose river basin projects. Accordingly, the capacity of the Joint Rivers Commission (JRC) could be strengthened for that. Partnership and coordination of JRC with BWDB and WARPO can be useful. Possible multi-purpose technical options and cost–benefit analysis of these options will need to be developed as background research to dialogue with India. Routine dialogue involving exchange of conversations and general purpose concerns or complaints and information sharing can give way to a more strategic dialogue aimed at problem solving and finding solutions.

11.5.4 Creating a Delta Information and Knowledge Portal

Under BDP 2100 and SIBDP, an Information and Knowledge Portal was formed to collect, store and develop the relevant BDP 2100 data and knowledge which is needed as collective memory and basis for adaptive management execution and updating tasks. The existence of knowledge gaps and a knowledge agenda for research and development are already mentioned and will be further discussed in Chapter 12. Research in connection to the knowledge agenda can be out-sourced to universities and local research institutions. Following the institutional framework, the GED/ Delta Wing performs tasks on maintaining and developing or renewing the Information and Knowledge Portal, under responsibility of the DGC. Besides, the Information and Knowledge Portal with open access, a dedicated part can serve as platform for information sharing and other interactions, e.g., supportive to the Focal Points, between the with BDP 2100 involved ministries and agencies. Following approval by the Delta Governance Council of the knowledge work program and budget, implementation of such additional functions can proceed, e.g., in conjunction with the Bangladesh Bureau of Statistics and other delta-related ministries and other institutions with linkages to their data bases.

11.5.5 Developing an Adequate Monitoring and Evaluation System

Monitoring and Evaluation (M&E) are important governance aspects in the policy and planning cycle. Policy making and planning incorporate goal setting and providing a way of generating feedback to the policy makers on how far goals have been achieved and on possible issues which hamper goal achievement. The M&E issues are discussed in Chapter 14.

The Investment Plan prepared under BDP consists of a total of 80 projects. Among them 65 are physical projects, and 15 are institutional and knowledge development projects. Important is to know what is actually happening with these measures in terms of outcomes in the physical systems in relation to achievement of delta goals. It is equally important to monitor and evaluate the climate change and other relevant developments with respect to the in BDP 2100 identified strategies and measures in the Investment Plan and to how the physical systems are developing. Here lies the essence of an adequate monitoring and evaluation framework in that as much as it is required to constantly keep an eye on execution of projects and measures and the results they generate, expected and unexpected outcomes. Reflective evaluation means that the outcomes are discussed against expected achievement of goals and vision and in relation to the actual development of climate change and other relevant developments. Reconsideration of strategies, measures or priorities can follow on this reflection in order to adapt effectively. The practice of conducting monitoring and reflective evaluation of government policies and programs is an item that can be improved in the Bangladesh policy and planning approaches.

With respect to governance, the responsibility for M&E is interesting. So far as investment projects and their progress in execution are concerned, M&E lies with the implementing agencies. Within GoB, M&E of sectoral policies, plans and programs rests with the concerned sectoral ministry and IMED. Regarding M&E of the BDP 2100, this is one of the responsibilities of the Delta Governance Council and with respect to actual tasks, of the GED/Delta Wing. As in the case of implementation of the knowledge agenda, the Delta Governance Council and Delta Wing can draw on capabilities of its implementing partners, involving inter-ministerial agencies dealing with delta issues and the IMED.

References

Alam, S., De Heer, J. M., & Choudhury, G. A. (2018a). *Bangladesh Delta Plan 2100: Baseline studies on governance and institutional development* (Vol. 6). General Economics Division, Planning Commission, Government of Bangladesh. https://plancomm.gov.bd/

Alam, S., De Heer, J. M., & Choudhury, G. A. (2018b). *Bangladesh Delta Plan 2100: Baseline studies on disaster and environment management* (Vol. 2). General Economics Division, Planning Commission, Government of Bangladesh. https://plancomm.gov.bd/

Alam, S., De Heer, J. M., & Choudhury, G. A. (2018c). *Bangladesh Delta Plan 2100: Baseline studies on water resources management* (Vol. 1). General Economics Division, Planning Commission, Government of Bangladesh. https://plancomm.gov.bd/

Barkat, A., Khan, S. H., Majumder, S., Badiuzzaman, M., Sabina, N., Ahamed, K., Abdullah, Md. (2015). *Local governance and decentralization in Bangladesh. Politics and Economics*. Pathak Shamabesh, Dhaka.

Government of Bangladesh, Water Act (2013)

De Heer, J. M., Choudhury G. A., Islam, R. (2016). *Study Report on Governance*, BanDuDeltAS.

Ministry of Planning, GED, BDP 2100 Formulation Project. (2016). *Sixty years of water resources development in Bangladesh: Lessons Learnt.*
Ministry of Planning. (2018) *Bangladesh Delta Plan 2100.* https://plancomm.gov.bd/
Ministry of Water Resources. (2015). *Rule on participatory water management.*
Muhith, A. M. A. (2007). *Agenda for good governance.* From lawlessness and corruption to a caring and prosperous democracy.

Chapter 12
Need of up to Date Data and Knowledge

Abstract Adaptation to climate change implies that concerned strategies and plans need to be checked and modified frequently. This means that up to date data and knowledge must be gathered and related to already existing data and knowledge in order to find out what is still valid and what should be adjusted. It becomes clear in this chapter that a specific form of knowledge management is required. The availability of local and regional as well as world wide data and analysis of these and of developing patterns with respect to the various climate change indicators are of crucial importance for considering new measures and investments. In this respect, new questions may arise and additional research can be required to answer these and to clarify upcoming problem situations, also in terms of complex adaptive decision making and investments. New knowledge is often followed by innovations which could deliver alternative insights and solutions. Adaptive management expects a large investment in knowledge, research, and innovation, all of which are interconnected. For addressing the items on the Delta Plan knowledge agenda, the vast body of technology, practical experiences and knowledge capabilities of universities, research institutions, and business sector enterprises can be utilized. The necessity of institutional and interministerial research and knowledge coordination is yet another challenge. A knowledge management method must be anchored and coordinated in the adaptation process, as the BDP 2100 is meant to be continually scientific and knowledge driven. In light of this, the goal of this chapter is to define the key concerns and components for a knowledge development strategy and knowledge agenda, as well as to emphasize the context and contemporary knowledge management techniques and factors pertinent to adaptive management.

Keywords Knowledge management · Knowledge strategy · Knowledge agenda · Knowledge institution · Knowledge needs · Knowledge generation · Knowledge accumulation · Knowledge availability · Knowledge application · Knowledge community · Technology · Innovation · Data · Information · Decision support · Delta atelier

J. M. de Heer and G. A. Choudhury, *Strategies for Adaptation to Climate Change in a Transformative Approach*, Springer Climate,
https://doi.org/10.1007/978-3-031-88884-7_12

12.1 Setting of BDP 2100 Knowledge Management

The significance of data, information and knowledge management, as well as the necessity of an over longer time structured approach, is underscored by the extensive knowledge domains of climate, delta and related issues. Adaptive management is preferably built on managing knowledge in a process of acquiring, creating, preserving, sharing and utilizing data and knowledge. The process of knowledge updating and creation is from societal and adaptation points of view need based, following important questions of uncertain climate change and ways to deal with it in the involved actors practices, interactions, and learning process.

Uncertainty can for a large extent be interpreted as lack of (reliable) data, information, knowledge. Ambiguity is related to this, as the meaning of a statement or situation cannot be definitely resolved due to uncertainty and allows several interpretations with different options for choices and behavior. Development, implementation and renewing of the Delta Plan can be seen in this perspective as continuing concern, in various data gathering and research forms with analysis and results accessible to stakeholders. As stated in Chapters 1 and 2, ambiguity on climate change is wide spread and serious. Generating and structuring reliable data and knowledge management are therefore important mechanisms of the Delta Plan, serving decision makers and participating stakeholders in adaptive decision making concerning measures, priorities and investments. A specific Information and Knowledge Portal has been created for BDP 2100 as part of the ongoing knowledge management process, initially in order to (a) preserve the relevant knowledge and (b) to make it accessible (Ministry of Planning, 2018). Keeping the data and information up to date is also an important task, while new issues may require generation or collection of new knowledge. The fundamental principle is knowledge production, viewed as the generation of new ideas through the interplay of explicit and tacit information inside individual cognitive processes. It comprises socializing (tacit to tacit), externalization (tacit to explicit), combination (explicit to explicit), and internalization (explicit to tacit).[1] Knowledge generation involves the ongoing transfer, combination, and transformation of various knowledge types as users engage, interact, and learn. Data is here characterized as unprocessed facts, information is regarded as a structured collection of data, and knowledge is understood as significant information (Bhatt, 2001). Knowledge is derived from information, which in turn originates from data (Davenport & Prusak, 2000).

Regarding BDP 2100, large quantities of data and information are generated following regular tasks, e.g., on databases and through numerous initiatives and studies by different government entities, see Table 12.1.

The National Water Resources Database (NWRD) has been created by WARPO, to assist in the development of the National Water Management Plan (NWMP). This database, which has been regularly updated, contains national-level information necessary for water resources planning in accordance with the NWMP. The Integrated Coastal Resources Database (ICRD) was also developed by WARPO to provide

[1] Nonaka & Toyama, 2015.

Table 12.1 Important databases and data collecting institutions

Data bases/Government institutions	
- National Water Resources Database (NWRD) - Integrated Coastal Resources Database (ICRD) - Integrated Haors & Wetlands Resources Database (IHWRD) - Bangladesh Water Development Board (BWDB) - Bangladesh Meteorological Department (BMD) - Bangladesh Bureau of Statistics (BBS) - Directorate of Land Record and Survey (DLRS)	- Survey of Bangladesh (SOB) - Local Government Engineering Department (LGED) - Roads and Highways Department (RHD) - Department of Environment (DoE) - Soil Resource Development Institute (SRDI) - Bangladesh Agricultural Research Council's (BARC)

the necessary information for Integrated Coastal Zone Management and Planning. This database comprises information regarding the coastal region at the local level. During the development of the Haor Master Plan, the Bangladesh Haor and Wetlands Development Board has established the Integrated Haors and Wetlands Resources Database (IHWRD) which is enhanced with information from various sectors to assist and direct planners in the development of the Master Plan and the harmonization of Haor management activities by the various line agencies. With active collaboration of the Bangladesh Agriculture Research Council (BARC), CEGIS has developed a computer-based Drought Assessment framework (DRAS) that employs GIS and other analytical tools. There are more institutions in Bangladesh involved in the distinction between data and information gathering and knowledge creation. BUET, CEGIS, DoE, and IWM are among the foremost knowledge producers on water and climate change. The Ministry of Agriculture, along with its affiliated institutes and research bodies, is responsible for technical research on the aspects of agricultural productivity under varying soil, climate, and water conditions. There is however minimal or no coordination among the research organizations in the context of delta management. Consequently, the data or knowledge regarding the impact of sea level rise, e.g., on farm productivity, income, and employment are not readily accessible. Another issue is that a significant amount of data is outdated, as derived from designs or studies that were conducted many years ago. Furthermore, there is a lack of a systematic, well-coordinated national climate change research agenda to which the knowledge producers are contributing. These factors imply, that there is a substantial knowledge deficit and outdated data in nearly every aspect of delta planning and management. Therefore, the knowledge base must be expanded over time through systematic research, planning, and decision-making that are aligned with the Delta Plan's requirements. In that regard, it is imperative to cultivate a Delta Knowledge Community which is able and willing to facilitate the exchange of knowledge and perspectives from academia, policy-making, civil society, and the private sector. Approximately 100 experts from a variety of water-related disciplines participated in a BDP 2100 Knowledge Agenda Round Table on March 1, 2016, to investigate a draft Knowledge Agenda. This was the first step in this direction. During the Round table and the in knowledge conferences that followed the initial one, knowledge gaps and challenges, including the most important research questions and burning issues, were identified in the context of the BDP 2100.

12.2 Reflection on Knowledge Questions or Issues

Managing data and information on BDP 2100 formulation and implementation deals besides generating knowledge also with knowledge sharing, e.g., to implementing agencies, knowledge institutions and planning institutions.. More difficult to manage is taking care of understanding and application of shared knowledge by users, and getting it into practices of executing agencies or for renewing BDP 2100. Renewal of data and information is a difficult issue because tasks and especially budgets of the involved agencies are too limited while it is often also not clear who should do what. The national statistical institution (BBS) could provide climate change and sectoral impact related indicators useful to BDP 2100 renewal. In practice, the climate change related consequences for the national and regional economy, socio-economic development and damage and loss in various sectors are based on assessments. Stocktaking and data gathering are also very important for proper monitoring and evaluation and for related decisions on budget allocation.

Although the BDP 2100 Information and Knowledge Portal will primarily address information sharing, the institutional duties, finance and coordination among the several players engaged with the knowledge management and agenda could be clarified. Furthermore major obstacles include lack of necessary information or availability of information from source organizations in outdated formats and often problems of data distribution and sharing with others. While some databases in Bangladesh have GIS maps that can zoom in and out, most of the ones accessible to the general public are lists and PDFs and not interactive searchable. A limited number of organizations keep relational database systems to handle their data and records. Sustainability of systems and the use of various digital formats or in hard copy makes formation of a relational BDP 2100 database difficult. A requirement of the National ICT Policy (2018) is to properly share generated knowledge as good as possible. The BDP 2100 Information and Knowledge Portal aims to contribute to this requirement for the sake of effective use and renewal of data and information for climate change adaptation purposes. It is clear that many knowledge aspects still can be improved.

12.3 BDP 2100 Knowledge Management Approach

For its knowledge management approach, BDP 2100 builds further on earlier experiences and publications on knowledge management,[2] following an approach that handles the main elements.[3]

Collecting existing data and information and making it accessible are important elements to start with. As said, a lot of data, e.g., on rivers or coast are outdated and collection and renewal of data is costly. Besides, the sequence and performance of storage systems and issues about the continuity and accuracy of information and whereabouts still provide many challenges. The internet and intranets constitute and serve a shared environment for knowledge access, supplies and applications. Despite strong search engines, there are many difficulties concerning reliability of data, ambiguity of information and the challenge of rapidly locating required, meaningful knowledge and information. Examining, categorizing and structuring current knowledge and information assets as well as generating meta-level information (that is, knowledge about information) helps with this.

Examination of new needs for data, information and knowledge is another element. In a scientific and techno-economic context, the knowledge and data-driven strategy of adaptive management amid a rapidly evolving knowledge-based world results in a constant evolution of the demand for information, validated data, and associated competencies. This pertains to BDP 2100, focusing on enhancing current communication methods and identifying novel avenues for knowledge-intensive collaboration among delta-related countries, authorities, academic institutions, and corporate sector entities. The earlier mentioned Tracking of strategies and the Baseline Studies are applied practices which provided data and information related to first the knowledge needs on BDP 2100 formulation and for identification of knowledge gaps. The knowledge conferences and formation of a knowledge agenda show other activities in this area.

The next element of knowledge management concerns generating new data, information and knowledge. Face-to-face interaction, casual settings, focus and diversity of information, and interdisciplinary research and teaching methods are all conducive to the generation of new knowledge. Universities and other knowledge institutions both domestically and internationally produce a significant portion of new scientific studies and discoveries. But also frequent data collection by government agencies may show new patterns of certain system behavior. Data collection also forms the basis for advanced analysis or modeling. Moreover, pilot projects and experiential learning are major ways for knowledge generation and accumulation in human networks and in daily actions. Without in-person interactions or more structured information sharing, a significant portion of developed knowledge is tacit and thus difficult to convey to students, coworkers, new staff or officials. Knowledge creation and exchange are also influenced by international exchanges, and developments by the many virtual communities, and artificial intelligence applications. Although there

[2] Kautto-Koivula, 1998; Wiig (1995).

[3] See also: Kautto-Koivula (1999a, 1999b). Main operations for societal Knowledge Management.

is a vast amount of knowledge available, its quality is frequently unclear or even questionable which makes verification of sources really important. Renewal of research methods and educational reform are ongoing and taking place as a result of the digitalization and modernization of applications, working procedures and tools in many sectors. The developments concerning knowledge generation have implications for adaptive management, with new forms of networking and knowledge sharing with the various stakeholders, between knowledge institutions. New ways of adaptive programming and design as well as decision support for officials come forward and expect further capacity building.

The utilization of knowledge and innovations in practice is a crucial element. With respect to BDP 2100, this can be interpreted as occurring within a knowledge community that is connected by relationships in which the creation, application, and exploitation of new knowledge is discussed and taken care of. Point of attention is the support of users, especially small agencies and companies, with respect to their capacities, operational mode and the acquisition of the necessary skills to modernize and implement new knowledge or working methods. Another point of attention form government institutions, as new data and information do not lead inevitably to renewal of policies, plans or rules. Dedicated knowledge transfer and training and sometimes reform will be needed to achieve updating or modernization. It is possible in this way to accomplish results for BDP 2100 implementation or renewal, also by emphasizing BDP 2100 vision and goals. Keeping such an approach sustainable requires frequent interaction between knowledge users and providers. Collaboration of the for BDP 2100 responsible institutions with the knowledge community surrounding BDP 2100 and implementing agencies and private companies will be very productive.

It is the objective of the BDP 2100 Knowledge Management Strategy to facilitate the attainment of the Delta vision and goals by making use of these four elements (Ministry of Planning, GED BDP 2100 Formulation Project, 2018). Figure 12.1 shows the elements in a logical order in more detail. Strategies and proposals can be developed with the assistance of a vast array of technology and knowledge. The Knowledge Agenda must prioritize and resolve fundamental research questions, burning issues, and knowledge gaps related to the goals and strategies. However, funding the knowledge activities and systems is an important aspect of successful knowledge management which needs to be arranged in a structural way based on an earlier decision that 5% of the Delta Fund would be allocated for research and development and to coordinate the knowledge agenda.

The Delta Plan's Knowledge Agenda originates from the formulation of knowledge requirements and gaps related to the delta issues and challenges or to the proposed interventions. The knowledge gaps found in the Baseline Studies and suggested research are the focus of this Knowledge Agenda. The 26 themes covered in the Bangladesh Delta Plan 2100 Formulation Project were considered the starting point for the selection of subjects to be addressed in the knowledge agenda. A selection of research questions was formulated during two preparatory meetings and a large Round Table Conference that included experts from the various thematic study groups, as well as external experts and officials. The Baseline Studies have been prepared as a first step in stocktaking and the acquisition of knowledge. These

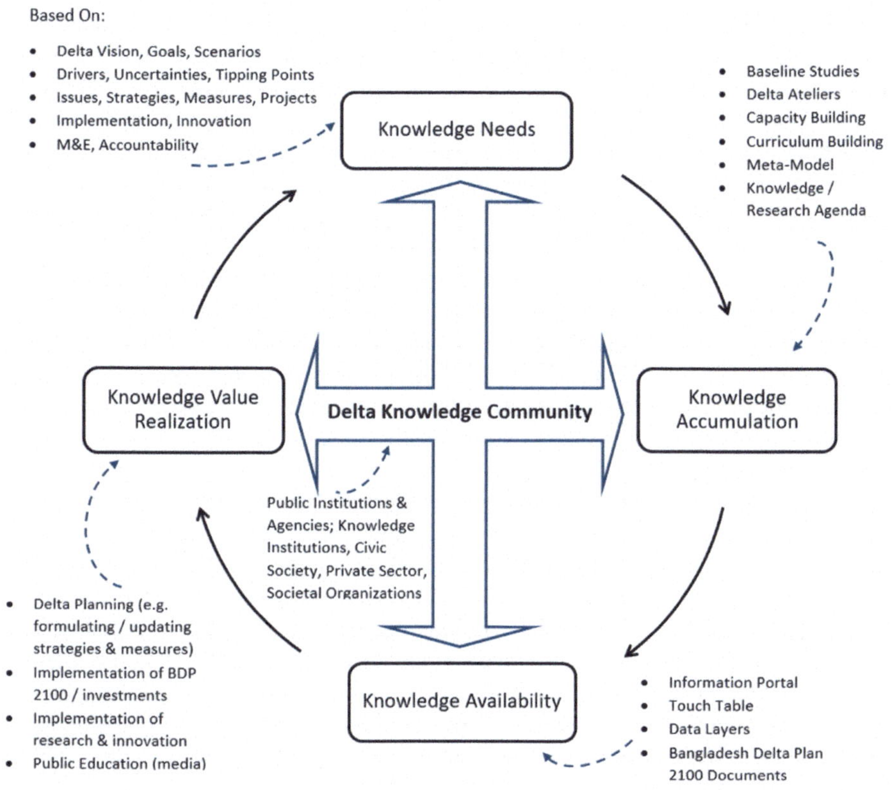

Fig. 12.1 Elements in BDP 2100 knowledge management approach. *Source* BanDuDeltAS, 2016

studies are comprehensive resources for learning and gathering information. In this regard, the broad Delta Atelier rounds were used to collect field knowledge. Since the themes-related knowledge and data must be updated, particularly on actual water systems and field-level data, this knowledge can be further developed in collaboration with the stakeholders. Developing partnerships with climate and delta knowledge centers is a promising avenue for this and for knowledge and data handling, e.g., with respect to modeling.

Through a web-based information portal, the Delta Plan disseminates the knowledge it has accumulated, including research proposals and results, draft policy documents, consultation reports, and geographical data layers. Beginning with the Baseline Studies, a great deal of information and scientific data was collected and produced during the BDP 2100 Formulation Project. The Climate Atlas and the creation of the Delta Vision, Mission and Goals, the scenarios, Delta governance framework, numerous measures and implementation programs are all based on data and information. For the purpose of creating and implementing the Information and Knowledge Portal, a specific Baseline Study of the subject has been prepared. Here, also information and knowledge management tools or services of government agencies and other stakeholders are described. These sectoral knowledge and information management systems, like the National Water Resources Database, foster the organization and archiving of information and support planning and decision-making at both the local and national levels.

Delta Knowledge can be used to create value, e.g., to inform actual planning decision making with the creation of the Climate Atlas, the Bangladesh Delta Plan 2100, the Investment Plan, or the Five Year Plans, or it can be used to inform related policies based on scientific findings and existing knowledge (e.g., SDGs). Additionally, feasibility studies will be carried out and projects and innovations will be implemented utilizing and thereby realizing the value of delta information. Additionally, people profit from the public's increased awareness of delta knowledge (for example, through the media) by making better judgments. When developing the content of Project Concept Notes, a lot of knowledge has been employed. The creation of Project Concept Notes and their subsequent implementation are covered in the Investment Plan.

For flexible and efficient knowledge development, coordination and application, an alliance between the public and private sectors can be established. Research on water, agriculture, the economy, the climate, and environment is already conducted by specialized, semi-autonomous institutes or government institutions in Bangladesh. It would be possible to establish or broaden a partnership with other public and government organizations that produce knowledge related to BDP 2100. This can be expanded with private knowledge institutions that specialize in feasibility studies, project preparation, management and organization, economics, and financial analysis, and also with businesses who handle various aspects of the delta plan's execution.

Adaptive management and delta knowledge have been introduced into an active delta knowledge community. To update stakeholders on developments and results and engage them in pertinent delta problems, a delta conference was organized in 2022 and would be held on a regular basis. The primary objective of establishing a delta community and an institutional framework for it is to foster innovation, increase ownership, and deeper involvement in the creation, accumulation, and storage of information as well as activities related to accessibility and application.

12.3.1 Development of a Decision Support System

A 'meta-modeling initiative' is initially a pilot activity of the Bangladesh Delta Plan 2100 project. The main purpose of the pilot is to demonstrate the potential of meta-modeling which is based on important underlying models, e.g., regarding water resources, agriculture, climate change, fisheries, to support the decision making process on delta planning and investments in Bangladesh. Decision-making for the Delta Plan includes key water aspects, food security aspects and socio-economic data. Integrated impact analysis of proposed measures is generated with the metamodel to support decision making. The pilot initiative is also aimed at supporting decision-making under conditions of uncertainty (linked to the scenarios) and developing an adaptive plan. The measures included in the BDP 2100 are a mix of pre-feasibility studies, concept notes and rationalization programs. The metamodel aims to supports the: (i) assessment of impacts of combinations of measures on a consistent set of policy indicators; (ii) assessment of the synergy, competition or overlaps across measures; and (iii) prioritization of measures from a list of multiple possible measures. The metamodel can be further developed with for adaptive management essential components or models as, e.g., sedimentation, water quality etc. The metamodel was further developed with support of EKN and used in the SIBDP basin wise adaptive programming.

References

BanDuDeltAS. (2016). Baseline Study on Governance – Knowledge Management.

Bhatt, G. D. (2001). Knowledge management in organizations: Examining the interaction between technologies, techniques, and people. *Journal of Knowledge Management, 5*, 68–75.

BDP 2100 Knowledge Portal. http://bdp2100kp.gov.bd

Davenport, T. H., & Prusak, L. (2000). *Working knowledge: How organizations manage what they know* (pp. 1–7). Harvard Business School Press.

Kautto-Koivula, K. (1999). *Can societies learn from global organization*? EGEDL99 Conference. http://www.uta.fi/EGEDL/outline/kauttokoivula.htm

Kautto-Koivula, K. (1998). The Pitfals of Knowledge. Information Strategy.

Kautto-Koivula, K. (1999). *Main operations for societal Knowledge Management.*

Ministry of Planning. (2018) *Bangladesh Delta Plan 2100.* https://plancomm.gov.bd/

Ministry of Planning, GED, BDP 2100 Formulation Project. (2018). *Baseline study on Governance.*

Ministry of Science and Information & Communication Technology. (2018). *National ICT policy.*

Nonaka, I., & Toyama, R. (2015). The knowledge-creating theory revisited: Knowledge creation as a synthesizing process. In: *The essentials of knowledge management* (pp. 95–110)

Wiig, K. (1995). *Knowledge management methods.*

Chapter 13
Realization of Measures and Interventions

Abstract The realization approach of the Delta Plan concerns ongoing realization of measures and interventions in interplay with crucial actors in decision making, funding and execution tasks. This chapter highlights the involved aspects in the realization process. To steer this, integration and mainstreaming of BDP 2100 is needed at high government levels to incorporate BDP 2100 measures and actions in sectoral programs and planning. GoB decided to assign these tasks to GED which has direct access to the highest levels of policy making of relevant Ministries and it plays a central role in negotiating and preparing the successive Five-Year Plans (FYP), that constitute an important vehicle to coordinate and integrate sectoral plans. Multiple ministries are involved in the BDP 2100 strategies, investment planning and realization. In addition, implementing agencies, local government institutions, knowledge institutions and other stakeholders are involved in planning and implementation in line with their mandates and tasks resources and interests, both in the private and public sectors. The Investment Plan with 80 selected programs and projects forms another important coordination mechanism. During SIBDP, regional and basin-wide programs have been prepared in line with BDP 2100 to facilitate its implementation and to identify new opportunities which led to an additional climate change adaptation investment plan. However, funding of the BDP 2100 Investment Plan and the additional one is a huge challenge in an environment of constrained fiscal space and economic and societal challenges. Coordination at the implementation level is important to effectively execute and realize the BDP 2100 and Investment Plan in the right manner. To contribute to coordination and execution, a network of Focal Points is active in implementing agencies and build the capacity of this network on project preparation, execution and M&E. As said, responsibility for coordination, facilitation as well as M&E of BDP 2100 implementation have been assigned to GED with a dedicated Delta Wing. The other main institutional reforms with impact on realization concern: (a) formation of a Delta Governance Council (DGC); (b) formation of a Projects/Programs Selection Committee (PPSC); (c) establishment of a Bangladesh Delta Fund; (d) establishment of Focal Points and at subsequent stages; (e) establishment of regional/local Water bodies.

J. M. de Heer and G. A. Choudhury, *Strategies for Adaptation to Climate Change in a Transformative Approach*, Springer Climate,
https://doi.org/10.1007/978-3-031-88884-7_13

Keywords Implementation · Implementation readiness · Implementation constraints · Readiness · Realization · Measures · Interventions · Execution · Executing agencies · Capacity · Capacity building · Administrative capacity · Connective and organization capacity · Knowledge and learning capacity · Funding capacity · Physical system adaptive capacity · Development results framework · Training · Basin-wise program · Additional investment program

13.1 BDP 2100 Implementation Strategy

13.1.1 Ongoing Process of Strategizing, Programming and Execution

Realization will be an ongoing process characterized by reflection on strategies and measures, by adaptive programming and adaptive design while the strategies, measures and interventions are again frequently reviewed, adjusted or intensified. In such ongoing adaptive process, there is no clear distinction between policy, planning and implementation as, e.g., described in classic public management literature.

As the BDP 2100 follows a holistic multi-sectoral approach, many ministries and agencies are involved in implementation, showing inter-linkages and interactions related to BDP 2100 goal achievement (Ministry of Planning, 2018). With respect to effective and efficient implementation and insight in goal achievement, SIBDP contributes to methodology and tools of result-based monitoring and evaluation which is needed to create frequent overviews and understanding of progress or delays in the BDP 2100 related performance of involved ministries and agencies. The BDP goals, strategies and targets constitute the Development Results Framework of BDP 2100 and form the focus of BDP 2100 M&E.

Closely related to this are knowledge management to support implementation. As BDP 2100 is continuously science and knowledge driven, an adequate and practical knowledge base is essential as said before, for effective and realistic planning and execution of the work. Knowledge drives the integration of sectoral priorities and stakeholder interests into workable programs, it is needed to create funding and highly supports coordination and execution of tasks. The Delta Plan utilizes Knowledge and Information Portal to enhance access to information for the main stakeholders.

Aspects of adaptive management as mentioned in Chapter 4 are also relevant here. The practical implications of adaptive management during implementation of BDP 2100 play a meaningful role in the following implementation and executive actions:

- Managing institutional developments and activities to enable further strategy operationalization and implementation, e.g., by streamlining procedures, shortening approval processes.

- Capacity Building and learning what adaptation means and requires regarding implementation, especially with respect to adaptive programming and adaptive design.
- Seeking adaptation options or pathways consisting of effective measures and interventions to realize effective adaptation with optimal and state of the art solutions.
- Implementing any enabling actions to ensure future long-term options remain open while lock inn situation are avoided.
- Managing knowledge and data according to implementation needs and innovation opportunities.
- Mainstreaming BDP 2100 input and frequent updates in the Five-Year Planning Cycle and or other national plans.
- Result Based monitoring and reflective evaluating of results or outcomes with respect to effective adaptation in light of vision, mission, goals and strategies.
- Review again of the main drivers of change (as climate change, transboundary issues, economic development), assessing uncertainties (gaps, unknowns) and looking for approaching tipping points in the developments.
- Updating or renewing the BDP 2100 and strategies frequently.

13.1.2 Specifications in Investment Plan and Execution Strategy

Besides the above mentioned aspects and along with developing the delta vision, mission and strategy, Bangladesh has also developed a BDP 2100 Investment Plan. As part of BDP 2100 Investment Plan development process, different delta-related line ministries and agencies were invited to provide delta project ideas and subsequently 500 project ideas were obtained from these ministries and government agencies. Subsequently, these 500 projects have been reviewed on relevance from BDP 2100 points of view and stakeholder consultations were arranged to obtain further insights. This led to 134 preferred proposals and subsequently, 80 projects have been included in the BDP 2100 Investment Plan based on the best fit of issues and challenges, delta vision, goals, future scenarios, and preferred strategies.

For analytical reasons explained earlier, and later for ensuring effective implementation of the Bangladesh Delta Plan, the entire country has been divided into six Hotspots (see Chapter 9) and each of these Hotspots include a separate set of the Investment Plan. Then, while the implementation of these 80 projects are important for Bangladesh delta, the obvious question is which project will be executed first and what could be a possible sequence of project implementation. Within each Hotspot program of the BDP investment plan, projects were prioritized and sequenced, so that investments are neither ‘too much, too soon’, nor ‘too little, too late’, but instead deliver ‘enough, in time’. The proposed projects are divided into three clusters: from Cluster 0+ (Improvement of current system), Cluster 1 (Change in current approach), and Cluster 2 (Change of the water system’s behavior). In Cluster 0+, projects which can enhance the effectiveness of already existing infrastructure or other structures are selected. For example, all projects that involve knowledge or institutional capacity building are Cluster 0+ projects. Other types require a change from the current approach, but do not change the configuration of the entire water system. Subsequently, projects that require a change in current approach are included in Cluster 1. For example, building new embankments to provide flood protection to new areas are Cluster 1 projects. On the other hand, projects that require change of the water system's current approach are included in cluster 2. These are often high-risk projects and require high levels of preparation and execution capacity. For example, construction of storm surge barriers, or abandoning tracts of cultivated or inhabited land etc. are Cluster 2 projects. However, depending on the climatic conditions and socio-economic situation, Bangladesh may need to implement these cluster 2 projects, even in the next decade.

Overall, the delta investments can be gradually implemented by the delta ministries and agencies observing established practices followed by the government. GED prepares the Delta Investment Plan in cooperation with the line ministries and sector divisions of the planning commission. The 7th and 8th Five-Year Plan already contain a large number of BDP 2100 investments which get priority in the approval process by the Planning Commission. For ensuring clearly identified responsibilities for delta plan implementation, Bangladesh Delta Plan also developed a mapping document. This provides a guideline to the ministries/divisions/agencies describing their respective sectoral involvement in implementing their investment programs. For facilitating this implementation, a committee named as Delta Plan Projects/ Programs Selection Committee (PPSC) has already been developed. The PPSC is primarily responsible for selection of specific projects and program and will facilitate critical scrutiny of only delta-related program/projects in consultation with the Planning Commission and relevant Ministries and agencies. However, each of the selected projects needs to go through the full range of cost–benefit analysis and project feasibility analysis that is a part of the government's project approval process. In addition, the delta plan's Investment Plan is a dynamic one, it needs updating as time passes some projects may be redundant while some new priority projects can get entry into it.

13.1.3 Towards a Basin-Wise Approach

For the implementation of the BDP 2100, a basin approach is followed considering the Bangladesh Water Act of 2013 and following principles of Integrated Water Resource Management (IWRM) and BDP 2100. Adopting a basin approach entails that the analysis of key issues and opportunities as well as implementation, consider the interaction and integration of delta and water management inside the basin, and between the basin and adjacent rivers, sea or estuary. By adopting the basin approach, Bangladesh wants to overcome some of the challenges to implementation discussed in the BDP 2100; summarized here as: (1) weak capacity of water and water-related institutions; (2) absence of water management organizations throughout the country; (3) lack of integration of water issues.

The basin-wide water resources planning is indeed a legal obligation under the National Water Policy. This requirement mandates that water management must be conducted at the basin level, reflecting the natural flow and interconnections of water systems rather than conforming to administrative boundaries alone. This legal obligation ensures a comprehensive approach, where planning considers the full scope of each basin's hydrological, environmental, and socio-economic characteristics. This basin-centric framework helps manage water resources in an integrated manner, addressing issues like water availability, flood control, ecosystem health, and land use in a cohesive plan. By fulfilling this legal obligation, initiatives such as BDP 2100 can prioritize projects that align with both national strategies and specific regional needs, leading to sustainable and effective water management across Bangladesh's delta regions.

The National Water Policy (1999) has highlighted the need for river basin management and cited that "basin planning provides the most rational basis of development of water resources under the influence of one or more major rivers." At the same time, it has been recognized that the international river basins, such as the Ganges basin, the Brahmaputra basin, and the Meghna basin, present unique problems. Due to its location as the lower-most riparian, Bangladesh has no control over the rivers entering through its borders. The adverse effects of this are floods and water scarcity, which occur frequently. Although the 1996 Treaty on Sharing of the Ganges Waters with India has brought some relief to the drought-prone area of the southwest, the water shortage problem during the dry season is likely to aggravate in the Ganges and other basins with rising demands of the increasing population. The policy also recognized that working out of joint plans for different river basins with other co-riparian countries would take considerable effort and time. Therefore, as a long-term measure, it recommends to "undertake essential steps for realizing basin-wise planning for the development of the resources of the rivers entering its borders."

As an initial step in basin-wide water resources planning, the Water Resources Planning Organization (WARPO) created detailed maps marking the catchment areas, both within and beyond Bangladesh's borders, for all rivers flowing through the country. This mapping effort was undertaken as part of the formulation of the National Water Management Plan (NWMP). By delineating hydrological boundaries that align with Bangladesh's unique hydro-morphological characteristics, WARPO established a foundation for effective basin-wide planning. These delineated boundaries support integrated planning and ensure that water-related initiatives are consistent with the natural flow patterns and characteristics of each watershed.

The Bangladesh Water Act (BWA) of 2013 was enacted to support comprehensive and integrated management of the country's water resources, covering aspects such as development, extraction, distribution, use, protection, and conservation. Under this Act, all water within Bangladesh's territory is vested in the State on behalf of the people, superseding any conflicting laws. This centralization of water rights emphasizes the State's role in ensuring sustainable and equitable access to water resources. The BWA formally recognizes the National Water Policy, which remains in effect provided it aligns with the Act's provisions. Specifically, the Act mandates basin-wise development planning (Section 15(2)(10)) for water resources, reinforcing the necessity of managing water resources by natural river basin boundaries. This directive acknowledges river basins as practical and effective units for water management, as they inherently account for upstream–downstream hydrological processes and interactions. By utilizing basin boundaries, this approach allows for a coordinated, holistic strategy that reflects the interconnected nature of water systems across regions, fostering sustainable development and management of Bangladesh's water resources.

However, experience in practice of the BDP 2100 learns an important point: institutions and staff as well as the broader public seem not yet fully aware or convinced of climate change and possible serious impacts and are not ready to work on them while usual daily worries are overwhelming and require full attention. The Readiness to Implement needs to be improved, which given the discussed aspects, boils down for government institutions to developing and applying five required capacities to deliver effective implementation.

13.2 Building Capacities for Implementation and Adaptation

The outcomes of the approval process are key to the implementation of the Plan, with respect to the strategies and investment plan, the institutional setting and financial ambitions. The BDP 2100 Executive Summary indicates (pg. vii; draft BDP 2100 April 2018) that Bangladesh has 60 years of experience in managing water resources. "Despite this rich history, effectiveness of implementation of integrated water management is weak." The reasons are (pg. vii): (1) weak capacity of water and water-related institutions; (2) absence of water management organizations throughout the country; (3) lack of integration of water issues with climate change, environment, land management and other delta-related challenges owing to inadequate institutional coordination. The monitoring and evaluation of policy implementation is also weak. In relation to this and with respect to substantial new investments and adequate financing arrangements, "a substantial overhaul of the Delta governance and the institutional arrangements is needed to implement the Delta Plan." (Executive Summary, pg. vii). The 'Summary of Delta Governance and Institutional Reforms' provides the series of proposed reforms and required actions. In our approach, we follow this overview (Executive Summary, pg. viii–xi). Chapters 11 and 12 of the Delta Plan elaborate on investments, implementation and these mentioned governance and institutional matters.

In connection with the above analysis and approach and based on experiences with strategy making and implementation, SIBDP 2100 has been established to deliver support to implementation from 2018 till December 2024. The organization has been presented in the next Fig. 13.1.

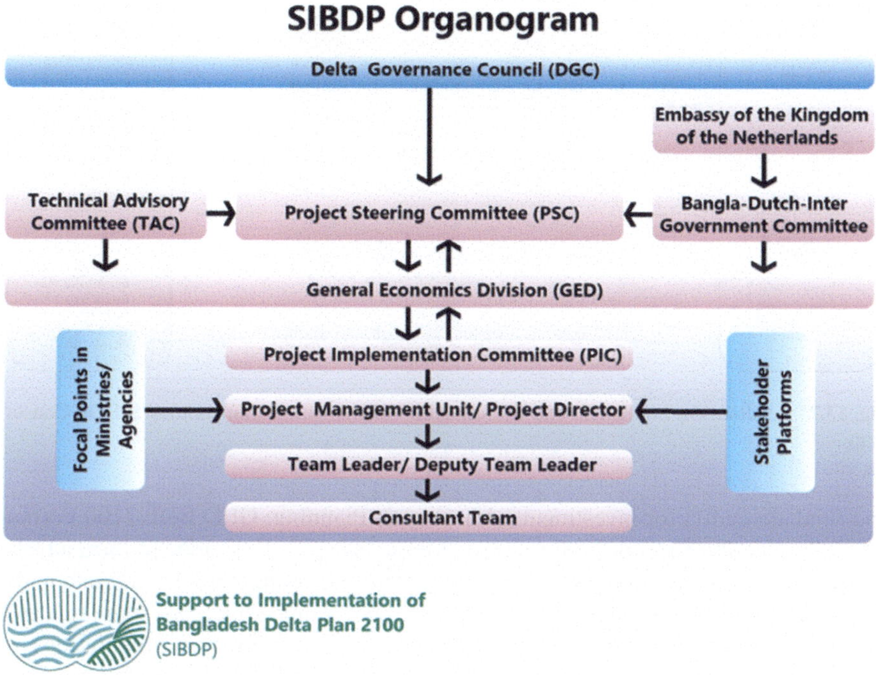

Fig. 13.1 SIBDP 2100 project organization and consortium brand

SIBDP analyzed and focused in its support to BDP 2100 implementation on developing five capacities for effective BDP 2100 implementation which have to be addressed in order to achieve expected results.[1] These five capacities are shown in Fig. 13.2 and are explained below.

13.2.1 Administrative Capacity

Implementation of a huge and significant plan like BDP 2100 requires an adequate institutional framework to make further operationalization and execution legal and possible. Integration of BDP 2100 into the legal framework, governance rules, regulations and national planning cycle mainstreams in fact BDP 2100 within policies and plans like the National Five-Year Plans, the (draft) Perspective Plan 2041 and the yearly budget allocation process. Furthermore, a legal and strong institutional basis is required for coordination of the multi-sectoral tasks, for monitoring and evaluation and for cyclical updating and mainstreaming as well as for decision making

[1] See e.g. on Readiness to implement the Center for Effective Services (2022): https://thecenterforimplementation.com/toolbox/tag/Readiness+and+context.

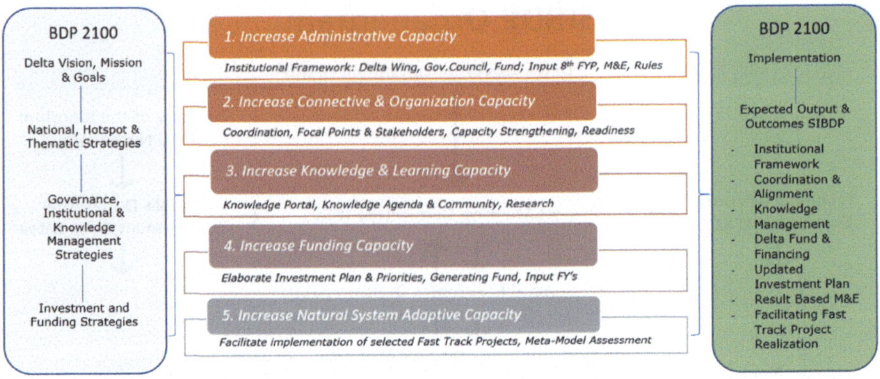

Fig. 13.2 Readiness to implement and five capacities of implementation. *Source* SIBDP Consortium

and fund allocation for investments (Ministry of Planning, GED BDP 2100 Formulation Project, 2018a). For performance of these tasks, GoB decided to form a Delta Governance Council, a Projects/Program Selection Committee, a Delta Fund and a GED—Delta Wing. Besides, the network of Focal Points which are supported in a number of cases by a dedicated Delta Cell supports the coordination, preparations for execution of projects and monitoring. By performing and completing this overhaul, the administrative capacity for BDP 2100 implementation and follow up will increase considerably. The transformation with respect to Water Institutions and legislature of the Delta Act needs more time to further elaborate, for approval and introduction.

13.2.2 Connective and Organization Capacity

Prerequisites to implement a multi-sectoral plan are interaction and coordination as well as 'readiness' of involved implementing agencies, knowledge institutions and main stakeholders. Readiness means being prepared to implement: (1) knowing and being trained on the BDP 2100 content with its methodology and strategies, (2) bringing involved agencies in good shape (e.g. mandates, rules, structure and tasks, processes, human and financial resources, other capacities), (3) contributing in preparation and execution of projects. BDP 2100 implementation requires strong emphasis on coordination and interaction, creating an implementation network with pro-active Focal Points and effective procedures for micro planning, preparation, decision-making, funding and project execution. These elements and training will be part of capacity strengthening indicated in the project document.

13.2.3 Knowledge and Learning Capacity

Implementation can only take place when necessary data, information, maps and knowledge of enough quality are available. Knowledge management, research and learning from outcomes as well as innovation are therefore of crucial importance in implementing an adaptive delta plan. Data, information and understanding are needed to deal with the drivers of change and development in a pragmatic way, supporting and managing interventions in the natural system as well as in institutional, organization and management and financial matters. This will build on the already developed Knowledge Agenda (with a possible research program with knowledge institutions), the created Knowledge and Information Portal and data layers. Additionally, a periodic review of BDP 2100 implementation will offer systematic insight in progress of implementation, natural system behavior and goal achievement.

13.2.4 Funding Capacity

Proper investment planning, funding and financial mechanisms are a must to any implementation. Especially for the implementation, BDP 2100 committed a provision of 2.5% of the GDP annually. Nevertheless, additional financial resources still need to be allocated by the GoB, development partners, IFI's with a linkage to, e.g., the Green Climate Fund and Public–Private Partnerships. Updating the Investment Plan and building trust and relationships with development partners and IFI's to this are important elements. Especially funding of preparation of selected projects is of crucial importance as (pre) feasibility studies take considerable time and budget. The transformation from usual traditional design to adaptive design is also of great importance here. Experience show that proposals for GCF and PPP go through really time consuming processes. An important SIBDP advice is to shorten project preparation and reduce preparation costs for Rationalization Programs. As said, rationalization means that already existing embankments, other infrastructure and buildings go through a process of serious maintenance, where needed renewal and also climate change adaptation. Because of the fact (1) Rationalization Programs are no regret programs and (2) that concerned infrastructures and buildings already exist, it is logical that the approval process for rationalization can be shortened. The relevant approval procedure (DPP) can be followed but streamlined leading to reduction of preparation time and costs.

13.2.5 Natural System Adaptive Capacity

BDP 2100 implementation builds on more detailed planning and adaptive programming exercises. So, investment priorities, e.g., per Hotspot project with already a feasibility study prepared (see Investment Plan, belonging to the 0 + category), which after final approval and financing can be executed. Implementation requires a well-considered approach, a systematic sequence and involvement of stakeholders. To facilitate this, Hotspot strategies have been elaborated into basin wise river basin programs, as presented in the next paragraph, developed in consultation with stakeholders. The physical system requires upgrading in terms of safety standards and risk reduction, see, e.g., Fig. 13.2. This means that many no regret measures can be taken which need to be done anyway, many interventions will be effective in terms of contributions to achieving the delta goals for quite some time. However, given the ongoing climate change and transboundary activities, effectiveness of measures and possible needs for transformational change of the physical system still are important criteria for monitoring and reflective evaluation.

13.3 Preparing Regional Implementation Programs

BDP 2100 has been elaborated during SIBDP into regional programs, anticipating a Basin Wide approach. Six Basin Management Plans have been prepared for the six Hotspots. These are: (i) North-West Basin Management Plan for Barind and Drought-Prone Areas Hotspot, (ii) Extended Coastal Zone Basin Management Plan for Coastal Zone Hotspot, (iii) North-East Basin Management Plan for the Haor and Flash Flood Areas Hotspot, (iv) Urban Areas Implementation Program, (v) Chattogram Hill Tracts Implementation Program and (vi) Major River Systems and Estuaries Implementation Program.

The Extended Coastal Zone program is designed to extend the existing coastal zone management efforts under the Bangladesh Delta Plan, see, e.g., Fig. 13.3. This program aims at restoring and conserving coastal ecosystems to enhance resilience and protect biodiversity. For achieving this, the development process of this program ensured the participation of local communities in planning and decision-making for sustainable development.

For the ECZ basin, the following strategies and programs have been developed, to contribute in the BDP delta goals and BDP 2100national and thematic strategies, see Fig. 13.4.

The Chattogram Hill Tracts is an ethnically diverse region of Bangladesh and it's river systems plays an important role in the biodiversity of Bangladesh. Sustainable development planning is crucial to support the economic activities of different indigenous communities of this region while preserving the environment. Recognizing this issue, the following strategies and seven distinct programs have been developed for Chattogram Hill Tracts to contribute to the BDP delta goals and BDP 2100 thematic strategies, see Fig. 13.5.

North-East Haors are characterized by their seasonal flooding, which occurs during the monsoon season when rivers overflow and fill the depressions, creating vast wetlands. During the dry season, these areas become fertile agricultural lands. The fertile soils of the Haors basin support rice cultivation and other agricultural

Fig. 13.3 Coastal protection works. *Photo* J.M. de Heer

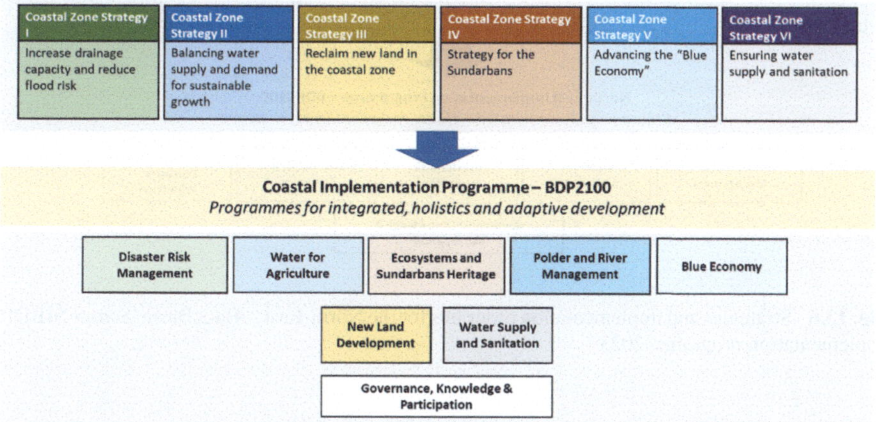

Fig. 13.4 The suggested implementation programs for the extended coastal zone. *Source* SIBDP, implementation programs, 2023

activities during the dry season, contributing to the livelihoods of local communities. Haors act as natural water reservoirs during the monsoon, helping to regulate water flows and prevent downstream flooding. The flooded Haor areas provide breeding and feeding grounds for various fish species, supporting the local fishing industry. Considering the significant importance of this area, the BDP 2100 developed the following strategies and six distinct programs, see Fig. 13.6.

The North-West Barind region is a semi-arid drought-prone zone of Bangladesh. Due to limited water availability and frequent droughts, the region faces challenges

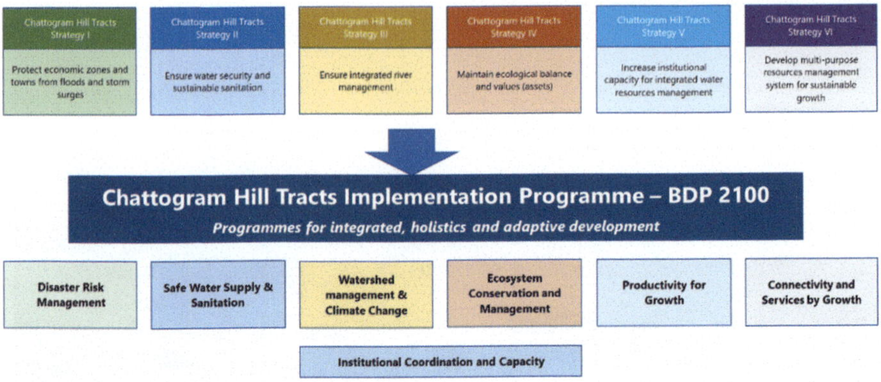

Fig. 13.5 The suggested implementation programs for the Chattogram Hill Tracts. *Source* SIBDP, implementation programs, 2023

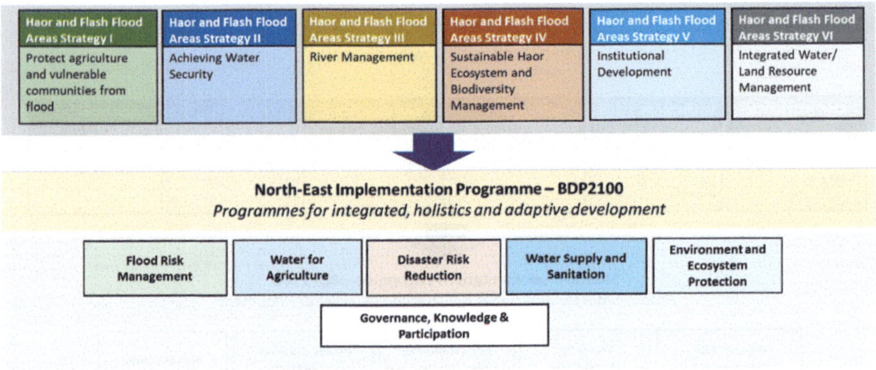

Fig. 13.6 Strategies and implementation programs for the North-East—Haor Basin. *Source* SIBDP, implementation programs, 2023

related to agriculture, water scarcity, and livelihoods. Effective water management is essential to address the water scarcity issues in the region. This includes both surface water and groundwater management, as well as the construction of water storage and distribution infrastructure. For addressing these issues, four distinct programs have been developed for the North-West Barind basin, see Fig. 13.7.

Bangladesh is crisscrossed by a network of major rivers and their tributaries, including the Padma, Meghna, Jamuna, Brahmaputra, and more. These rivers and their associated estuaries play a crucial role in the country's geography, environment, and economy. The rivers serve as a primary source of freshwater for domestic, agricultural, industrial, and other uses. These rivers and their waterways are the main modes of transportation in many areas of Bangladesh. They enable the movement of goods and people, particularly in regions with limited road infrastructure. Estuaries, where freshwater meets the sea, are rich in biodiversity and support various

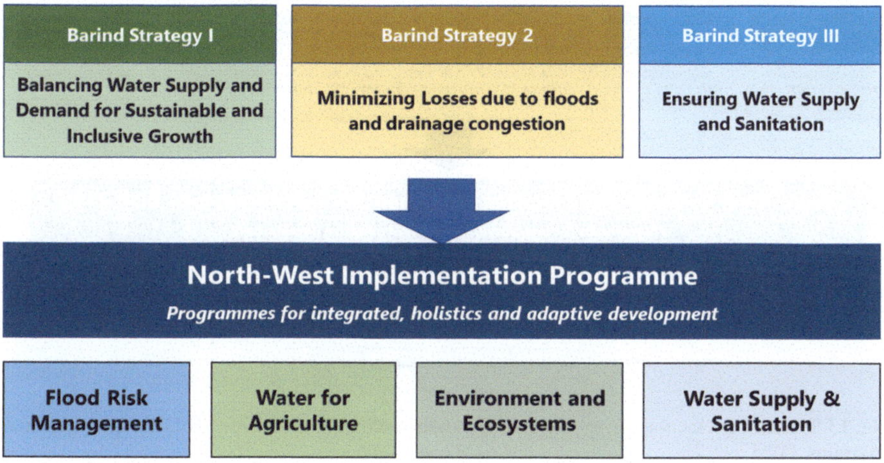

Fig. 13.7 Strategies and implementation programs for the North-West Barind and drought-prone areas. *Source* SIBDP, implementation programs, 2023

ecosystems including mangroves and wetlands. Embankments need to be improved in many areas to provide protection against flooding and erosion, see Fig. 13.8.

Recognizing the importance of these major rivers and their tributaries, seven programs have been designed for the Major River Systems and Estuaries to contribute

Fig. 13.8 Various solutions for erosion protection. *Photo* J. M. de Heer

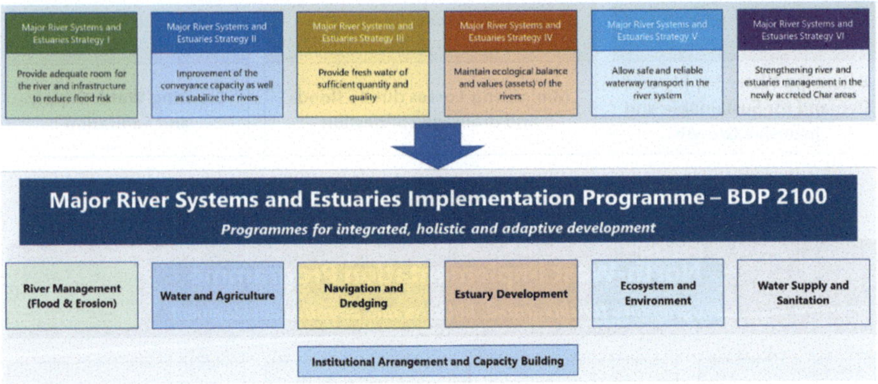

Fig. 13.9 The implementation programs for the main river systems. *Source* SIBDP, implementation programs, 2023

to the six BDP Delta Goals and the BDP 2100 national and hotspot plans, see Fig. 13.9.

Urban areas in Bangladesh have been growing rapidly due to factors such as population growth, rural-to-urban migration, and economic development. This rapid urbanization also presents various challenges, including infrastructure deficits, environmental degradation, inadequate services, and social inequalities. Given the challenges and opportunities posed by urbanization, an "Urban Areas Program" within the Bangladesh Delta Plan and 8 distinct programs have been developed. These programs intend to make urban areas more resilient to the impacts of climate change, including sea-level rise, increased flooding, and extreme weather events. In addition, these programs aims to play a key role in ensuring proper drainage and wastewater management systems to mitigate flooding and waterlogging in urban areas, see Fig. 13.10.

By focusing on basins, the Bangladesh Delta Plan ensured effective resource allocation based on the specific needs and priorities of each basin. This led to the development of interventions that are tailored to the local conditions and challenges. Much more can be said about the Programs with the basin characteristics and development of strategies, elaborated in measures and adaptation pathways as well as investment proposals which led to an additional Investment Plan.[2] This basin-wise approach also matches the principles of the Global Water Partnership on Integrated Water Resource Management (IWRM),[3] a principle adopted by Bangladesh in the 2013 Water Act. This coherence ensures that the previous planning process and Bangladesh Delta Plan reinforce each other.

[2] See further De Heer, et al. (2023a; 2023b; 2023c; 2023d; 2023e; 2023f).

[3] Global Water Partnership on IWRM and Tools (2023) etc. https://www.gwp.org/

Fig. 13.10 Urban implementation program. *Source* SIBDP, implementation programs, 2023

13.3.1 Additional Investment Program

In addition to the BDP 2100 Investment Plan with the 80 selected projects, this Basin-Wise Investment Program presents the outcomes of further regional elaboration of the BDP 2100. The National, Hotspot and Thematic strategies have been elaborated in a Basin-Wise Implementation Program approach. The initial Hotspots are still relevant in terms of analysis of issues and challenges. However, for planning purposes a clear planning unit needs to be the basis for regional and local planning. A river basin clearly shows its own characteristics physically, in natural resources but also in terms of involved people and businesses. Each delta plan intervention should for planning and execution reasons be placed into its context for which river basins have been chosen. The identification of the basins has been explained; it is argued that they logically follow out of earlier studies and approved government documents.

BDP 2100 Investment Plan is a dynamic document and as already mentioned would be updated after every five-year (Ministry of Planning, GED BDP 2100 Formulation Project, 2018b). About 5 years have already been passed since adoption of this plan and initiative has been made to update the current BDP 2100 IP. The additional Investment Plan report is a part of this initiative of updating BDP 2100 IP. The main objective of the additional IP report is to provide an outline towards updating of the BDP 2100 Investment Plan. Specific Objectives are:

- Recognizing the challenges and prospects within the Bangladesh delta region.
- Proposing comprehensive basin-wide programs in response to these challenges.
- Establishing connections between these programs and the BDP's vision, mission, goals, strategies, and delta scenarios.
- Suggesting potential financing options for implementing the investment plan.
- Offering recommendation on the implementation arrangements.

- Defining a results-oriented Monitoring and Evaluation (M&E) system, encompassing information and knowledge management considerations.
- Identify and define linkages with the other relevant policy and plans.

The Basin-Wise Investment Program report has been organized into nine chapters, offering besides the basin characteristics an introduction on developments and an assessment of the current implementation status. Selection and prioritization of projects was needed because during the making of the six Basin-Wise Implementation Programs many new project ideas and potential new projects were generated. For pragmatic reasons, the number of projects has been reduced in multi criteria based expert sessions into a shorter list of realistic projects which were during detailed expert discussions reduced to the presented 68 preferred projects. The selection methodology incorporates the Delta Vision related goals and indicators also with respect to climate change. The Basin-Wise Investment Program needs further discussion with involved ministries and agencies as well as consultation, e.g., on prioritization of Delta projects and finally approval.

Attention for worsening climatic conditions is essential with respect to final prioritization and funding decisions. An important consideration for the current BDP 2100 Investment Plan and this Basin-Wise Investment Program is, that climatic conditions seem to get worse and that also the financial-economic situation of the country is not as expected during the formulation of BDP 2100. The temperature is approaching + 1.5 °C. Impacts of the rising temperature have been seen worldwide in extreme weather conditions, with, e.g., heavy rains, flooding, cyclones, droughts, fires, warm ocean water. Besides economic conditions are less positive due to global disturbances, inflation, interest rates and less economic growth. This combined means, that a shift in the presented scenarios could be ongoing from scenario 'Productive' towards scenario 'Active'. So, as urgency of climate related action increases, selection and prioritization of projects needs to be considered in that way. Less lead time seems to be available for adequate adaptation. The climatic but also food production and socio-economic impacts need to be viewed more prominently.

Given the increasing climatic, safety and food security based urgency to implement delta plan projects, the GoB implementing organizations could better make time and resources available to prepare projects according to the requirements. Within the formal framework, seeking a 'good enough' approach of project preparation would be sufficient as many projects are already existing in some form and state and do not involve totally new development. This means looking for optimization of the preparations with the normal approval procedure. The advantage is that more project preparation can be achieved with less man power and financial costs.

13.4 Dedicated Basin-Wise Elaboration

After being approved by the Government of Bangladesh in September 2018, the implementation of the BDP 2100 is structured and elaborated in distinct implementation programs at the National and Basin levels. While previous water sectoral plans focused only on national-level issues, the Bangladesh Delta Plan focused on both national and basin-levels. This focus on basin-level issues is relevant as the Bangladesh delta is formed by the confluence of multiple major rivers Ganges, Brahmaputra, and Meghna and each area in the country belongs to a river basin. The basins of the major rivers are of course very large, covering multiple countries; here the focus is on the Bangladeshi parts of basins. This approach is interesting for the GoB as well as for stakeholders, as a basin provides a logical planning entity and practical scale. Besides, interventions need to be seen in their direct context with characteristics of the basin and adjoining rivers, areas and cities or villages. Interventions alter a basin, which has socio-economic and physical but also ecological consequences, see Fig. 13.11.

Moreover, a BDP 2100 intervention relates to other interventions or development activities in the basin. A coherent approach in good interaction with stakeholders and the ecological features of the basin is important for sustainable adaptation.

As per directive of the Bangladesh Water Act, the BDP 2100 addresses the limitations posed by the transboundary nature of major river basins by implementing a basin planning concept within the country's territorial boundaries, as per the National Water Policy. Recognizing that Bangladesh lacks jurisdiction over the full extent of its main river basins, BDP 2100 focuses on managing water resources at a domestic

Fig. 13.11 People losing their land due to dramatic river bank erosion along the Jamuna. *Photo* courtesy of Amirul Hossain

basin level. This approach facilitates a comprehensive, integrated framework that considers water-related challenges alongside other critical sectors.

By establishing basin-level "hotspots," BDP 2100 links national and thematic strategies with regional contexts, effectively translating high-level goals into region-specific action plans. This adaptive framework enables more targeted responses to local challenges, making national strategies actionable and relevant to each basin's unique needs. However, BDP 2100 acknowledges that significant governance and institutional gaps hinder the integration of water management with other delta-related issues and the achievement of broader development goals. Key gaps include:

- Integration Challenges: Water management is often inadequately coordinated with climate resilience, environmental preservation, and land management efforts.
- Governance Weaknesses: Many institutions responsible for water resources lack the capacity for effective and sustainable management.
- Insufficient Regional Presence: The absence of dedicated regional and local water and environmental agencies limits effective implementation at these levels.
- Stakeholder Exclusion: Key stakeholders, including community beneficiaries, are frequently excluded from decision-making processes.
- Overlapping Roles: Redundant roles and responsibilities across institutions can lead to conflicts and inefficiencies.

Addressing these governance and institutional challenges is crucial for BDP 2100's success. By focusing on the regional basin level, the plan aims to prioritize projects more effectively, aligning initiatives with each basin's specific objectives and thereby enhancing overall water management across the Bangladesh delta.

BDP 2100 has organized the country's water resources into distinct hydrological regions, each considered as a unique river basin for targeted planning and management. BDP2100 hotspots focus most dominant issue in a hydrological vis-à-vis a basin. These hydrological regions, adopted as river basins, enable focused interventions tailored to each area's geographic and hydrological characteristics.

The steps of the basin approach in relation to the BDP 2100 are visualized in Fig. 13.12. Each basin program consist of a Characterization Report and a Basin Implementation Program including existing and new investment ideas and proposals.

The Basin-Wise Program approach recognizes the interconnectedness of the rivers and aims to address the challenges of water management and development in an integrated and coordinated manner. It also considers the interaction and integration of delta and water management inside and between the basin and adjacent rivers, seas, or estuaries. The entire country has been divided into six basins, a map of these basins is shown in Fig. 13.13.

At the start of this basin-wise approach, the Bangladesh Delta Plan arranged separate stakeholder workshops for each of the basins. During the workshops, basin-level stakeholders shared the different water management challenges in their basin and suggested possible solutions and programs for addressing these challenges. These workshops also provided a platform for discussing the conflicts of interest among different water users and contributed to resolving conflicts among different water users through dialogue and negotiation. This ensured that decisions were equitable

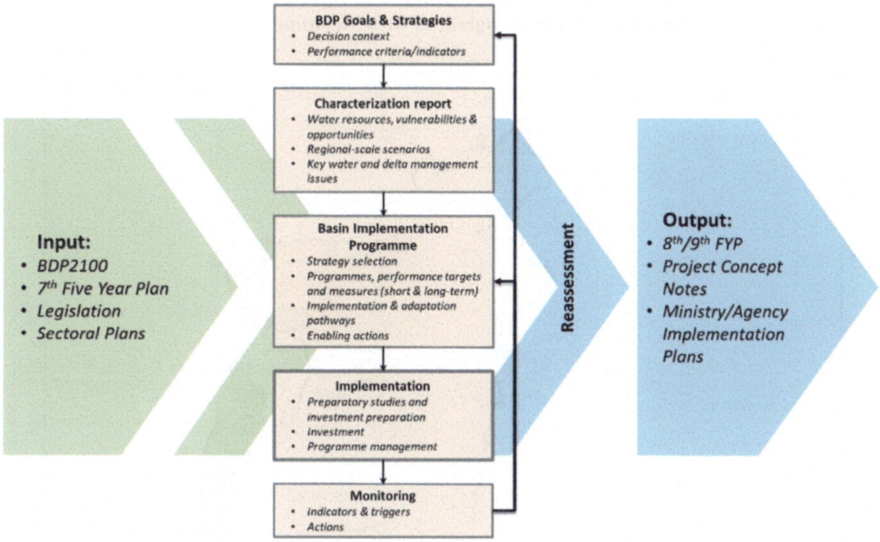

Fig. 13.12 Steps in the basin-wise programming approach. *Source* Basin-Wise Programs, SIBDP Consortium, 2023

and balanced. Subsequently, programs and projects have been developed to address the basin-level challenges.

The hydrological regions adapted to basin are: (i) north-west hydrological region as the Hurasagor Basin; (ii) north-central hydrological region as the Dhaleswari basin; (iii) north-east hydrological region as the Upper Meghna Basin; (iv) south-west hydrological region as the Gorai-Passur Basin; (v) south-central hydrological region as the Baleswar-Tentulia Basin; (vi) south-east hydrological region as the Gumti-Muhuri Basin; (vii) eastern hilly hydrological region as the Hilly Rivers Basin; (viii) major rivers and estuary hydraulic region as Major Rivers and Estuary Basin. Because of the coastal issues are more or less similar, the Gorai-Passur Basin; the Baleswar-Tentulia Basin; the Gumti-Muhuri Basin and the coastal plains of the Hilly river basin have been grouped under the Extended Coastal Zone (See Fig. 13.13).

By structuring the country's river systems into these hydrological regions, BDP 2100 facilitates basin-specific strategies that address unique challenges and opportunities, from upstream–downstream interactions to coastal and estuarine management. This approach supports more targeted and effective water resource management across Bangladesh's deltaic landscape.

BDP2100 Investment Plan comprises 80 projects in total, with 65 physical projects focusing on infrastructure development and 15 projects dedicated to institutional and knowledge development. Together, these projects represent a capital investment of BDT 2978 billion (approximately $38 billion). The Investment Plan aims to build both physical resilience—such as embankments, water infrastructure, and urban drainage systems—and institutional capacity for managing Bangladesh's unique

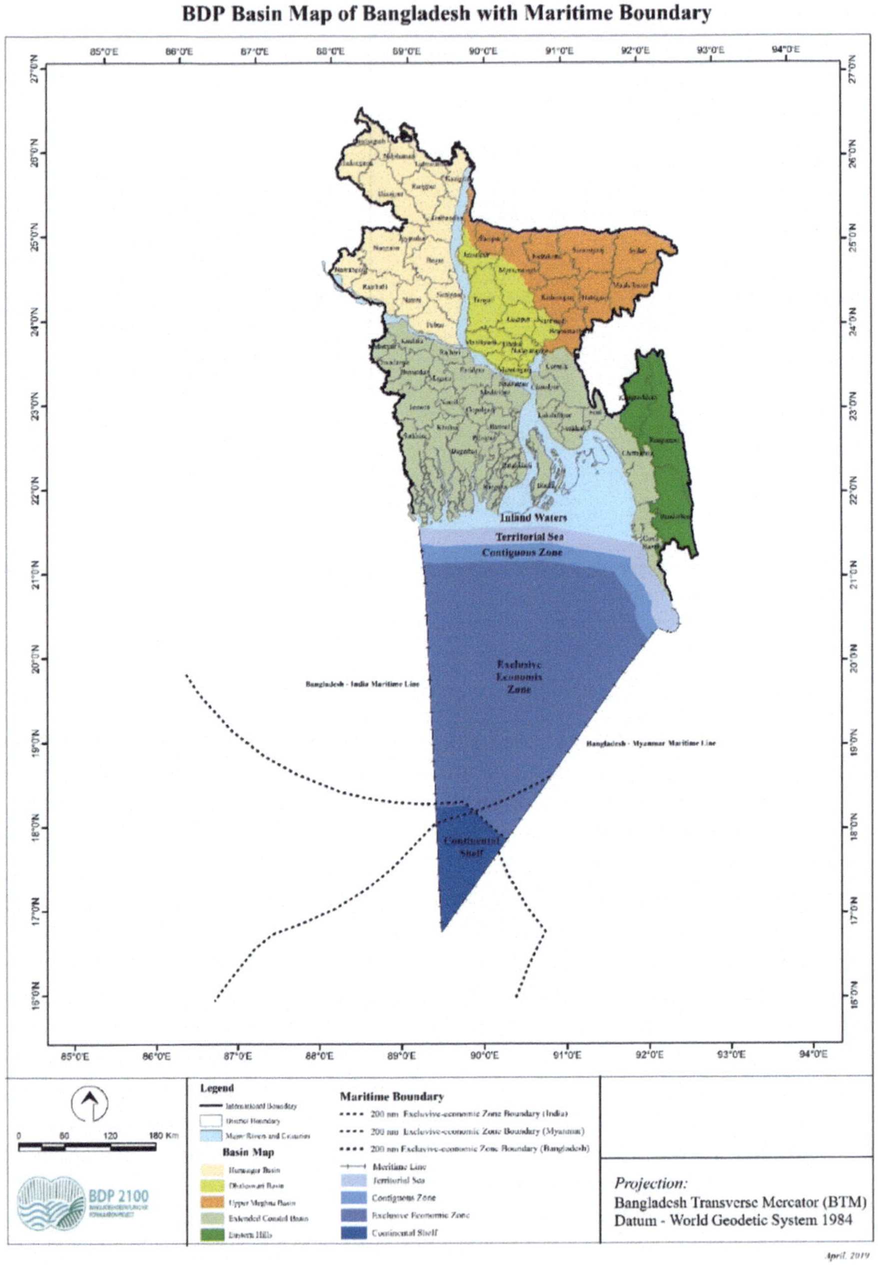

Fig. 13.13 Basin management map of Bangladesh. *Source* SIBDP, Basin-wise implementation programs, 2023

deltaic challenges in the long term. BDP 2100 has organized its 80 projects basin-wise in conformity to seven strategic "hotspots" as well as hydrological regions. Some of the notable programs/projects are described below.

- The major water management programs/projects are: North Rajshahi Irrigation Projects, Kurigram Irrigation Projects, the waters of the Brahmaputra River etc.
- The major flood control and drainage programs/projects are: Revitalization of Khals all over the country Revitalization and Restoration of Hurasagor and Atrai Rivers etc.
- The notable river management programs/projects are: Integrated Jamuna-Padma Rivers Stabilization and Land Reclamation Project, River Bank Improvement Project, Sustainable Restoration of Connectivity of Major Navigation Routes etc.
- The important urban development projects include: Dhaka Integrated Flood Control Embankment cum Eastern Bypass Road, Improvement of Drainage Congestion and Flood Control for Chattogram City Corporation Area, Improvement of Drainage Network, Flood Control and Solid Waste Management for Khulna City, Improvement Storm Water Drainage in the City Corporation Areas, Improvement of Urban Drainage in Districts and Upazilas of Bangladesh, Water Supply Project for City Corporations Area etc.
- The important Extended Coastal Zone programs/projects are: Padma (Ganges) Project, West Gopalganj Integrated Water Management Project, Improved Drainage in the Bhabadha Area, Char Development and Settlement Project, Development of Water Management Infrastructure in Bhola Island; Integrated Land Reclamation Project in the Meghna Estuary; Southern Agriculture Improvement Project etc.

13.4.1 Importance of Rationalization Programs

Over the past 60 years, BWDB has implemented an impressive array of water management projects. Through the implementation of 910 projects, BWDB has developed significant water infrastructure and assets, which have been critical in managing FCD, FCDI, and riverbank protection. These projects have played a crucial role in protecting agricultural lands from flooding, ensuring irrigation supply for crops, and stabilizing riverbanks to reduce erosion and protect communities from river-related hazards. These investments have helped establish a strong foundation for water resources management, significantly contributing to the country's agricultural development, flood control, and overall water security. Effective operation and maintenance (O&M) of existing water infrastructure is essential for ensuring the proper functioning and sustainability of vast water management systems. Proper O&M ensures that these infrastructures continue to deliver critical services such as flood protection, irrigation, and drainage, which are vital for agriculture and resilience to climate hazards. However, maintaining these systems presents significant challenges for BWDB.

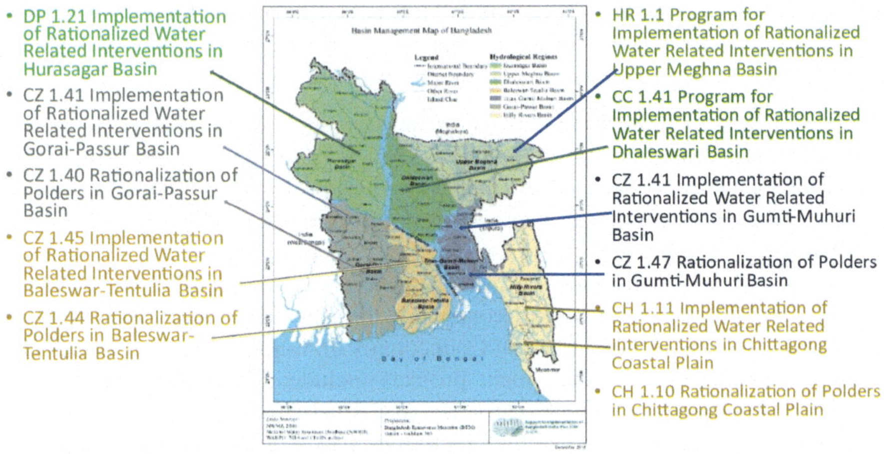

Fig. 13.14 Overview of rationalization programs

BDP 2100 Investment Plan incorporates the rationalization programs of existing water resources management infrastructure, emphasizing the need for updates and improvements to ensure effective and efficient management of water resources as well as to meet the growing challenges of water management, particularly under changing climate conditions. A key aspect of the plan focuses on rationalizing, large maintenance and or re-designing, or updating existing water management structures. Climate change factors such as more erratic rainfall, increased storm surges, rising sea levels, and the intensification of cyclones need to be incorporated into the design parameters. The rationalization of infrastructure will enhance drainage capacity, improve flood control, and increase water retention to ensure that infrastructure is resilient ensure and adaptable to future climate challenges.

BDP 2100 IP includes 11 rationalization programs spreading over the whole of the country. These rationalization programs are (See Fig. 13.14):

- Rationalization of Water-Related Interventions in Hurasagar Basin.
- Rationalization of Water-Related Interventions in Dhaleswari Basin.
- Rationalization of Water-Related Interventions in Upper Meghna Basin.
- Rationalization of Water-Related Interventions in Gorai–Passur Basin.
- Rationalization of Polders in Gorai–Passur Basin.
- Rationalization Water-Related Interventions in Baleswar–Tentulia Basin.
- Rationalization of Polders in Baleswar–Tentulia Basin.
- Rationalization of Water-Related Interventions in Gumti–Muhuri Basin.
- Rationalization of Polders in Gumti–Muhuri Basin.
- Rationalization of Water-Related Interventions in Chittagong Coastal Plain Basin.
- Rationalization of Polders in Chittagong Coastal Plain.

These projects focus primarily on updating and rationalizing existing water management infrastructure to improve its effectiveness and resilience in the face of climate change. This approach ensures that Bangladesh's water management systems remain robust, adaptive, and capable of addressing both current and future challenges.

13.5 Implementation: Ongoing Transformation and Project Realization

13.5.1 The Early Stage of BDP 2100 Implementation

Support to the Implementation of the Bangladesh Delta Plan 2100 (SIBDP), is one of the projects directly related to the tasks of the General Economics Division (GED) of the Bangladesh Planning Commission within the Ministry of Planning. The role of GED/ad hoc Delta Wing concerning BDP 2100 implementation and updating consists of (1) coordination of the 11 with BDP 2100 involved ministries, (2) facilitating data and information exchange for decision making in the BDP 2100 institutional framework on planning and investments and (3) monitoring and evaluating the progress of BDP 2100 implementation, resulting in feedback to the Ministry of Planning and Cabinet on goal achievement, recent updates on climate change and impacts with possible need to modify strategies and measures.

Because of the multi-sectoral approach of the Delta Plan, the considerable number of ministries and their implementing agencies require coordination and support e.g., for setting implementation priorities and selection of projects, assessing feasibility and impacts as well as preparation of projects and for investment decision making. It is recognized that the strategic planning concept and adaptive approach of BDP 2100 imply a paradigm shift in the national planning approach, accompanied by institutional reforms and changes of working procedures. Most of these reforms and changes were already indicated in BDP 2100 and endorsed with the final approval of the delta plan in 2018. Most of these transformations have been initiated by GED. The Technical Assistance (TA) Team (SIBDP Consortium consisting of 3 Bangladeshi and 3 Dutch companies) assisted GED. The SIBDP TA Team was appointed by the Government of the Netherlands in 2018 and finalized its tasks in December 2023. SIBDP focused on developing the five earlier mentioned capacities by performing studies, analysis and providing training, but also provides relevant content, additional documents, procedures and frequent communication, participation and consultation of many institutions and stakeholders. The GED with the TA Team were guided by a Project Implementation Committee, a Technical Advisory Committee and an overall high-level Steering Committee. Besides, an Inter-Governmental Committee consisting of a Bangladeshi and Dutch delegation, provides advice on a yearly basis regarding implementation issues. The period of 2020 until 2022 was, due to Covid19, dominated by a hampered process with absence of team members and officials, closed offices, travel restrictions, limited face to face interactions etcetera. However, numerous online meetings and adjusted working procedures resulted in continuity and progress.

13.5.2 Activating the Institutional Framework

Implementation of a comprehensive and significant long-term plan like BDP 2100 with its focus on adaptation to climate change, contribution to food security as well as socio-economic development, requires an adequate institutional framework for further operationalization of strategies and measures and for decision making on investments. The institutional framework as approved with the endorsement of BDP 2100 in 2018, is incorporated in the core governance of Bangladesh in two ways. First, the approved BDP 2100 has been integrated into the legal framework, governance rules, regulations and national five-year planning cycle, to mainstream BDP 2100 within policies and plans like the National Five-Year Plans, the Perspective Plan 2041 and the yearly budget programming and allocation plan. BDP 2100 related projects have been included as priority one in the Annual Development Program (ADP) formulation guidelines. Therefore, BDP 2100 projects get priority for inclusion in, and having allocation of ADP. BDP 2100 has already been added to the relevant items of the Development Project Proposal (DPP) format with other plans like Perspective Plan, Five-Year Plan, and SDGs action plan.

Second, institutional reform was carried out as follows. A strong institutional basis has been elaborated for performance of the (horizontal) coordination, facilitation and monitoring tasks of GED, with respect to the multi-sectoral tasks and investments of the involved ministries and for the (vertical) coordination of agencies within those ministries and of regional/local government. To perform these tasks at GED, a dedicated Delta Wing (with about 45 officials) has been designed and in is in the final stage of approval while a temporally Delta Wing has been established. Moreover, a new dedicated political body, called Delta Governance Council (DGC) has been established and officially announced in 2021. This is the political decision-making body consisting of eleven ministers, chaired by the Prime Minister, with the aim of providing BDP 2100 guidance and decision making on BDP 2100 matters and investments. The first meeting of the DGC was held on 22nd of May 2022. The meeting focused on the progress of BDP 2100 implementation and made a number of policy decisions. The DGC has, among others, decided to take the initiative for formulation of a Delta Act and for establishment of the Project/Program Selection Committee (PPSC). The meeting also decided on actions for creating a Delta Wing in the relevant ministries. The Project/Program Selection Committee has been formed in 2022 at the level of Secretary of the involved ministries, chaired by the Member (Secretary) of GED. This PPSC was established under the Delta Governance Council to coordinate selection and preparing decision-making on programs and projects to be funded and executed. Clearly, the institutional framework requires adequate coordination by GED and involved ministries. That's why a Focal Point network was formed and trained. Each involved ministry is with its Focal Point linked to GED/Delta Wing on a daily basis. A number of Ministries and agencies formed a dedicated Delta Wing or Cell for this Focal Point, to take care of BDP 2100 matters. A draft Delta Act (still in process) has been developed to provide a legal basis and proper mandates for this BDP 2100 institutional framework.

13.5.3 Improving Connective and Organization Capacity

Prerequisites to implementation of BDP 2100 are interaction and coordination, as well as 'readiness' of involved implementing agencies, knowledge institutions and main stakeholders. 'Readiness to implement' is a well-known concept covering many factors for being prepared to implement: (1) knowing and being trained on the BDP 2100 content with its methodology and strategies, (2) bringing involved agencies in good shape (e.g., decision making and coordination, structure and tasks, rules and procedures, human and financial resources, other capacities), (3) contributing to execution of projects. BDP 2100 implementation requires strong emphasis on coordination and interaction, creating an implementation network with pro-active focal points and effective procedures for planning at meso and micro level, decision-making, funding and project execution.

Building capacity of the GED and other ministries, divisions and agencies related to BDP 2100 implementation is crucial. With the support of SIBDP assistance team, GED has implemented a massive capacity development program for its own officials and officials of Delta-related ministries, divisions, agencies and institutions. A large number of officials were appraised about BDP 2100 vision, mission, goals, challenges of hotspot areas, strategies and measures, investments, funding, M&E framework and knowledge management. A series of seminar and workshops were held on the investment plan, study reports, policy briefs and knowledge products of the SIBDP 2100 project. As of November 2022, 10 seminars/workshops on BDP 2100 related issues, 3 Workshop-cum training and 19 short training courses (3–4 days) were organized with the support of SIBDP. A total of 615 officials has been trained on the various mentioned items. These activities are also supported by public relations and communication actions with many parties and stakeholders, supported by communication material as brochures, video documentary and Knowledge Portal.

The 1st International Conference on the 'Bangladesh Delta Plan 2100: Issues and Challenges of Implementation' took place on 26–27 May 2022 in Dhaka with a series of events and sessions. The conference was arranged by GED under the auspices of the Delta Governance Council (DGC). GED was assisted by SIBDP TA team, the Embassy of the Kingdom of the Netherlands, and the World Bank. The officials opened the Conference and emphasized the importance of BDP 2100 for the safety and socio-economic development of Bangladesh. A number of events were arranged during this international conference; they were one Plenary session on Institutional challenges and opportunities with international experiences; 4 breakout sessions on Coastal Zone, River Systems, Urban Areas and Agriculture Transformation along with 2 side events on Valuing Water and Youth Panel Dialogue, one Round Table Discussion on improving and accelerating BDP 2100 Implementation and a concluding session. The aim is to repeat such conference frequently.

Another important item is to strengthen the connective capacity through participation and consultation by the formation of a new Water User Institution. Several institutions in Bangladesh are responsible for the management of water resources and related aspects of the delta plan. Within and around the ministries and agencies, there are many stakeholders with diverse and often conflicting interests. There is competition for obtaining government resources or assignments and most water-related services including operation and maintenance (O&M) can be improved. Consequently, many BDP 2100 issues are managed more as an individual service in each entity rather than an integrated service. In the absence of basic institutional arrangements for the proper use of water and infrastructure in terms of ownership, responsibilities and cost sharing, new infrastructure will not sustain up to their economic lifetime. Here, sustainability is the key issue. Investments in water services and infrastructure are rationalized when these are durable and deliver benefits to a larger section of the population to the required extent. Coordination at the macro level and community participation at the micro level are key to success. The question is how to integrate and mainstream delta governance and integrated water management in a manner that would facilitate its transformation from temporary ad hoc project-driven efforts to long-term self-sustained programs with ownership and responsibility of the community. BDP 2100 envisages effective approaches that go beyond specific projects and instead give local government institutions and associated communities and representatives from the private sector a voice in water decisions at the national and regional level. In a nutshell, the formation of Water User Institutions addresses three major issues of concern. These are: coordination among implementing organizations, stakeholder participation at all levels and financing of O&M with, besides the GOB funding, contributions based on the 'beneficiaries pay' principle. The proposed approach has been elaborated and will have follow up within the government.

13.5.4 Developing Knowledge Management and Learning

Implementation can only take place when necessary information, data, maps and knowledge of enough quality are available. Knowledge management, feasibility studies including assessments of environmental impacts and socio-economic development, research and learning are therefore of crucial importance. These include e.g., flood management, sediment, river and polder management, food security, water quality in urban areas. Integrated assessment modeling with the 'Bangladesh Meta-Model' offers support for taking decisions on interventions and investments. Data, information and understanding are needed to deal with the drivers of change and developments in a pragmatic way, supporting and managing interventions in the natural system as well as in institutional, organization and management and financial matters. This will build on the already developed Knowledge Agenda, the created Knowledge and Information Portal and data layers in many different data bases and models. Additionally, a frequent review of BDP 2100 implementation will offer systematic insight in progress of implementation, natural system behavior and goal achievement.

Table 13.1 Projects under study/preparation. (*Source* GED, SIBDP, Ministry of Planning, 2023)

No.	Project name
1	Jamuna River Economic Corridor Development Project
2	Revitalization and Restoration of Chalan Beel
3	Dhaka Rivers Ecological Restoration Project
4	Climate-Smart Agriculture and Water Management Project

13.5.5 Support to Funding of Investments

Proper investment planning, funding and financial mechanisms are a must to any implementation. Especially for the implementation, BDP 2100 committed in 2018 a provision of 2.5% of the GDP annually. The allocation to this provision is gradually increasing while additional financial resources still need to be allocated by the GoB, development partners, IFI's with a linkage to, e.g., the Green Climate Fund and Public–Private Partnerships. Updating the Investment Plan and building trust and relationships with development partners and IFI's to this are important elements.

The Investment Plan (IP) of the BDP 2100 comprises 80 projects to be undertaken by the year 2030 i.e., during 8th and 9th Five-Year Plan period. However, implementation of these projects will take longer time beyond 2030. It may be noted that most of the projects of BDP 2100 IP are multi-sectoral and programs in nature. Out of these 80 BDP IP projects/program, 17 are at different stages of preparation and implementation. Table 10.2 (par. 10.1) presents the BDP 2100 IP projects under implementation. Table 13.1 depicts BDP 2100 IP projects which are (end of 2023) at preparation stage.

A total of 315 projects, directly or indirectly linked to BDP 2100 strategic objectives, are included in the 2023 Annual Development Program (ADP). It is to be noted that, the delta-related annual public investment in pre-Delta Plan stage was 0.8% of GDP, which has gradually increased to about 1.1% of GDP in 2023. Many of the BDP 2100 IP program/projects are good candidates for funding from the Green Climate Fund (GCF). Project Concept Notes for six BDP IP projects have been developed and submitted following the prescribed format of GCF.

13.5.6 Area/ Basin-Wise Programming for Natural System Adaptation

BDP 2100 implementation builds on the previous planning exercise, which delivered project preparations and investment priorities per Hotspots projects with feasibility study (see Chapter 11, BDP 2100) belonging to the so called "0 + category", which after final approval and financing can be implemented. Projects will be selected with stakeholders, tested for coherence and goal achievement, linked to the overall investment portfolio for each Hotspot/river basin, subsequently improved where needed and programmed. Implementation requires a well-considered approach, a systematic sequence and participation of with the intervention involved stakeholders. To facilitate this, six programs have been developed for basin-wise implementation of strategies and river basin investment plans, which will be further developed in close interaction with stakeholders and key decision-maker/implementing agencies. The Meta Model assessment has already partly been integrated into support of decision-making by planning and implementing agencies.

13.6 Implementation Constraints

Practices of preparation and execution of a strategic long-term plan as BDP 2100 are quite different from the practices applied during the BDP 2100 formulation stage which was characterized by initiation, base line studies and diagnostic search in terms of issues, challenges and Hotspots, addressing uncertainties in scenario writing, a comprehensive strategy process, lots of consultations, financing strategy, investment planning and finally decision making. Now, execution of measures 'simply' means putting strategy into action, realizing the goals and ambitions by taking measures, preparing and executing projects on the ground.

The constraints can be focused on two problem areas, closely related to the implementation problems described in Chapter 11. and also regarding the 'readiness to implement factors' which refer to the extent individuals or organizations are willing and capable of jointly delivering an intervention.[4] Readiness factors include, e.g., general or system capacity (staff, rules, IT, organizational processes), intervention specific capacity (dedicated knowledge and training, specific policies and resources), leadership (giving direction, setting priorities, tasks, instructions), psychological and behavioral readiness (awareness, information, motivation and acceptance) by individuals and teams of staff and stakeholders, access to resources needed to implementing tasks, tools and materials.

In fact, the usual way of doing things, the normal working procedures are most probably confronted or at least linked with the new actions, bringing change and transformation as the institutional framework indicates. An important principle is, that implementing ministries and agencies would take up BDP 2100 implementation in their sector and that the Planning Ministry/GED would coordinate the multi sectoral collaboration and execution. So, in practice, GED intends to coordinate and facilitates but also monitors actions which ten ministries and agencies are used to do themselves. First point is that GED was not equipped to do that in terms of leadership, staff and behavior as well as resources and secondly, ministries and agencies tend to follow their usual course of action, aren't aware of how to do their execution work differently and with which resources.

[4] See: (Weiner, 2009; Barwick, 2011; The World Bank Group, 2023).

Preparation of projects, e.g., by conducting feasibility studies requires budgets which need to be proposed by the implementing ministries and agencies to the Planning Ministry. Each implementing organization has however its own wish and priority list for realization which initially does not or only partially correspond with the investment plan of BDP 2100 and indicated priorities in the Five-Year Plan(s). This needs to be coordinated and streamlined which takes a lot of time. GoB already funds a number of projects. A consideration with Development Partners is, that Bangladesh will become a middle income country which brings changes, in fact a decline in the development cooperation compared to what it used to be. Then, in this regard there are international financial institutions (IFI's) like the World Bank Group, Asian Development Bank, Asian Infrastructure Investment Bank (AIIB), European Investment Bank (EIB). Their main focus is on what they call bankable or investment ready projects. The constraint for Bangladesh is that projects need to be prepared first, including adaptive design, economic impact assessment, environmental impact assessment and other requirements. These IFI's are not funding the huge related costs of preparation. IFI's keep on asking for bankable projects but don't take up the already by GoB followed strategies and selected projects which only need preparations as indicated. Funding constraints are also seen with the six proposals submitted for funding by the Green Climate Fund and comparable funds. Procedures are time consuming, funding highly complicated. Public–Private Partnership is in principle a promising mechanism to increase investments by private companies against returns on investments in the BDP 2100 investment areas. However, due to development of Covid19, global unrest and increasing inflation and interest rates, risks to invest also increase and is hardly taking place in line with the PPP-ambitions.

References

Barwick, M. A. (2011). *Checklist to assess organizational readiness.*

BDP 2100 Knowledge Portal. http://bdp2100kp.gov.bd

Centre for Effective Services. (2022). *An introductory guide to implementation.* https://thecenterforimplementation.com/toolbox/tag/Readiness+and+context

De Heer, J. M., Choudhury, G. A., Oliemans, W., Ahmed, F., Tahsin, A., Hossain, J., Slager, K., Islam, T., Ara Parveen, I., Sadiq, S., Haque, Z., Rashid, A., Schasfoort, F., Van de Vat, M., Rahman, M.A., Bani, A., & Ahmed, S. (2023a). *Implementation program for the north-west region (Hurasagar Basin).* SIBDP.

De Heer, J.M., Choudhury, G.A., Hossain, Md A., Ahmed, S., Haque Khan, Z., Hossain, Md J., Islam, T., Ara Pervin, I., Chandra Mridha, G., Alam, R., Kanti Magumdar, T., Uddin, Md S., Salam Sikder, Md A., & Tahsin, A. (2023b). *North-East (Haor) basin management plan: Implementation program.* SIBDP.

De Heer, J.M., Choudhury, G.A., Ahmed, F., & Ahmed, S. et al. (2023c). *Implementation program for the urban areas.* SIBDP.

De Heer, J.M., Choudhury, G.A., Islam, R., Hossain, Md A., & Ahmed, S. et al. (2023d). *Chattogram hill tracts implementation program.* SIBDP.

De Heer, J.M., Choudhury, G.A., Oliemans, W., Ahmed, F., Rahman, Md M., Sadik, Md S., Haque, Z., Rashid, A., Rahman, Md A., Banik, A., Hossain, Md J., Islam, T., Ara Pervin, I., Begum,

M., Slager, K., Schasfoort, F., Tahsin, A., Hossain, Md A., Ahmed, S. et al. (2023e). *Extended coastal zone basin management plan.* SIBDP.

De Heer, J.M., Choudhury, G.A., Akter, J., Tahsin, A., Haque Sarker, M., Hossain, M.A. Ahmed, S. et al. (2023f). *Major river systems and estuaries: Implementation program.* SIBDP.

Global Water Partnership on IWRM and Tools (2023). etc. https://www.gwp.org/

Ministry of Planning. (2018) *Bangladesh Delta Plan 2100.* 2018. https://plancomm.gov.bd/

Ministry of Planning, GED, BDP 2100 Formulation Project. (2018a). *Baseline study on Governance.* https://plancomm.gov.bd/

Ministry of Planning, GED, BDP 2100 Formulation Project. (2018b). *Investment plan.* https://plancomm.gov.bd/

World Bank Group. (2023). *ODRA assessment approach.* https://opendatatoolkit.worldbank.org/en/data/opendatatoolkit/odra

Weiner, B. J. (2009). A theory of organizational readiness for change. *Implementation Science, 4*, 67.

Chapter 14
Reflective Evaluation Key to Adaptation

Abstract Monitoring provides policymakers with the information about the current status of a policy, program or project relative to expected targets and outcomes, whereas evaluation gives insight in whether targets and outcomes are being achieved or not. However, in the case of adaptation to climate change, targets and outcomes are not fixed, they are moving targets and outcomes are depending on actual change of the climate, the impacts and progress with adaptation of the related systems. The primary purpose of M&E is therefore twofold: to help the government to measure the quantity, quality and targeting of outputs and to quantify how the outputs affect the lives of the people against continuous evaluation of climate change taking place. Intended policy outcomes are not easily attained in the public domain. Outcomes may be unexpected or discarded. Monitoring of implementation processes and critical or reflective evaluation offer insights to make adjustments or take additional measures in time. Monitoring is therefore also concerned with defining goal or target related indicators and getting information on how far and sometimes also in what way goals are realized. This information can be used to evaluate a program or project and reflect on its impact, possibly leading to lessons learned or to modifications for the future. It can help policy makers to choose applicable measures among a set of alternatives or even to check the design of a plan or program. Reflective evaluation is the systematic and objective assessment and critical review of a plan, program or project in relation to its impact, efficacy and goal achievement. As a whole, monitoring and evaluation promote transparency of what is going on in implementation and on accountability in public spending, To achieve this, public institutions need to have the right financial and human resources capacities as well as adequate data collection and processing systems and provisions. In case of a results-based M&E system, results and outcomes in terms of societal impact are emphasized. It can show delay, cost overrun, failure or other problems of implementation but also the efficiency and effectiveness of problem solving by these measures or the requirement of additional or new measures. Result-based monitoring and reflective evaluation are strong instruments for good governance including overall goal achievement of 'adaptation by development'.

J. M. de Heer and G. A. Choudhury, *Strategies for Adaptation to Climate Change in a Transformative Approach*, Springer Climate,
https://doi.org/10.1007/978-3-031-88884-7_14

Keywords Monitoring · Progress · Evaluation · Reflective evaluation · Development results framework · Updating · Adaptation · Corrective · Additional · New measures · Adjustment · Monitoring and evaluation · Result-based M&E

14.1 Importance of Monitoring and Reflective Evaluation

Monitoring is an important activity to determine the progress of implementation of a program, plan or project. It is a continuous process in a systematic and routine way for collecting information and data on the progress indicators. Monitoring is checking the real progress against intended execution of program, plan or project. Thus, monitoring helps the concerned stakeholders and decision makers by providing information on the state of implementation and achievements till a certain point in time. While comparing the outcomes of a specific program with a set of targets, monitoring indicates whether the efforts are on the right track. Monitoring can help during the implementation process by showing the right information in improving the design and implementation of a program or project.

Evaluation is an impartial assessment of a phase of an ongoing action or of a completed project, program or policy with a view to determine the relevance, effectiveness and fulfillment of objectives, impact, and sustainability of that project, program, or policy. Information gathered in relation to these aspects during the monitoring process provides the basis for the evaluative analysis and conclusions. In such analysis, evaluation delves deeper into the relationships between the results of a program or project, the effects they produced and the overall impact a program or project have in a policy sector or in society or, e.g., with respect to adaptation to climate change. Such an assessment is usually undertaken at design, implementation, and results stages. An evaluation usually provides information which is credible and useful, enabling the incorporation of lessons learnt into the adaptive decision-making process concerning follow up or future action. In general, monitoring is integral to evaluation. As a whole, Monitoring and Evaluation (M&E) generate information on and evaluate the initial values of chosen indicators and desired outcomes. M&E often plays an important role in the dialogue on development and its progress between involved stakeholders. Monitoring and evaluation promote transparency and accountability in spending resources, and ensure that resources are injected in the right direction, being efficiently utilized for attaining agreed goals.

BDP 2100 Investment Plan shows a collection of potential projects and programs concerning adaptation to climate change and water resources management, with related sectors and subsectors of the economy (Ministry of Planning, 2018a). These multi-sectoral programs and projects have been selected out of a large set of proposals through multi-criteria analysis and are closely aligned with the BDP 2100 vision and goals, strategies and sub-strategies. These projects and programs have been taken up at the hotspot, thematic and national level to address the prevailing challenges and utilizing opportunities. Proper and timely intervention is crucial to face the adverse effects of climate threat and other delta related challenges. The BDP2100 Investment

Plan, therefore, strongly pursues 'no regret measures' in the first place to ensure that the right investments are made at the right time (Ministry of Planning, 2018b). It aims to avoid 'too little, too late' and 'too much, too early' investments. An effective M&E system is crucial to monitor and evaluate the implementation of projects and programs of the BDP 2100 Investment Plan. It can play a vital role in resources allocation and spending in an efficient manner in projects and programs which are fully consistent with the BDP 2100 priorities. Moreover, BDP 2100 is an adaptive long-term plan which requires frequent updating of climate change and other problem situations. Adjustment of strategies and measures based on M&E information and assessments may be needed accordingly.

14.2 M&E Practices

In Bangladesh, M&E is at implementing agency level mostly concerned with monitoring the financial and physical progress of a project or program. The Implementation Monitoring and Evaluation Division (IMED) of the Planning Ministry performs M&E of the projects and programs as a whole. The key stakeholders for the IMED are the ministries, divisions, agencies as well as other autonomous bodies. Monitoring primarily oversees the implementation process, identifies the challenges in terms of quality, time and costs and provides recommendations for improvement. IMED is involved at various stages of the project cycle, from project preparation, project completion to post-project impact evaluation. In the pre-project phase, IMED offers ideas to increase efficiency, efficacy and adjustment, while it in later phases focuses on progress, time, budget and quality. Data collection is elaborated in reports and discussed in monthly coordination and review meetings and special meetings with the Project Directors. The performance of implementing institutions is next to monthly meetings and reporting also shown in quarterly and annual documents.

In the framework of the national planning, the General Economics Division introduced the concept of results-based monitoring and evaluation (RBM&E) in the 6th Five Year Plan (FYP). This was further elaborated in the BDP 2100 project with a Development Results Framework (DRF) which indicates results to be achieved over relevant time in relation to national priorities. The DRF also provides the sources of data and mentions the responsible institution for realizing the given target. Evaluation is made to assess to what extent the results are achieved. GED performs evaluation of the FYP implementation at two stages; the first one is done in the middle of the plan period (Midterm evaluation) and the second one is carried out after completion of the plan period (End Evaluation).

14.3 Integrated BDP 2100 M&E Approach

As earlier seen, BDP2100 includes a strategic vision, long-term goals and four scenarios, adaptive strategies, measures, and interventions for the long term as well as investment options for the short-term (up to 2030) and medium-term (2030–50). The BDP2100 Investment Plan includes in total 80 programs and projects, out of which 65 are investment programs and projects and the other 15 are knowledge management and institutional programs and projects.

BDP 2100 suggested to incorporate the elements of adaptive management including the holistic water resources approach in the design of each project. This means, that M&E needs to focus on the status of realization of these elements in the project documents. In this respect, the adaptive management and holistic water resources approach include: (i) water security, (ii) climate change and environmental sustainability, (iii) food security and livelihood, (iv) economic growth, (v) social development, and (vi) institutional and knowledge development. Each of these themes have sub-themes as described below (De Heer et al., 2021):

- Water security includes: (i) fresh water resources management, (ii) flood risk management, (iii) drought risk management, (iv) disaster risk management, (v) climate change impact management and public health, water supply and sanitation.
- Climate change and environmental sustainability includes: (i) biodiversity, (ii) forestry, (iii) environmental pollution, (iv) climate change adaptation, (v) climate change mitigation and (vi) Disaster risk reduction management.
- Food security and livelihood includes: (i) crop production, (ii) livestock production, (iii) fisheries production.
- Economic growth includes: (i) integrated transportation system (road, river and rail), (ii) market linkage, and (iii) energy and power.
- Social development includes: (i) poverty reduction and (ii) gender development.
- Institutional and knowledge development includes: (i) water management organizations, (ii) land management, and (iii) knowledge development and (iv) research.

14.3.1 Monitoring Indicators

BDP 2100 adopted a holistic adaptive delta management approach in the BDP 2100 investment programs and projects. With respect to the Result Based M&E framework, BDP 2100 has identified delta goal-wise indicators and sub-indicators as follows (Alam et al., 2018a).

BDP Goal 1: This goal is about ensuring safety from floods and climate change related disasters. The indicators are identified as: (i) risk free zones from natural disasters and (ii) population vulnerable to natural disasters. The sub-indicators for risk free zones from natural disasters indicators are reduction of: (i) average flood affected area, (ii) catastrophic affected area, (iii) extreme drought free area, (iv) storm surge affected area, (v) dry season salinity intrusion free area and water logging free area. The sub-indicators for natural disasters vulnerability are the reduction of: (i) flood vulnerable people, (ii) cyclone vulnerable people, (iii) erosion vulnerable people, and (iv) water logging vulnerable people.

BDP Goal 2: This is to ensure water security and efficiency of water usages. The indictors are identified as: (i) dry season flow availability (January–May), (ii) dry season irrigation coverage, (iii) irrigation water efficiency, (iv) urban domestic water efficiency, (v) rural population with safe drinking water access, (vi) rural population with safe sanitation, (vii) surface water polluted by industrial wastes, and (viii) surface water sources polluted by other wastes.

BDP Goal 3: This goal is about ensuring sustainable and integrated river systems and estuaries management. The identified indicators are (i) erosion along major rivers and (ii) area of reclaimed lands. The sub-indicators for erosion reduction focus on the area eroded along the Jamuna, the Ganges, the Padma, the upper Meghna and the Lower Meghna.

BDP Goal 4: It is about conservation and preservation of wetlands and ecosystems and their wise use. The monitoring indicators are identified as (i) permanent wetland with connectivity, (ii) seasonal wetland with connectivity, (iii) habitat protection, and (iv) harnessing of ecosystem services and goods. The sub-indicators for habitat protection are increase of: (i) perennial aquatic habitat area, (ii) seasonal aquatic habitat area, and (iii) marine habitat area.

BDP Goal 5: This goal promotes development of effective institutions and equitable governance for in-country and transboundary water resources management. The monitoring indicators are (i) rural people with adequate capacity for WRM, (ii) equitable share of water among users, (iii) Adequate monitoring mechanism, and (iv) O&M budgeting.

BDP Goal 6: This goal states about achieving optimal and integrated use of land and water resources. The monitoring indicators are (i) spatial zoning of integrated land and water uses, (ii) spatial standardization of drainage density, (iii) flood control, drainage and irrigation capacity, (iv) sectoral use of water, and (v) navigation capacity. The sub-indicators for flood control, drainage and irrigation capacity are the increase of area under (i) irrigation projects, (ii) drainage projects, (iii) irrigation and drainage projects, (iv) FCD, (v) FCDI, (vi) coastal FCDI. The sub-indicators for sectoral use of water are: (i) surface water used for irrigation, (ii) ground water used for irrigation, (iii) domestic water and (iv) industrial water use. The sub-indicators for navigation capacity are: (i) wet season navigation course and (ii) dry season navigation course. The indicators for socio-economic development are: (i) poverty reduction, (ii) gender development, (iii) land use and safe housing, and (iv) socio-economic impact. The sub-indicators for socio-economic impact are: (i) food security, (ii) households affected by droughts, floods and salinity, (iii) displaced people due to disasters, (iv) access to safe drinking water, (v) access to sanitation, and health. The indicators for institutional development are: (i) water management organizations, (ii) land acquisition, (iii) land settlement, and (iv) knowledge development and sharing.

Apart from the above goal-wise indicators, there are some cross-cutting indicators for the overall evaluation of the socio-economic status, evaluating the social and economic conditions for all over the country. These concern: (a) economic growth with indicators: (i) integrated transportation system (road, river and rail), (ii) market linkage, and (iii) energy and power. (b) Economic impact and indicators are: (i) flood damage, (ii) crop production, (iii) fish production, (iv) river navigability, (v) inequality (e.g. due to flood, salinity and drought damages), (vi) employment–short term, (vii) employment–long term, (viii) economic growth and (ix) poverty. (c) Social impact with indicators: (i) food security, (ii) households affected by droughts, floods and salinity, (iii) displaced people due to disasters, (iv) access to safe drinking water, (v) access to sanitation, and health. See also Annex 2 for a brief overview.

14.3.2 Results-Based M&E for the BDP 2100

A number of ministries, divisions and agencies implement the BDP2100 Investment Plan and provide a range of services to the public. These services contribute to achieve the targets of BDP2100 Delta Result Framework (DRF). However, not all actions and programs are directly linked to this framework. Over 40 different minis-eries/divisions and their corresponding agencies are involved directly or indirectly in BDP2100 implementation. Besides the general ministries as of Planning and Finance, key organizations involved in the BDP 2100 IP implementation include the following ministries and their implementing agencies: Ministries of Water Resources; Agriculture; Environment, Forests and Climate Change; Disaster Management and Relief; Local Government, Rural Development and Cooperative; Shipping; Fisheries and Livestock; Chattogram Hill Tracts affairs; Land; Housing and Public Works; Power, Energy and Mineral Resources (Alam et al., 2018b).

BDP 2100 advocates integrated and adaptive management of the Bangladesh Delta, following long-term development goals. However, there are considerable uncertainties about the future path of the climate variables and of policy responses of upstream countries related to transboundary water management. The quality, relevance and adequacy of the current BDP 2100 interventions and the need for changes in light of new evidence on climate variables and transboundary river interventions can only be assessed properly, if there is a well-developed M&E system that monitors the implementation of BDP 2100 and shows progress on expected results over time. Critical evaluation by the key stakeholders is then required together with adaptive decision making on whether or not climate change adaptation and other challenges are addressed correctly or that corrective, additional or completely new interventions will be needed.

BDP 2100 goals and result based targets constitute the DRF for BDP 2100, which becomes the main focus of the BDP 2100 M&E. The actions and programs of the line ministries and divisions are all of direct relevance to the DRF. This primacy of the concerned ministry and division to the DRF of BDP 2100 makes it the primary focus of the M&E effort. The evaluation of the selected indicators is anticipated to reveal how different actions or inaction can degrade or improve conditions that lead to water safety and sustainability. The Framework is built around both statements of intent (e.g., objectives) and themes (e.g., water quality). A crucial priority for BDP 2100 is to establish a proper DRF that adequately captures progress in its implementation. The DRF should be treated as an adaptive framework that can be modified and refined as more, better and updated information is available. In the DRF, targets have been set at two levels for the year 2030 and 2050 with baseline information for 2016; all are indicative and will have to be updated frequently as implementation of the Delta Plan proceeds. A total of 25 indicators and 30 sub-indicators spreading over the 6 BDP 2100 goals have been incorporated in BDP 2100 DRF.

As usual, IMED will continue M&E of BDP 2100 development projects. But it will be the responsibility of the GED to perform Results-Based M&E of the entire BDP2100 implementation. GED will perform this function with the support and cooperation of the delta related ministries, divisions and agencies. M&E is needed to be made consultatively on the basis of the analysis of the monitoring and evaluation at the project and sectoral level. For doing this, GED will also closely interact with IMED and statistical organizations including BBS.

The quality and usefulness of the M&E will mostly depend on the availability and adequacy of data and sufficient budget and means to collect the data. The BDP 2100 M&E system is aimed at generating information that can be used in achieving these objectives. The M&E system is envisioned to be a living process and would need to remain flexible and robust to respond effectively to unanticipated events. So, there is scope to further improve the existing DRF provided in the BDP 2100 Strategy Report to make it as comprehensive and detailed is needed.

Monitoring and evaluation of the Investment Plan will supply hotspot and nationwide information on the physical and financial progress of projects and for the Investment Plan and adaptation as a whole. A comprehensive M&E Web Portal is useful to be developed for M&E, while a working protocol and template may be used by the GED for monitoring the physical and financial progress of implementation of projects and programs. This Result Based M&E-system may be developed also to provide support to a participatory and interactive planning process for ensuring adaptive management of the Bangladesh delta. The M&E web portal will allow users to access all generated and collected information and show how far adaptation has progressed. This will also provide community based adaptation with information for adaptation activities at the local level.

14.4 Key Points of Concern

There are some key points of concerns to check through M&E of BDP 2100 Investment Plan. These are briefly narrated as follows. There is in the first place the state of readiness of investment projects (in terms of approved feasibility study; prepared project documents; arranged budget etc.) listed in the Investment Plan. Development of the right project proposals is a key for processing approval of the public sector projects. Experience shows, that the implementation of identified projects is often delayed due to a lack of timely preparation of project documents supported by a complete feasibility study. During M&E, it is needed to check to what extent the project documents are timely prepared, and processed for approval and inclusion in the Annual Development Plan (ADP) budget according to the priority and order mentioned in the BDP Investment Plan and/or Five Year Plan.

Secondly, allocation and spending of resources are very important. Timely allocation of adequate resources is vital for implementation of a project within the given time frame. Physical progress of a project mostly depends on the financial progress, meaning the way and timeframe the project authority needs to spend the allocated money timely and appropriately. There are examples that in a good number of projects the implementing authority could not utilize the allocation during the given time frame, e.g., because of the monsoon season or execution problems. Therefore, it is needed to monitor whether allocations are sufficient enough and the executing agency is able to spend 100% of the allocated fund or not. Spending capacity of the project implementing agencies is important for timely implementation of a project. Regular M&E can help to accelerate the pace of BDP 2100 IP implementation by identifying the spending bottlenecks.

Third, BDP 2100 recommended for gradual increasing the allocation of resources to a level of 2% of the GDP for financing the BDP 2100 Investment Plan. The status of resource allocation through the ADP to BDP 2100 investment projects is needed to be monitored to check whether the allocation is in right track of meeting the 2% of the GDP.

Fourth, a substantial Result Based M&E focused capacity building is required to be coordinated between BDP 2100 ministries and agencies. Use of the DRF with data collection and management activities and systems as well as the evaluations are crucial for adaptive decisions on follow up measures. The knowledge base must match with the expanded scope of multi-sectoral M&E according to BDP 2100. Good quality M&E will allow implementing agencies to execute the project in the right direction, being flexible to changing circumstances. A sound knowledge base will also provide implementing agencies with the best information to develop new projects in line with BDP 2100. The implementing agencies also need capacity building to implement the projects successfully and develop their knowledge and data management as well.

So, there are many challenges in applying Result-Based M&E of BDP 2100 implementation for tracking the progress and achievement of the core targets. The major challenges are M&E capacity constraints and lack of tools, data and knowledge for performing Result Based M&E in a systematic manner. Quality of the staff responsible for M&E and adequacy of tools and data are prerequisites for reliable M&E. Hence, the M&E organizations require institutional strengthening including working procedures, tools and human resources development. The existing human and financial resources of the data collecting, processing and storing agencies are often inadequate to perform satisfactorily. Access to data is also a challenge as many of the organizations do not want to share data, they capture. The GED could equip itself with adequate qualified human resources for effective M&E of the BDP 2100 IP implementation.

14.5 Reflection, Updating, and Adjustment

The core of scenario-based adaptation is to be able to deal with uncertain developments over time. This means that M&E is not only about the progress of implementation but also of the developments taking place, which is of utmost importance. Consequently, that the drivers of change will be studied, analyzed and compared to the chosen adaptation approach. Signposts on crucial indicators can be defined and involved in this. Frequent updating or renewal of the BDP 2100 and Investment Plan with the Delta vision and goals in mind is important and should incorporate the findings of the above indicated analysis. If change in speed of implementation or of the overall approach is necessary, strategies could be reviewed and where needed modified.

The formulated delta vision and goals will be achieved by adaptation and transformation through frequent updated planning, implementation and modification of BDP 2100. Its implementation demands, among others, institutional reforms, capacity and awareness building, knowledge generation and database management, regional (basin-wise) programming and execution with adequate financing of selected and prepared projects as well as thematic approaches, e.g., on navigation and Blue Economy. With the adoption of this dynamic adaptive plan, Bangladesh experiences a paradigm shift in its planning and implementation system. This paradigm shift consists of adaptive management with emphasis on climate change related adaptive planning and decision making, a holistic and multi-sectoral approach, a new institutional framework concerning decision making, coordination, and funding as well as organization of decentral water or delta institutions. Consequently, significant institutional and policy implications, financing challenges and tasks needed to be developed and practiced which is indicated as transformation. By working on these implications, challenges and tasks, The government is going through a transformation process, aimed at climate change adaptation and related delta goals. Transformations take time and also need to be monitored and evaluated to generate feedback and adjustments in the desired direction of achieving goals and vision.

The by BDP 2100 triggered transformation process is ongoing and requires support to realize optimal performance, e.g., in (1) the Institutional Frame Work on decision making (Delta Governance Council) with the role of the Planning Commission; (2) in (horizontal) coordination with the involved sectors and (3) (vertical) coordination with (partly new) regional or local institutions; (4) performance with the Delta Wing; (5) involvement of the Private Sector; (6) The government's strategic positioning towards financial challenges with International Financial Institutions, Development Partners and utilization of PPP's and the Green Climate Fund. On this basis, updating and, where needed, modification of strategies and interventions can take place, which belong to important characteristics of adaptation and possible additional transformation. The signs of worsening climate change and impacts are already there and need to be taken seriously, which is a major reason for updating of the Delta Plan.

References

Alam, S., De Heer, J. M., & Choudhury, G. A. (Eds.) (2018a). *Bangladesh Delta Plan 2100: Baseline Studies on Water Resources Management. Volume 1*. General Economics Division, Planning Commission, Government of Bangladesh. https://plancomm.gov.bd/

Alam, S., De Heer, J. M., & Choudhury, G. A. (Eds.) (2018b), *Bangladesh Delta Plan 2100: Baseline Studies on Governance and Institutional Development. Volume 6*. General Economics Division, Planning Commission, Government of Bangladesh. https://plancomm.gov.bd/

De Heer, J. M., Choudhury, G. A., Khan, F. A., Haq, M. M., Huq Mou, B., & Ahmed, S. (2021). *Monitoring and Evaluation Report*. SIBDP.

Ministry of Planning. (2018a). *Bangladesh Delta Plan 2100*. https://plancomm.gov.bd/

Ministry of Planning. (2018b). *GED, BDP 2100 Formulation Project. Investment Plan*. https://plancomm.gov.bd/

Part V
Adaptation and Transformation Experiences

Findings of Adaptation in a Transformative Approach with the Crucial Role of Applied Practices

Chapter 15
Lessons Learned from the Bangladesh Experience

Abstract Based on the Bangladesh experience, this chapter provides reflection and lessons learned from preparatory study up to realization start of BDP 2100. These lessons will be combined with briefly mentioned other international findings, e.g., from the Netherlands and Vietnam. The lessons learned from the Bangladesh experience show in the first place the necessity for optimal exchange and collaboration of government institutions and involved other parties or countries. Trust and respect as well as the willingness to collaborate and share many valuable assets as knowledge, manpower and financial resources are basic requirements for a fruitful joint process. Besides, attention for specific aspects of cross cultural collaboration are important to come to mutual gain and strategy activities fitting in administrative culture and societal expectations. In the course of time from the start early 2014 till 2024, a number of episodes with their own characteristic focus and activities have been identified and indicated as adaptation practices. Managing a process of adaptation and transformation means in fact leading professional practitioners in their performance of a sequence of deliberate practices. An adaptive process is recurring although there will be differences in how elaborate practices need to be performed after the first round. Keeping a collective memory alive is important for continuation. A well-known factor is the potential break or gap between practices of strategy formulation and practices of execution also because these practices take place by different practitioners in completely different places. Knowledge and reasons behind choices and priorities may evaporate quickly. Even the sense of urgency can easily be lost in busy daily routine affairs when, e.g., the main strategizing organization transfers measures for realization to executing agencies. In the approach presented here, realization of transformations (institutional, financial, capacity building, technical) and start of execution of measures started already during the strategy process and application of practices. Because preparation of measures and interventions needs the experience and technology of executing agencies, they are from the beginning involved in the strategy process and capacity building. To increase effectiveness, they can be even more involved in the strategy process and in the preparation of execution of, e.g., no regret measures. Another aspect is that, although Bangladesh is a centralized unitary state, execution of measures and interventions could for reasons of effectiveness and involvement be stronger embedded in regional and local institutions next to the

J. M. de Heer and G. A. Choudhury, *Strategies for Adaptation to Climate Change in a Transformative Approach*, Springer Climate,
https://doi.org/10.1007/978-3-031-88884-7_15

national level. Utilizing community-based adaptation also in addition to national actions, is essential to mobilize local people to contribute in the adaptive activities with their capacities and technologies.

Keywords Strategic planning · Strategy as practice · Practices · Practitioners · Diagnostic search · Strategy process · Interactive strategy making · Institutional framework · Adaptive measures · Adaptive programming · Adaptive design · Delta atelier · Water institutions · Community-based adaptation

15.1 Strategic Planning in a Sequence of Applied Practices

Managing adaptation and related transformation can be achieved by performing a number of well chosen, dedicated practices. The strategy as practice literature shows the importance of the relationships between practitioners and their perceptions, sensemaking and analysis of the broader context in relation to the strategy praxis and their choices for performing professional practices in order to generate appropriate and fitting strategies and measures. In doing so, the practitioners directly interact in the presented approach with stakeholders from government agencies at various levels, from academia, civic society, financial institutions, the private sector, local people etc. This interaction is of crucial importance for a fruitful process and for shaping of not only professional but also meaningful strategic content in the actual context. Here is the essence of a transformative approach, generating changing mindsets, performative action and real change already during strategizing, in choices and decision making, in preparatory work and in execution of measures.

The experience from Bangladesh shows that, with an idea of a delta plan in mind and experiences in the team, a sequence of professional practices can be applied to generate a strategic adaptive long-term plan with first reforms and measures in execution right after approval of the BDP 2100 by the highest authorities. The point for the transformative approach is, that each applied practice brings change with it in governmental activities and services, in a broad range of stakeholders activities and societal processes accordingly and in certain areas also activities in the physical systems. Selection and utilization of for the strategic planning process useful practices requires awareness and discussion of the consequences of applying those practices, although consequences cannot always be known on beforehand nor directly visible during the process. It can be expected from practitioners that they act upon effects of their practices in an adequate way.

The consideration for starting a long-term strategic adaptive process should include awareness and sensemaking not only of the necessity but also of the methods and intricacies of strategizing for adaptation over a long period of time. The advantage is that many suitable practices have been developed in strategic management and other disciplines or can be taken from studies of strategy applications in situations with comparable characteristics. Searching for the right adaptation strategy practices and valuable experiences in and support by other countries is therefore a logical government activity, which may be accompanied by exploring the feasibility of such strategy process. Specific practices are involved for the preparation of a strategic planning process for adaptation, e.g., on strategy process agreements like an MOU and the contracting of an assistance team as well as a decision on process design. Once this is done, practices of mobilization and management of a specific project for strategy preparation and diagnostic search need to be performed, together with identifying issues and challenges. The practice of visioning, managing uncertainties and strategy formulation follow next with, e.g., choices like 'adaptation by development'. Normally a government is not used to and equipped for a long-term adaptive process so, developing a dedicated institutional framework with required transformation and tools is needed. In line with this and with the adaptive measures, investment planning and regional or basin wise adaptive programming utilizes specific practices. Regular or routine practices, sometimes also implying transformation, can be performed by implementing agencies on execution of measures and interventions, e.g., including adaptive design of physical system interventions and innovation of working methods and tools. Funding practices are also specific within the existing budget process but also in relation to various financial sources. Because adaptation will be ongoing over a long period, the practices of monitoring and reflective evaluation, updating and adaptive adjustment form a crucial action for continuity and achievement of results as formulated in goals and vision.

The lessons learned from the Bangladesh Experience require against the above attention and have been summarized in the following practices as follows in Fig. 15.1.

The practices have been elaborated as follows.

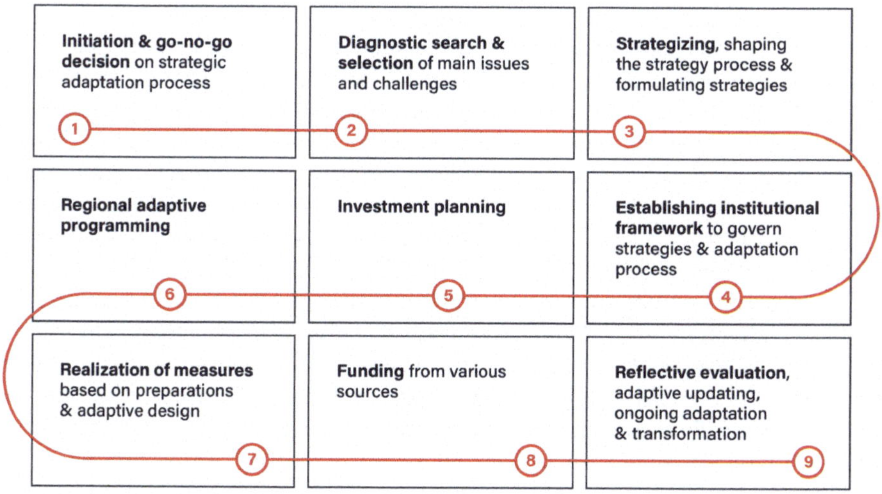

Fig. 15.1 Repertoire of adaptation practices

15.1.1 Strategy Initiation, Process Agreements and Design and Start-Up

Initial awareness rose in an early stage as Bangladesh is located in a vulnerable and dynamic delta. So processes threatening its development were picked up from international reporting on climate changes, knowledge exchanges during international conferences etc. GoB was also aware of strategic delta planning in the Netherlands and Vietnam as well as in the USA concerning impacts and planning after hurricane Katrina. So, awareness led to sense making in terms of taking the initiative for an own delta plan. The Dutch government and private sector were in that time quite active with warnings about implications of climate change and the necessity to adapt in time. GoN provided GoB with ideas and assistance for preparing a delta plan. A preparatory study followed, which was highly useful. With inspiration of the Dutch and Vietnamese delta plans in mind, conditions in Bangladesh were studied and the team concluded in a positive way about the need for a Bangladesh delta Plan (Choudhury et al., 2012). This was followed by agreements on cooperation as an MOU and an Administrative Arrangement between Bangladesh and the Netherlands.

A Terms of Reference has been developed by GED and EKN to prepare requirements and basic ideas of a delta plan (Ministry of Planning, 2018). The ToR was incorporated in the tender documents as project document and was of crucial importance to communicate what type of delta plan GoB wanted to develop. Consortia of companies were formed on that basis, which proved to be effective. A Proposal was chosen and rewarded by GED and EKN, leading to a contract between EKN and consortium BanDuDeltAS. Here, learning to adapt and planning as learning orientations were starting points, together with inspirations of Adaptive Delta Management. This approach was applied in the Netherlands and a number of its principles were used or were viewed as points for inspiration to organize and prepare the BDP 2100.

Mobilization followed to establish the project organization and assistance team as well as GoB guiding committees and support. Mobilization means detailed formation of the assistance team, finding an office, purchasing equipment, furniture and vehicles etc. to start working. Besides EKN and GoB/ GED prepared themselves and contacts and first meetings with the assistance team were established. Apart from some practicalities mobilization was a swift but important period.

The inception activities are meant for confronting the Terms of Reference (De Heer and Choudhury et al., 2014) and Project Proposal with the situation in reality. The Inception Report shows the findings of this confrontation in order to modify the Terms of Reference, Proposal and or project budget (De Heer et al., 2014). Besides a Work Plan contains the actions and activities to deliver the Delta Plan combined with planning of necessary human resources per action. Many atelier type of working sessions including prototyping proved to be very useful to start a demanding project. The assistance team in several occasions together with GED, forms during a prototyping session an image of what it has to prepare, discuss the characteristics and product types to be delivered. This is a very useful way to find direction, share the joint ambition and focus as well as defining the scope and dividing the work packages. Frequent meetings with officials and stakeholders are of great importance already in an early phase, to learn about expectations, share and discuss or correct first orientations on the strategy process, problems and outcomes. This all contributes to learn to know each other and learn about the cultural differences and cross cultural collaboration.

Both governments have formulated a joint Terms of Reference in which characteristics of a strategic delta plan and the strategy process are given. The assistance team further elaborated this in a workable approach in the Inception Report. Besides, the Government of Bangladesh installed guiding committees and an expert committee which were also described in the Inception Report. The function of the report is initially, besides the inception function as such, especially a vehicle for communication and interaction and thereby for sensemaking and sensegiving about the diagnostic search and strategic planning process ahead.

15.1.2 Diagnostic Search: Identifying Characteristics, Issues and Challenges

Tracking strategies: the historic background of water resources management has been utilized, learning from the rich history, many aspects were in the mind of the team and utilized in diagnostic search, strategy making and later more explicitly in 60 years history report. Project proposals from earlier plans which were not executed were reconsidered against the new challenges and fitted into the strategies.

The Baseline Studies provided a means to start with extensive ways to understand the delta in an integrated way, 26 Baseline Studies carried out and discussed also in an integrated way. Many sectors have their own technology, which needed to be arranged by adding specialists to the team. Stakeholders proved to be very important to discuss the interactions with stakeholders were of crucial importance. Initially during consultation in the Inception phase but also with respect to outcomes of the Baseline studies, with scenario development. During a roadshow many places all over the country were visited in the study phase where stakeholders participated in so called 'Delta Ateliers'. The 'Ateliers' or 'Studios' are dedicated workshops organized around analytic and strategy practices, where after brief introduction stakeholders focused on analyzing their existing situations in their areas or basins including impact of climate change, see Fig. 15.2. Next, in the constructing part stakeholders formed ideas what to do about it for the content of strategies, e.g., with respect to flooding, fresh water and many specific items relevant to them.

Fig. 15.2 Diagnostic search in dedicated session. *Photo* J.M. de Heer

15.1.3 Interactive Scenario-Based Strategy Formulation, Adaptation by Development and Related Adaptive Measures

An interactive way of working is also characteristic for the strategy process. Many workshops informal and formal were organized to share knowledge and opinions. Managing uncertainty of climate change and, e.g., transboundary issues led to the choice of the following timeframe: 2100 as long-term strategic orientation with the delta vision in mind, 2050 as medium term and 2030 for the short term. Thereby working with a methodology of: (1) adaptive strategies and measures as response on the selected issues and challenges; (2) scenario-based interactive strategic planning and adaptive delta management ideas on strategy formulation and (3) a mixed scanning approach for selecting the strategies and measures and for detailed implementation programming. Selection of strategies was needed because of the huge quantity of during the above mentioned roadshow and Delta Ateliers developed ideas and actions. The selection sequence in a funneling process by experts was from all possible measures to realistic measures to preferred more detailed strategies, sub-strategies and measures.

Scenario building for the strategic planning took place in two large (about 110 participants) workshop sessions, with a number of team members explaining and guiding the sessions. Major developments and their drivers were selected and discussed as well as the impact and uncertainty matrix. Discussion took place on which major drivers of developments the scenarios would build. High and Low Economic Development issues and challenges became the vertical axis, while water variability including climate change and trans boundary issues and challenges became the second axis.

Strategies were formulated and selected in sessions with experts and officials on the national level, Hot Spot level and on thematic issues and challenges. Most of the national and Hot Spot focused strategies deal with respect to sea level rise with 'protection' (both open or closed) and along the major rivers and coastal zone 'advancing' with plans for land reclamation an development of protected new land. Responses as 'doing nothing', 'accommodate' or 'retreat' are discussed but not included yet in the BDP 2100. In practice, these responses do to some extend happen.

Focus on uncertainty and scenarios brings forward the question of 'what if'? To be flexible the notion 'Adaptive Policy Path Ways' and Tipping Points' were introduced to underline the possible necessity of shifting attention to priorities or accelerating implementation and adaptation when needed. In terms of monitoring, a monitoring and evaluation mechanism was developed and introduced to show progress on the one hand and to frequently adjusting BDP 2100 including pace of climate change and of other developments.

Right from the start of BDP 2100 formulation, GoB gave the instruction to include food security and socio-economic development next to water safety and security. This clearly indicated, the government wanted to use the Delta Plan for development purposes, which became more explicit later in the three so called 'higher level goals'. This makes the Bangladesh Delta Plan a unique plan compared to other delta plans which are much more focused on water and climate change adaptation.

15.1.4 Developing a Dedicated Institutional Framework and Tools

The strategic planning process has been hosted by GED as part of the Planning Commission, Ministry of Planning. Initially, there was a struggle because the Ministry of water Resources wanted to do the delta plan development. However, due to the long-term and multi-sectoral planning, GoB decided to bring the delta planning process under responsibility of the Ministry of Planning which already had experience with the 20 year Perspective Plans and others. This experience was of great help for the formulation process of BDP 2100. The responsible GED-official was of the high level of 'Senior Secretary' who showed leadership, provided feedback and guided the final delta plan to decision making at the highest level.

Inclusion of the BDP 2100 strategic planning process and its outcomes into the Five Year Planning Cycle proved to be of great importance. By doing this, the BDP 2100 is in the by the Cabinet approved FYP mainstreamed, becomes a national plan with status. Initially the 7th FYP was utilized in 2016 already to announce the plan and provide overall information about the plan and importance of it. In the 8th FYP of 2021, the outcomes of the BDP including investments up to $ 19.7 bln were included.

Because of the importance of the delta vision and goals with multi-sectoral scope and impact, BDP 2100 has been developed as one of the top national plans and was incorporated or referred to in many important national documents as the Five Year Plans, the Perspective Plan 2041, National Adaptation Plan, a dedicated part of the Annual Development Budget while a Delta Act is foreseen to cover the whole institutional framework. This framework consists of Delta Governance Council, Project/Program Selection Committee, Delta Wing, Focal Point Network of the 11 officially involved ministries, Delta Fund and BDP 2100. This dedicated framework started functioning in terms of guidance and decision making, which is a transformation in itself.

Collaboration between different levels of government agencies involved in the delta planning and execution process is crucial for successful implementation of the Bangladesh Delta Plan (BDP). While the Delta Government Council and Delta Wing have already been established, it is essential to coordinate their work with national level agencies like WARPO. It is also essential to focus on the development of Delta Act and mobilization of delta fund to ensure proper implementation of Bangladesh Delta Plan.

In terms of horizontal coordination, currently 11 ministries are contributing to the implementation process of Delta Plan. Usually, these ministries have their own set of responsibilities, budgetary constraints, and decision-making processes. Delta Plan also deals with key sectors like water resources management, infrastructure development, agriculture, environment, and disaster management. These sectors are typically overseen by different government agencies. Vertical coordination among different ministries and agencies are important for facilitating the integration of policies and actions across sectors, reducing conflicts and optimizing resource allocation. In this regard, setting up clear communication channels between different government agencies are important to ensure that information flows smoothly, and issues can be addressed promptly.

Knowledge management is seen as absolute requirement for the long-term adaptive planning. As said, a delta plan is and will always be science and knowledge driven. The whole knowledge management process has been utilized. From setting information and data available for the planning process, to generating knowledge, e.g., in the 26 Baseline Studies and BDO 2100 itself. This and other knowledge is available and accessible on the BDP 2100 knowledge portal. Besides, knowledge gaps relevant to adaptation have been identified and included in the knowledge agenda. Frequent updating of BDP 2100 requires new insights in information and data of progress, trends and developments. Tools have been utilized to deal with knowledge management, as usual computers, the mentioned knowledge portal but also computer aided Touch Tables utilized in many interactive sessions. Development of a meta model system which was developed in a different project has been co-developed and is being used for decision support on interventions.

15.1.5 Investment Planning and Decision Making

GoB indicated during the formulation stage, that the Delta Plan needed to be elaborated in an investment plan, to streamline the upcoming debate where and what should be done against which investment costs. Given the comprehensive package of multi-sectoral strategies, this was a huge additional task and was because of that carried out together with a team of the World Bank Group. The Investment Plan preparation is important to make really clear what the exact projects are and what belongs to them. This is in fact the operationalization of the strategies in real terms of measures and projects including locations, estimates of designs for problem solving. Stakeholders, environmental and socio-economic impacts, stakeholders and costing.

While Bangladesh Delta Plan have 80 investment projects, many projects in the plan are interdependent. Deciding where to start becomes more complicated when some projects rely on others for success. This mainly happens as various stakeholders, including local communities, government agencies, and private sector entities, have different priorities and interests concerning specific projects. Balancing these interests and aligning them with the overall goals of the plan can be challenging (see e.g. De Heer, 2009). In addition, some projects require significant environmental and social impacts, and these need to be carefully assessed before implementation. Choosing the wrong starting point could hinder the effectiveness of subsequent projects.

15.1.6 Regional Programming, Preparation of Measures, Adaptive Design of Interventions

The readiness of implementing agencies is a crucial aspect of the successful implementation of the Bangladesh Delta Plan (BDP). The readiness of these agencies directly impacts the efficiency, effectiveness, and timely completion of delta projects. It concerns especially the right amount of people, skills and funding, including necessary rules and regulations.

Because of the impact of climate change, implementing agencies will encounter unforeseen challenges and changing circumstances during the delta plan implementation process. As Bangladesh is one of the most active deltas in the world, the implementation agencies need to have access to most recent climate and project data. For ensuring climate change adaptation, implementing agencies will need to constantly train themselves. For ensuring readiness of implementing agencies, technical training, knowledge-sharing workshops, and exposure field visit to similar projects in other countries need to be arranged.

15.1.7 Further Preparation and Realization of Interventions

Realization starts with project preparation according to the regional or basin wise programming. Feasibility studies are mostly quite comprehensive, including all related research and detailed design activities as well as societal, economic and environmental assessments.

Project management capabilities, technical know-how, and financial capacity are important because of the upcoming project preparations and execution in the field. Assessing the capacity of implementing agencies is essential to determine if they have the necessary expertise, technical skills, and resources to undertake the projects. Localization of execution works is important in terms of adequate local preparations of interventions and stakeholder involvement.

15.1.8 Utilizing Funding from Various Sources

Funding has been arranged along the lines of normal project preparation by the implementing agencies while the BDP 2100 team expected significant steps and short tracks based on the gradually developed sense of urgency with respect to climate change adaptation.

While Bangladesh has achieved a remarkable growth in the last decade, in recent times the country is going through some macroeconomic challenges. The Government of Bangladesh has recently taken some austerity measures and currently focusing on the implementation of existing projects. Despite macroeconomic challenges, the Bangladesh government is trying to prioritize the funding of BDP projects in its annual budget. At the same time, Government of Bangladesh needs to explore various innovative sources of funding like Reducing Emissions from Deforestation and Forest Degradation (REDD +), Global Environment Facility, Mitigation Action Facility, Adaptation Fund (AF) for delta plan implementation.

REDD + is a UNFCCC-developed framework to provide financial incentives to reduce greenhouse gas emissions. Through REDD + , it is possible to implement reforestation, water conservation and afforestation initiatives. While this financing mechanism is not suitable for large scale hard infrastructure projects, REDD + can be an effective mechanism for promoting nature-based solutions for implementing Bangladesh Delta Plan. REDD + offers results-based payments to communities that successfully reduce deforestation and forest degradation emissions. These payments are based on the amount of carbon emissions that are reduced or avoided.

The Global Environment Facility (GEF) is a multilateral fund that helps confront various environmental issues, including biodiversity, climate change (mitigation and adaptation), international waters, waste management, and land degradation. While Bangladesh has an active GEF project portfolio of 27 projects, the country can still access additional funding from GEF for implementing Bangladesh Delta Plan. Integrated water resource management to address water security, droughts, and flooding is one of the key adaptation themes for GEF. One unique aspect of GEF is the mobilization of mainly grant financing mechanism, which is not the case for other climate funds like GCF.

The Mitigation Action Facility, provides technical support and climate finance for mitigation projects. Renewable energy is one of the thematic strategies of Bangladesh Delta Plan and for implementing renewable energy projects Bangladesh can utilize the funding support from Mitigation Action Facility. Like GEF, the Mitigation Action Facility mainly uses grant financing mechanism. While this facility has already supported 47 projects with total funding commitment of EUR 668 million, till now Bangladesh have not received any funding from the Mitigation Action Facility. The Adaptation Fund (AF) provides adaptation finance to developing countries. Though AF started on a small scale, the fund size has grown with time. As of November 2020, the fund has received approximately USD 1.05 billion in financial contributions. While Bangladesh is currently facing climate change adaptation challenges, so far Bangladesh have received only one Adaptation fund supported project.

15.1.9 Monitoring, Reflective Evaluation and Conducting Adaptation as Ongoing Development and Transformation

The strategy preparation and making is really comprehensive and in a complicated and dynamic delta is of Bangladesh a huge endeavor. This is even more so in the government's delta vision containing a broad narrative of adaptation by development with the aim to contribute to reaching the middle income status. In this perspective, the government decided to develop a dedicated institutional framework in which decision making and coordination of the multi-sectoral approach can be streamlined. This indicates, that the GoB considers working with the BDP 2100 for a long time with periodic actions to monitor and evaluate progress of adaptive actions and investments as well as relevant developments. In order to make the delta plan up to date, with good insights of the state of the delta in the midst of climate change, monitoring and reflective evaluation with the related tools and indicators is of crucial importance. Tipping points in developments or their main drivers may imply revision of strategies and investments. An interesting aspects is whether or not the adaptation proposed measures with massive investments will be followed by politicians and policy makers given the competition among many issues also claiming budget.

Monitoring and evaluation have with respect to adaptation to climate change an important relationship with time and consistency. It's important to use the lead time climate change, e.g., with its impact on sea-level rise in a sense still has. Consistently moving towards achieving timely adaptation is however crucial for water safety and security, avoiding serious climate change impacts and disaster. A well-known factor is the potential break or gap between strategy formulation and implementation, in which knowledge and priorities can easily be lost. In the Bangladesh case, the government wanted to achieve continuity which resulted in a quick start-up of the implementation process. However, implementation needs the experience and technology of executing agencies. Lesson is that they should be much more closer involved in the implementation process preparations. Another lesson is that implementation not only should relate to the national level ministries and agencies but also to regional and local level institutions. Moreover, utilizing community-based adaptation is essential to mobilize local people to contribute in the adaptive activities with their capacities and technologies.

A crucial question with respect to monitoring of developments and reflective evaluation is: is a shift of scenario already taking place? Since the approval of BDP 2100 in 2018, changes have up to earl 2024 occurred on the two main axes of the scenario figure. The implications of ongoing climate change led to warmer sea water and atmosphere, which may have led in Bangladesh to heavy rains and resulting floodings, e.g., in the north of the country. At the same time, the economy showed constraints during the Covid-19 pandemic and due to changes as increasing inflation and interest rates. Besides, funding from international sources as the WB and GCF were disappointing under these circumstances. These developments could indicate the relevance of a less positive scenario and that adaptation strategies need to be considered and adjusted to the topical climatic and economic situation. This point needs to be further explored and considered in terms of data and impacts.

The Bangladesh Delta Plan 2100 has been developed as techno-economic, adaptive, long-term integrated science and knowledge driven plan. With the adoption of the BDP 2100, Bangladesh experienced a paradigm shift in its planning and implementation systems. This paradigm shift consists of (1) following an adaptive planning and implementation methodology with emphasis on adaptation to climate change, (2) contributing to food security and (3) to socio-economic development, with related cyclical instead of linear planning, a holistic and multi-sectoral instead of a sectoral (silo) approach, a new institutional framework concerning decision making, coordination and funding as well as organization of decentral water / delta institutions. Consequently, significant institutional and policy implications, vast financing challenges and many new managerial and organizational tasks needed to be developed and practiced. By working on these implications, challenges and tasks, GOB and involved sectors are going through a strategic reorientation process, aimed at readiness for climate change adaptation and realization of related delta vision, goals and investments.

The, by BDP 2100 triggered, strategy process consists of several transformations, e.g., in the GOB Institutional Frame Work on decision making (Delta Governance Council cs), the role of the Planning Commission with the upcoming Delta Wing in (horizontal) coordination with the involved sectors and (vertical) coordination with (partly new) regional institutions. Important is also to further utilize ways for involvement of the Private Sector, for adequate funding and utilization of the Green Climate Fund as already proposed and utilize other funds as well. The transformation process has not been finalized yet, partly due to the challenges of the transformation itself, partly due to the Covid-19 pandemic and partly due to the GOB administrative systems and culture. Changing the water governance related institutional framework, administrative systems and culture, also with respect to frequent updating of BDP 2100 and adaptive programming, sustainable project preparation and adaptive design, require pertinent choices and robust GOB action. These actions have been started and initial results have been achieved but they will take more time and budget for continuous learning and innovation as will be required for ongoing and adequate adaptation to climate change. Given the latest developments, extended support to BDP 2100 implementation by contributions of development partners, international financial institutions and climate funds can be expected shortly to gear up the preparation of adaptation projects and increase project execution. We don't know whether successful adaptation to future climate change and probably severe impacts will be possible in this complicated dynamic delta but by then, Bangladesh will have shown its experience with learning to adapt and increasing its resiliency.

15.2 Institutional Reforms Bring Transformation and Adaptation

Important lesson is that it was recognized during the strategy making of the multi-sectoral and adaptive BDP 2100, institutional adjustments needed to be made. The government was not used to multi-sectoral approaches, decision making and joint realization. In that sense, the setup to organize and prepare BDP 2100 was already exemplary of how a multi-sectoral set up could become. Besides, decision making and mainstreaming by the NEC and FYP facilitate coordination of multi-sectoral approach and multi-facetted realization of the delta plan.

Decision making development was required and took shape as proposed in BDP 2100 by arrangements of the Delta Governance Council, Project/ Program Selection Committee, Delta Wing of the Planning Commission. These institutions focus on adaptation by realization and updating of the delta plan, by investment planning based on an appraisal framework for prioritization within the strategy framework.

Adaptation by development is a typical choice for BDP 2100 as the delta vision incorporates not only water safety and security but also food security and socio-economic development. Although BDP 2100 is a long-term adaptive water centric strategic plan, it has a broad techno-economic scope. The government has chosen this formula of adaptation by development so that BDP 2100 could contribute in achieving a middle income position of Bangladesh in near future. The ambition is high as a senior official stated: 'development is also creative destruction' with respect to required institutional reforms and possible upcoming innovations as well as resistance to change. Sectoral and regional or local transformations mean doing things differently which is mostly not in everybody's interest and appreciated. Although ambitions for development are high, the intended Delta Fund and investment funding from other sources is still in a process of becoming. This can be seen in terms of changed economic and financial circumstances in the country. Besides the government wants to capture the mentioned institutional changes in a Delta Act which takes time because of the legal and political process to realize such Act.

The new adaptive and multi-sectoral planning approach with scenario-based long range planning is seen as ‘paradigm shift’, altering the existing linear and more or less trial and error planning approach. It transformed thinking and acting on adaptation in a learning approach, in which monitoring and reflective evaluation play an important role in combination with updating of climatic and other circumstances as dealing with transboundary issues. For the transboundary approach, BDP 2100 provided a water diplomacy approach in relation to climate change and impacts on rivers and drought situations.

Realization of delta vision and goals concerns to a large extent ‘rationalization’ programs. These are substantial programs for large maintenance and climate ‘proofing’ measures. Rationalization means capital maintenance of existing infrastructure and renewal when needed added by climate impact related measures. This was accompanied by the introduction of Asset Management as infrastructures, buildings and other facilities can be systematically maintained during a certain lifecycle. The Rationalization Programs are comprehensive no regret programs. Maintenance and or renewal of infrastructure needs to be done anyway, so, adaptation measures can be easily combined in the total program. The advantage is, that the infrastructure already exists. This means that the Rationalization programs could go through a shorter feasibility and preparation process than in case of development of new infrastructure. Asset Management provides the necessary management tools for systematic maintenance in a life cycle investment pattern. The purpose and costs of infrastructure are related to the quality of performance in its setting over a period of time. Reviews show whether or not infrastructure performs up to the mark or that maintenance and or climate adaptation measures are needed to keep it on the desired performance level.

15.3 Practices Performed by a Mix of Practitioners

The practices performed during the strategy and realization process are carried out by practitioners from various backgrounds as management consultants, water experts from knowledge institutions, engineers, economists, climate and environmental scientists, spatial development and ecology experts. During the strategy and realization process, dedicated teams have been formed to mobilize and setup the project, to conduct baseline study research on 26 themes requiring specific practitioners. The strategy development under different scenarios required specific knowledge and practices to analyze the main developments and drivers Consultation and participation processes during the roadshow and also later on demands special communication skills and tools utilized in the delta ateliers. Institutional specialists developed the institutional framework utilizing practices from public administration and organization and management sciences. Project proposals have been proposed and elaborated by technical specialists. Every discipline brings its own technology and practices developed in those disciplines or based on experiences elsewhere. The situation in Bangladesh and the scope of the BDP 2100 required in many cases rethinking and adjustment of the practices people were used to apply because of broad array of relevant factors in the holistic approach and attempts to integrate or at least inter-relate findings or actions on the various issues and challenges So the applied practices were partly recursive and adaptive in nature. Important is to relate the outcomes of the strategy and realization process to the sensemaking and sensegiving by practitioners in the application of their practices within the wider societal and physical context. Practitioners bring their knowledge and practices from their background into the strategy process in interaction with the dynamics they are in that process confronted with. Strategy is in this sense an expression of the politics of culture but also of in that setting performed practices (Jarzabkowski, 2004; Whittington, 2006, 2017).

At the start of the BDP 2100 strategy process, practitioners brought experience from Dutch delta planning with them. These practices were to some extent introduced as adaptive delta management. Against this backdrop of the Dutch delta planning, the Government of Bangladesh was interested in the experience with such approach with already developed professional planning practices, given the challenges Bangladesh was expected to be confronted with. In the tendering procedure for the Bangladesh Delta Plan, practitioners were required and reviewed to have vast experience with the Dutch delta planning. During the strategy process, they were challenged to expand that experience with the Bangladeshi demands to develop their unique BDP 2100. From a different theoretical angle, some consider this utilization of knowledge and experience from one country by another one as policy transfer or policy translation. Here it is seen as utilization by GoB of strategy practices from a trusted and experienced country. Practitioners who became part of the consultants team to develop BDP 2100 were in the joint tender documents explicitly requested to bring in proven experience with the Dutch delta planning. In organization and management praxis it is often seen that one party is interested in proven strategies or technology to have advantages of application of to some extend already developed practices.[1]

15.4 Regional Water Institutions and Community-Based Adaptation

BDP 2100 is not only meant as top down approach. The Delta Plan incorporates institutional reform by the formation of 'water user institutions'. First, this is a reaction to the fragmented development of local and regional participative organizations. The intention is to closely involve local and regional stakeholders in implementation activities and updating of the Delta Plan and Basin wise programs. One of the institutional sub-strategies is to develop local or regional water institutions, consisting of stakeholders directly related to foreseen adaptation or interventions in their river

[1] On progress e.g. Maroon, 2023.

basin or area. This in relation to local and regional institutions and, e.g., Water Management Organizations, their Associations and Federations. Second, there is an economic motive for formation of water user institutions. Taking care of water safety and security, food security and socio-economic development means in fact that many water related services (embankments, regulators, drinking water and sanitation, fresh water and groundwater for irrigation or other production functions, water purification etc.) are or will be provided to beneficiaries of these services. It is viewed as fair and logical to introduce service charges for delivered services. Water user institutions offer a way to arrange influence on features and quality of the service delivery and on decision making about them. Besides services and their quality or performance can be discussed to reach agreements on payment by stakeholders for service charges.

More attention should already have been given to community-based adaptation.[2] As climate change adaptation rises up the international policy agenda, matched by increasing funds and frameworks for action, there are mounting questions over how to ensure the needs of vulnerable people on the ground are met. Community-based adaptation (CBA) argues for tailored support at the local level to enable vulnerable people to identify and implement appropriate community-based responses to climate change themselves. Community-Based Adaptation to climate change explores the challenges for meeting the scale of the adaptation challenges on what can be done at the local level, using local knowledge and preferences. In BCA discussions and proposals, fundamental issues are addressed (a) like learning how to move from place-based projects towards more programmatic adaptation planning; (b) how community-based actions fit with larger scale adaptation policy and programs; c) what is the position of community-based activities related to institutions with tasks or actions that enable or undermine adaptive capacity and action. The same can be said and questioned about damage and loss due to climate change impacts. It can be observed, that BDP 2100 follows in its elaboration of large Rationalization Programs (see Investment Plan) and Basin Wise Programs a more programmatic approach instead of standalone projects. Also the intention to form Water User Institutions is a step towards the ideas about CBA. Transformation in this area should be further addressed and arranged accordingly. Another item concerns climate change migration with many questions surrounding it. Climate change migration is not considered as strategy in BDP 2100, which is based on protecting, accommodating, advancing and not retreat. So far, it's up to the people to stay or leave certain areas.

Adaptation also needs to be ecosystem based. Bangladesh is inhabited by large numbers of people, including farmers and fishers, depending on natural ecosystems throughout the country for their livelihoods. The BDP 2100 recognizes the threats and risks of climate change to the ecosystems, and takes actions to'protect it. However, current anthropogenic factors like industrialization, population growth, land use, pollution, pose a threat as well. So, ecosystem based adaptation is important and could be achieved by division of related tasks over existing executing agencies, to be addressed in an integrated way.

---

[2] See e.g. Schipper et al. (2020).

All the above aspects manifest themselves in the river basin and adjoining areas. The Basin Wise Implementation Programs and also the additional Investment Program provide the vehicle for incorporating Community-Based Adaptation interests, activities and funds. The Water User Institution can play an important role here to divide tasks and coordinate activities in relation to BDP 2100 project execution. The same counts for addressing damage and loss and ecosystem-based adaptation.

References

Choudhury, G. A., Terwisscha van Scheltinga, C., Van den Bergh, D., Chowdhury, F., De Heer, J. M., Hossain, M., & Karim, Z. (2012). *Preparations for the Bangladesh delta plan.* Alterra Report 2300, ISSN 1566-7197.

De Heer, J. M., Choudhury, G. A., et al. (2014). *Inception Report Bangladesh Delta Plan 2100 Formulation Project.*

De Heer, J. M. (2009). *Organizing under pressure: Strategy and change as expressions of the politics of culture.* Inaugural Address for the Chair 'Cultural Change and Strategy'. VU University Amsterdam.

Jarzabkowski, P. (2004). Strategy as Practice: Recursiveness, adaptation and strategic practices-in-use. *Organization Studies 25*(4)

Maroon, M. (2023). *From poverty to progress: Understanding humanity's greatest achievement.*

Ministry of Planning. (2018). *Bangladesh Delta Plan 2100.* plancomm. gov.bd/

Schipper, E., Ayers, J., Reid, H., Huq, S., Rahman, A. (2020). *Community-based adaptation to climate change.*

Whittington, R. (2006). Completing the practice turn in strategy research. *Organization Studies, 27*, 613.

Whittington, R. (2017). Greatness takes practice: On practice theory's relevance to "great strategy". *Strategy Science, 3*(1), 343–351.

Chapter 16
Adaptation in a Transformative Approach

Abstract This chapter broadens the scope by bringing the Bangladesh experience together with briefly described experiences in the Netherlands and Vietnam. International experience offers inspiration and show sometimes experiences in uncharted waters. This kind of experiences utilizes earlier thinking and acting and stimulates creative thinking for other situations rather than replication there. Bangladesh was highly interested in adaptive delta management ideas and experiences but asserted from the start that it wanted to develop its own Delta Plan, fitting in the Bangladeshi setting, context and also administrative culture as well as development position and challenges. Hence the focus on adaptation by development. There exist indeed of course important differences like in constitutional law, structure of the state and socio-political, economic and cultural characteristics. The adaptation and transformation aspects as elaborated in relation to the in Chap. 2 identified sensitizing concepts enrich the initial model with the field experiences. Based on these experiences, the adaptation process has been divided into a sequence of practices with characteristic activities. It is considered, that these practices offer a performative repertoire which can be addressed in every adaptation situation. However, not all will necessarily be as comprehensive as shown in the Bangladesh experience in this complicated delta. In different situations, choices can be made based on diagnostic search and identification of scope and challenges to be addressed. Moreover, the tracking of strategies will show if and how the existing strategic planning tradition and knowledge bases can sufficiently contribute in a certain situation to the needs for adaptation and the necessary practices, in line with the required or preferred time frame.

Keywords Transformation · Transformative approach · Sensitizing concepts · Practices · Sequence of practices · Praxis · Practitioners · Dutch delta program · Mekong delta plan · Institutionalization · Multi-sectoral approach · Adaptive programming · Adaptive design · Transformative technology management

J. M. de Heer and G. A. Choudhury, *Strategies for Adaptation to Climate Change in a Transformative Approach*, Springer Climate,
https://doi.org/10.1007/978-3-031-88884-7_16

16.1 Other Adaptation Experiences

Many countries are dealing with adaptation issues and challenges as about 33 major river deltas are existing worldwide. Throughout human history river deltas have attracted people. However, the inhabitants in these delta countries are confronted with climate change impacts in addition to their usual delta issues. Most delta countries are or will be confronted with additional, often threatening delta dynamics. The Ganges–Brahmaputra delta (Bangladesh) is considered to be the largest in the world. Other famous large deltas are the Chao Phraya (Thailand), Danube (Eastern Europe), Fly (New Guinea), Fraser (Canada), Godavari (India), Krishna (India), Lena (Siberia), Mackenzie (Canada), Mahakam (Kalimantan), Mekong (Vietnam), Mississippi (USA), Niger (Western Africa), Nile (Eastern Africa), Orinoco (Venezuela), Parana (Argentina), Po (Italy), Shatt-el-Arab (Iraq/Iran), Volga (Russia),Yangtze-Kiang (China), Yukon (Canada) and the Rhine-Meuse-Scheldt deltaic complex (Western Europe). Some 500 million people inhabit the delta areas of the world. According to studies of the 33 major river deltas, 24 are sinking due to flood-control efforts and other human-caused changes to the river systems (Syvitski et al., 2009) E.g., Jakarta is sinking about 10 cm a year and the Mississippi delta is losing land annually. Natural consolidation of delta areas combined with extensive groundwater extraction, and activities of flood control and use of sediments along the rivers cause serious problems of subsidence which are hard to overcome. The combination of sinking deltas and rising seas will increase the damages caused by natural hazards and flood events.

Deltas belong to the most densely populated areas in the world. Deltas normally provide plenty of food for their inhabitants because the soil is fertile to allow various forms of agriculture and the river and sea areas are easy accessible for fishing. Deltas often have large ports and harbors and therefore become hubs of industry and trade. On the other hand, deltas, with land hardly above sea level or often below it, are vulnerable to natural disasters. In future, as a result of climate change and rising sea levels, this vulnerability will most likely become more acute and cause severe impacts on the natural processes in deltas. The lives of the millions of people living there may be affected dramatically.

The Bangladesh Delta Plan 2100 shows in Part II the Bangladesh Experience and Chap. 12 summarized the lessons learned.

16.1.1 Aspects of the Dutch Delta Program

The Dutch Delta Program (2009) is the result of a strategic planning process as response on threatening experiences but also with an eye on future climate change developments. It is the second exercise in the Netherlands on preparation of a 'delta plan' for taking action, especially with respect to water safety, fresh water supply and to some extent spatial planning in relation to water management.

The Netherlands is located in the North-West of Europe in a delta area where the main rivers Rhine, Meuse, and Scheldt discharge on the western side of the Netherlands into the North Sea. This is a highly dynamic delta area, gradually shaped in the course of time. Many floods occurred in the delta area, caused by north-western storms, high river discharges from Rhine and Meuse, and sometime due to behavior of ice in melting rivers. The advantage of living in a delta area led to understanding of the water dynamics. This made construction of early protection works and new land development in coastal areas possible already before the year 1200. The technology developed and around 1530 reclaiming land of small lakes started to take place. The land reclamation became larger and larger with thanks to the windmill technology to pump water out of the lakes and discharge it via channels into the rivers. The need for reclamation was high because of shortage of land for food production and in the course of time for many other societal functions.

The location of the Netherlands in this delta area on a strategic position in North-West Europe made it interesting for trade and navy purposes to develop ports. Being used to shipping in this area and due to the availability of resources, the ship-building technology progressed quickly. With the intention to explore new possibilities and generate resources and wealth, trade relationships were established all over the world. It became possible to finance strategies aimed at larger protection works and advancing seaward or inland. The old Dutch word 'polre' means development of a pool. This led to the word 'polder' meaning that land in a pool or lake was reclaimed and surrounded by dykes, and provided by pumping (windmills or modern high capacity pumps) and channel provisions to drain the surplus of water into adjoining river systems or sea. In this way, people learned to live with water and created institutions to take care of related tasks and rules, in relation to the involved stakeholders. Nowadays, millions of people live in drained polder areas, in many cases even below sea level. However, maintenance of dykes, infrastructure and facilities was costly and frequently poor, causing dangerous situations. The political and financial priorities of the government after the second World War were focused on rebuilding the country and important signals about the poor condition of protection works were disregarded. A north-western storm pushed seawater to a high level in the funnel shaped North Sea causing together with strong wave attacks the devastating flood of 2 February 1953 when many dykes failed in the south-west. 'Never again' was the broad opinion because 1836 people died and the damage and loss were tremendous. The political mood changed and a decision was quickly taken to form the first Dutch Delta Commission as advisory Committee for the minister of Transport, Public Works and Water Management. The Commission prepared proposals for drastic improvements, the first Delta Plan, with strong focus on barriers, stronger protection works and closures. Shortening the coastline and thereby the required sea defenses is characteristic for the approach. A Delta Act was approved in 1958, to institutionalize the Delta Plan as well as a dedicated organization called Deltadienst, to indicate locations of the delta works and to increase safety standards. Besides reinforcing sea defenses and embankments, the delta Plan focused on protection of the coast by disaster risk reduction, reducing the risk of a disaster to a probability of one in 3000 years. The Delta Plan has been implemented over quite some time by the

Delta Works Program including research, engineering and innovative technology to realize the objectives of reinforcing and modernizing water defenses and shortening the coastline. The delta works are built following the Delta Act and a comprehensive program which started in 1958 and was completed in 1997, costing a total of US$ 15 bln. Transformations were reached politically with the changed political willingness and decision making, and institutionally with the formation of institutions as Delta Commission, Deltadienst (project organization), Delta Act and Delta Plan, and large dedicated private sector consortia, all leading to implementation of state of the art flood prevention infrastructures. Transformation of the physical systems is significant with considerable impact on safety and security but also on environment and water ecology.

Against the backdrop of the described earlier experiences, awareness of climate change and the possibility of higher sea levels and river discharges is increasing in the years 2000. In combination with gradual land subsidence, the risk of flooding was expected to increase. The interpretation of possible ongoing climate change with threatening impacts on the low lying delta area, especially in terms of a rising sea level, salt intrusion and higher river discharges, the government decided to appoint a second Delta Commission in 2008. This Commission presented its report in 2009. Based on its recommendations, the government prepared a second Delta Act in 2011, established a high level position of Delta Commissioner to form and manage a new Delta Program and set up the Delta Fund to provide and guarantee annual funding on a long-term basis. The Delta Commissioner reports to the minister of Infrastructure and Water Management and formed a support staff to assist in the tasks as per Delta Act. The new Delta Program is shaped as multi-level national program in which the national government, provinces, municipalities and water authorities work together and share financial resources. Besides, collaboration has been established with civil society organizations, the private sector and research or knowledge institutions. The Delta Program is aimed at maintaining the Netherlands as a safe and prosperous country. It focuses therefor on three National Programs: Fresh Water Supply; New Safety Standards on flood risks; Reconstruction and Development. These have been added by six Regional Programs: Coastal Zone; Wadden Sea Area; South-Western Delta (river Scheldt Delta); rivers Rhine and Meuse Delta; Lake IJssel Area and a Program for the two main rivers Rhine and Meuse. All Programs work on long-term scenario-based analyses and cover the period from 2015 to 2100. The Programs are expected to prepare so called Delta Decisions on their subjects. concerning issues and challenges of national scale or regional water management. (see GoN, 2018).

The institutional arrangements of Delta Act 2011, Delta Commissioner and Delta Fund together are basically meant for horizontal and vertical coordinating of the involved water institutions at various government sectors and levels. The delta governance functions, in line with principles of the Dutch decentralized unitary state, in a decentralized way and the Delta Program plays an essential coordination role in that, e.g., on research and modeling, long-term planning, policy and regulatory assistance and financial support. The preparations of investments and execution of interventions are taken care of by the decentralized institutions. Investments are funded accordingly, for a large part by resources of the decentralized institutions

(50% by regional water boards; 25% by municipalities; 5% by provinces and 20% by the national government). Investments projects of national importance are funded from the national budget. Next to the Delta Program, resources are spent on research and modeling, maintenance and upgrading of infrastructure. The Delta Act established a multi annual Delta Fund, to finance the needs of the Delta Program and ensure adequate funding of critical water programs, For the period till 2050 the Delta Fund, according to figures of 2024 provides € 27.4 bln.[1] This is about € 1,1 bln. a year, or about 0,1% of the Dutch GDP as it was in 2023. Spending priorities are agreed with participation of the mentioned water institutions, the national government and parliament. Besides national level investments and projects also co-financing could be possible of priority projects managed by the decentralized water institutions. Research and knowledge management by dedicated institutions, universities and private companies get high priority in the Delta Program to not only improve existing technology and infrastructure but also to innovate, experiment and develop delta related technology.

16.1.2 Aspects of the Mekong Delta Plan of Vietnam

The Mekong River stretches 4350 km, originating from its source in the Tibetan Plateau to delta streams discharging in the South China Sea. It is a lifeline for tens of millions of people in China, Laos, Myanmar, Thailand, Cambodia, and Vietnam who mostly rely on agriculture and fishing. The Mekong Delta shows significant agricultural modernization and development last years. Traditionally, the main focus in agriculture was on rice production made possible by the water management systems in this delta area. Recognizing the need for modernization and higher income generating possibilities, the Vietnamese government introduced in 2000 a policy on agricultural diversification which intended to shift production from the rice mono-culture to a more diversified system which also includes aquaculture, fruits and horticulture. The Mekong delta region is focused on the government development objectives in becoming a balanced multi-sector economy with increasing urbanization and industrialization. However, industrial development, e.g., related to food production in the Mekong Delta and actual developments in traditional agriculture are deviating from existing government policies aimed at high-tech agricultural development and full exploitation of the competitive advantages of the region. Against this background, with a strong emphasis on agricultural production, the Mekong Delta Plan has been prepared in 2010.

In its geographical situation, the Mekong Delta is likely to be severely affected by the adverse impacts of climate change. Peak flows and river floods are expected to increase in the wet season. Decreases in dry season flow may lead to serious fresh water shortages. The expectation is that sea level rise and saltwater intrusion will further increase, subjecting large areas of the coastal delta to a brackish environment.

---

[1] See: Delta Program 2023; 2024; https://english.deltaprogramma.nl/

Transboundary developments may further aggravate the stress on land and sufficient water resources. Hydropower dams on the mainstream of the Mekong River, including about 12 from China, pose risks of depleting the river's flow and reducing sediment transport to downstream areas. This contributes to worsening salinity intrusion and erosion in the Mekong Delta. Socio-economic developments also determine to a large extent the ever-increasing pressure on the delta's land and water resources. Developing an appropriate strategy to adapt to these changes has become crucial. The Government of Vietnam recognized the vital role of water as a key natural resource for the development of the Mekong Delta. In 2008, the National Target Program response to Climate Change (NTP) was issued with the main strategic objective to assess climate change impacts on sectors and regions, and to develop feasible action plans to effectively respond to climate change in the short-and longer-term. The premise for the Mekong Delta Plan is to contribute to sustainable management and use of water resources realizing and maintaining a prosperous delta, both economically and socially. At the same time adapted to changes in water resources due to climate change impacts. The delta plan indicates climate change uncertainties, analysis the water and agriculture related issues and challenges confronting the delta till 2050, and from 2050 towards 2100, and presents a clear long-term vision on agricultural transformation towards Agro-Business Industrialization as a promising future strategy. Taking into account the existing government policies and the institutional framework, it is recognized that the orientation for the agricultural sector in the delta needs to shift from a focus on production towards a focus on added value by improving the product-value chain, involving public–private partnerships and introducing market-based mechanisms. The Mekong Delta Plan includes 'no-regret' and priority measures, as well as measures that can be postponed towards the longer-term that Vietnam could adopt to ensure a safe, prosperous and both economically and environmentally sustainable and climate proof future for the delta region. The Mekong Delta Plan is presented as a reference document for Government agencies and organizations at all levels of the country. As such it provides a tool to support the review, coordination and integration of present and future master plans as well as a guideline towards implementation. Its findings are also important, indicating a course of action for the international development partners, universities and research institutes, potential domestic and foreign private investors as well as for a broader audience of individuals and organizations committed to the sustainable development of the Mekong Delta.

Brief comparison of the delta strategies and plans shows interesting differences in reasons for strategy making and focus of the strategies. Bangladesh wants to improve safety and security with respect to usual and worsening threats and climate change impacts, reduce economic setbacks and include food security and socio-economic development. Vietnam wants to diversify and therefore transform and develop its agricultural sector including adaptation to climate change impacts. The Netherlands focuses primarily on water safety, fresh water and to some extend on spatial aspects related to water. All three countries elaborate strategies for protection and accommodating while only Bangladesh so far expects a lot from seaward advancing and reclaiming and protecting land along the major rivers. No one choses strategies as 'doing nothing', or 'retreat'. The major differences are related to the specific delta locations, the dominant visions based on population size and growth, per capita income, phase in economic development, centralized or decentralized governance, development of competencies, administrative culture and institutional development over time. Given these differences, replication of an international model in original form is not a viable proposal. The Vietnam example shows a strong focus on agricultural transformation in relation to water and climate aspects. The governance approach is quite different due to its state form and procedures. The development and strategy making for the Mekong delta and recent achievements set interesting examples. Vietnam as well as Bangladesh follow an approach of adaptation by development, coupling important development initiatives to water management and more integrated delta governance. The Dutch water governance, as explained above, differs significantly from Bangladesh water governance especially in relation to the physical systems, the institutional framework, the strategies (national, hotspot and thematic) decision making, coordination with other sectoral agencies, in funding strategy, differences in attention and budget for operation and maintenance. In both countries, asset management is more utilized in combination with rehabilitation of infrastructure in combination to climate change adaptation. Lessons based on the Dutch experience can be utilized for developing the Bangladeshi water governance approach: principles of water governance and financing, structuring of institutional set up including water boards with a specific taxation regime, cost recovery in certain areas such as water supply, wastewater collection and treatment.

The three countries clearly show huge expectations from long term strategy making, to be prepared in their delta areas under uncertain conditions and the willingness and urgency to adapt to impacts of climate change and to transform systems in relation to that.

16.2 Typical Transformation Aspects

Transformation has been defined in Chap. 1. as process of meaningful transforming, a dramatic and marked change in form, nature or appearance. Some of the transformation are highlighted below in order to understand how they come forward in an adaptation process and learn more about them.

16.2.1 Deciding on an Adaptation Strategy Process is Already a Transformation

Deciding on the initiation of a strategic long term planning and adaptation strategy process means, that awareness about the need and possibility of such a process exist and that discussions and deliberations took place about it. This mindset and sensemaking by leading officials or even a Cabinet can be seen as meaningful for the country and marked starting point of change. In the course of time from the start in 2011 with the decision to install a preparatory team and especially since 2013 till 2024, episodes with practices including their characteristic focus and activities have been identified and indicated as sequence of adaptation practices. The result is BDP 2100 and outcomes and adaptation are forthcoming. The lesson is that in a delta as in Bangladesh combined with the government ambitions, proceeding in one go from problem identification to project execution on the ground is not possible. In the adaptation and transformation process concerning BDP 2100, nine practices have been developed and performed. The experiences and a number of these practices can be utilized for ongoing adaptation and renewal.

16.2.2 The Adaptive, Scenario-Based Strategy Process

The BDP 2100 strategy process composed by a sequence of applied practices, dealing with uncertainties and focused on adaptation brought a change in thinking and acting about planning. The new adaptive and multi-sectoral planning approach with scenario-based long range planning is, as said before, seen as positive 'paradigm shift', altering the existing linear and more or less trial and error planning approach. It transformed thinking and acting on adaptation in a learning approach, in which monitoring and reflective evaluation play an important role in combination with updating of climatic and other important circumstances as dealing with transboundary issues.

The Adaptive Pathways approach was an element of adaptive delta management but although attractive on first sight, not a very clear one for BDP 2100 development and implementation. The reason is probably that many fruitful no regret measures can be taken in the first place. Pathways could technically be 'designed' but tipping points, scenarios as well as development and choice of possible alternative measures require more in depth attention. Another issue was here, that experts defined the pathways but actual decision making takes place on different levels and in a different way. In the Basin Wise programming, pathways have been further explored and developed. Based on these experiences, review and further elaboration will be needed during the BDP 2100 updating or follow up.

16.2.3 Institutionalization of the Adaptation to Climate Change

A right institutional framework is basic for sustainable transformation and for climate change adaptation. In practice, many people think about content of strategies and measures. But as said earlier, these can only come into being through institutions and human choices and activities. If an adaptation action is not well understood or doesn't fit in working procedures so that people don't know what to do or when it doesn't follow the rules, institutional provisions and demands, it will be difficult to lead to some result. For BDP 2100, a dedicated Institutional Framework has been designed to take decisions and set priorities at the highest level and partly to serve implementation on the ground. The vertical coordination of decision making and support to it (by DGC, PPSC and Delta Wing) as well as horizontal coordination (by same institutions and Focal Points) is important for the multi-sectoral approach which was missing in earlier plans who had by consequence difficulty in implementation. The single sector plans were not mainstreamed, endorsed at central government level and consequently were not provided with required budgets.

The Five Year Plans and other national policies and plans incorporated BDP 2100 which in itself is a matter of mainstreaming and broad acceptance, also to be seen as transformation.

Adaptive Programming is programming for adaptation in a flexible way. As second part of the mixed scanning approach, the hotspot or basin wise programs have been developed like that, in a way where several adaptation path ways are included to change if needed the course of action on program level.

16.2.3.1 Preparing Adaptive Programming and Execution of Interventions

Many implementing ministries, departments and agencies have a qualified work force and division of the work incorporated in their organogram, based on traditional division of tasks, leadership and coordination. These are fixed and cannot be changed easily, causing a certain inertia to move quickly under changing circumstances. However, the BDP 2100 consists of new knowledge and another way of approaching the issues and challenges, namely traditional ones combined with modernization and climate change urgencies. The National level strategies are in this way combined with Hotspot and Thematic strategies and are furthermore elaborated in regional or basin wise adaptive programing. So the delta plan comes up with a work load of project preparation and execution activities, which does not fit in the existing organizations in terms of numbers of staff and, e.g., adaptive design knowledge and skills. Readiness factors are set to develop the capacities of involved agencies on five capacity areas by training, providing additional analysis and management tools and also trying to adjust the numbers of staff. Next to that, rules and regulations and especially funding of the measures and projects is of essential importance.

16.2.4 Preparation of Interventions and Adaptive Design

Adaptive design is important during intervention or project preparation. The idea is not to go for maximum construction solutions, but solutions which allow adaptation over time. Building with nature is such method, although the state of the delta with sometimes weak or low embankments or sea defenses and urgency to act may not allow much time for that. Adaptive design will require spatial reservations for dykes or embankments which are not easy to arrange.

A choice right from the start was that execution of adaptation works would be carried out by the already existing implementing ministries, departments and agencies. This requires coordination by the Planning Ministry and a Focal Point Network to make this work in an integrated way. This also means, that preparation and execution technology on adaptive design or other aspects can be shared instead of developed in separate silo approaches. Project preparation and execution could be more coordinated and experiences and technologies shared, e.g., in an 'adaptation project preparatory facility'.

16.2.5 Multi-Sectoral Approach Requiring Transformative Technology Management

As argued before, the chosen vision and goals cover many sectors of society with their problems, policies and technologies including solutions to deal with the problems. A holistic approach necessarily brings together the involved sectors and disciplines including their methods, tools and practices. Transformative technology management intends to utilize these technologies in order to address the issues multi-sector wise and optimize solutions. Especially to work on the readiness for transformation and implementation, this is useful, as has been done in the Bangladesh experience with, e.g., differing water issues, agriculture in various modalities, urban issues, navigation, Blue Economy.

Agriculture and water show in fact this need but also a joint option of an Agricultural Transformation Process (ATP), serving water security and food security goals. This is in essence a process of fundamental change with characteristic dynamics of multi-level change (societal/ regional, sectors, industry, organizations, individual). It is an adaptive process in terms of generating real change outcomes and impacts over time, able to adapt to changing circumstances or insights underway. Less water and space for traditional agriculture may lead to such transformation. This process depends on several key factors to be diagnosed and managed properly, among others with practices such as:

- Initiation: preparing the ground.
- Diagnostic Search: what is the state of affairs in core thematic areas? What are the real issues and challenges to be addressed and how; what about many items concerning readiness for transformation?
- Strategy Development: strategic / tactical and operational interaction, commitment and alignment of key stakeholders. Transformation requires broad ownership as well as interactive strategy formulation and elaboration, monitoring and evaluation and modern knowledge management. Action and show cases in pilot areas:

e.g., introducing higher value crops, business and working methods; water productivity, vertical farming, cold storage facilities, food processing, transportation and market facilities etc.
- Moving On: continuation of change actions with Programming and Execution of interventions while actively building required agricultural and water transformative capacities.
- Transformation: execution of programmatic actions and projects to achieve sustainable results and transformation goals.

Agricultural Transformation shows in most cases patterns of the above indicated transformation process and characteristic factors, each representing large knowledge areas or technologies, including modern water management and purification, professional agricultural practices and skills or tools as well as scientific knowledge and research. The emphasis will be on the use and implementation of knowledge and technology for ATP related interventions in the delta. Disciplines ('knowledge and technology area's') involved depend of the scope of the ATP. Here, the agriculture focus is predominant in an integrated food systems approach, which emphasizes a balanced approach in which agriculture is embedded in a setting of water, energy, transportation, socio-economic, ecology and environment, spatial planning and land use aspects, the following overview indicates main elements of the food systems approach.

- Pre-production: Improvement of land and water conditions; safe polders, improved protection and water management, land use, ecology.
- Upgrading agricultural production chain: aqua culture, livestock, horticulture, rice and grain production, water efficiency and productivity, sustainable production.
- Agro-Business Park value chain covering pre-production to end user markets, location choice, water and land infrastructure, buildings, food processing, logistics, business services inland and export.
- Governance, rules and regulations, ATP-process, business models and management, stakeholder process, smart program management.
- Feasibility studies, costs versus benefits, investment costs, business cases.
- Environmental and ecological development or protection.
- Technological applications, methods and practices.

Managing such a transformation process means basically managing the technologies to bring about the expected outcomes, e.g., in terms of modernization, food security, socio-economic wellbeing and poverty reduction.

16.2.6 Other Transformations by Creating Logical Innovative Combinations in the Philosophy of 'Adaptation by Development'

An advantage of the broad scope of BDP 2100 is that many more combinations can be formed, offering synergy to achieving the delta goals and adaptation. Examples are: sustainable water use by big water users as the garments industry, by methods of water productivity in combination with waste water treatment; piped drinking water development in combination with sustainable modern sanitation systems and waste water treatment; sustainable navigation via adequate water ways and inland and coastal port facilities all over the country as proposed in BDP 2100 in combination with multi-modality transportation systems; well protected multipurpose new land development in the coastal area and along rivers to provide space for urbanization, industrialization and, e.g., the mentioned ATP. New land development could be done in a large scale way to shorten existing sea defenses and reduce costs of adapting these defenses up to climate change standards.

16.3 A Transformative Approach of Adaptation

16.3.1 Building on the Sensitizing Concepts

In Chap. 2 the idea of searching for sensitizing concepts has been presented and briefly elaborated in a conceptual model in Fig. 1. Chapter 4, Sect. 4.4.2 showed the design of the strategy process with visualization in Fig. 4.2.

As argued, the theoretical approach of adaptation has been derived from strategic management and change as well as organization and public management sciences. Three levers of change have been mentioned and elaborated in the conceptual model. This also goes back on the question how a country can organize and adjust itself in the best possible ways to realize adaptation to climate change. On first sight, there is a strong emphasis in the model on strategy, organization and management aspects and one may wonder where the technology component is because strategies, measures and projects will most probably be technical in nature, e.g., concerning technologies on water safety and security or more specifically protection works, water supply, rehabilitation of areas to improve drainage and protection etc. The strategy and organization and in connection, the governance aspects have been intentionally highlighted because if there is no insight and understanding of these, adaptation will be problematic. Technology is however a cross cutting element of each conceptual model component. The technical content is reflected in the several streams of discourse, in elements of the narrative, the content of issues and challenge with related strategies and decision making, e.g., on infrastructure investments. Without use of technology, adaptation will be impossible. Emphasis in strategy making and implementation on transformative technology management is needed as it underlines the importance of utilization and innovation of technology as good as possible in combination with transformative and management dynamics.

Here interesting points will be highlighted to further develop the original conceptual model and to point to useful and doable activities in each practice.

16.3.1.1 Perception of and Interaction on the Urgency of Adaptation

The extent to which adaptation is found necessary depends largely on the tradition and culture regarding adaptation. First of all, there is the importance of a 'natural' awareness of increasing threats. Besides, the institutional arrangements and organizations are in place to act relative quickly. The nine lessons of Leary, 2008, are useful with respect to realizing and enacting climate change and adaptation as they indicate: (1) adapt now; (2) create conditions to enable adaptation; (3) integrate adaptation with development; (4) increase awareness and knowledge; (5) strengthen institutions; (6) protect natural resources; (7) provide financial assistance; (8) involve those at risk and (9) use place specific strategies. These lessons show that adaptation has been interpreted as urgent and fundamental process of change and intervention involving necessary decision making and taking appropriate measures swiftly and rapidly. However, whether adaptation to climate change will happen or not and how it happens depends to a large extend of the way the directly involved stakeholders make sense of the existing state of the delta (or other situation) and of the upcoming developments or even threats. Participation in the discourse in the discursive climate arena and generating narratives of climate and other changes, gives sense and initial direction to steps to be decided upon and managed. In this way of thinking, the conceptual model can be further elaborated in line with the experiences.

Sense making and sense giving in the midst of discourse, narratives and powerplay focus in the first place on interpretation of the trends and developments, secondly, on deciding to start the strategic planning process. Herewith, the praxis of strategizing on government side started and turns into a praxis of strategy decisions and approval which requires sensitivity for and knowledge of the rules of governance games. This aspect of adaptation is of crucial importance and is easily overlooked by experts or practitioners of the strategy making praxis. Results from this side need to be submitted to government institution where officials in their praxis come possibly based on consultation, to certain conclusions, modification proposals and or decisions. Practitioners in the strategy making praxis can of course try to anticipate an approval process but the question is whether real transformation can be achieved in that way or that such synchronization leads to more of the same solutions. Submitting a new proposed strategy or measure can cause curiosity about it or friction and non-acceptance. Here leadership of the practitioners of strategy making play a crucial role in establishing fruitful relations and trust as well as good will as representative of the strategy team. Explaining the professional background of performed practices and methods, emphasizing the consistency with earlier work and vision and ambitions of the process are important points to use.

Aware of knowledge of developments and risks, governments acting in countries with specific delta areas have taken actions to deal with these risks. These actions have emerged in the institutional framework and approaches related to traditions and the specific delta profile, issues and risks and the socio-economic environment. Many of the institutional experiences have evolved over a long period of time, in response to usual challenges in a way of learning by doing and also to emergencies often in an improvising way. Some of the principles and features can be summarized as follows. Depending on development stage and vision, the focus and scope of the strategic planning and transformative approaches differ. Climate change triggers strategic planning processes in various ways: the Netherlands focused explicitly on water safety and security with some regional planning as well; Bangladesh included next to water safety and security important ambitions concerning food security and socio-economic development; Vietnam focuses primarily on agricultural transformation and includes climate change related water resources management as part of that.

16.3.1.2 Tracking Strategies and Knowledge Management

This is highly useful in line with learning about the interventions taken before, also in terms of what worked well and what didn't work. Moreover, the knowledge of earlier plans and their background studies were used. This was important because additional research was not possible during the strategy process. Utilizing earlier plans and intervention was sometimes criticized, when expectations were that BDP 2100 would provide completely new strategies and interventions. However, it was felt that earlier solutions are not bad only because they were generated earlier. So, they were utilized after checking why implementation didn't take place.

Knowledge management in terms of basic scientific knowledge but also practical knowledge and practices with respect to the delta features, climate change development etc. is of utmost importance. Strategic and scenario-based delta planning in a transformative approach is science and knowledge driven. So, not only water resources are important but also adjoining areas of delta management with respect to food and socio-economic development. This planning and transformation require adequate monitoring and evaluation of trends, progress etc.

The Baseline Studies were of utmost importance in order to understand the existing natural system with its major rivers and existing complicated natural characteristics. Problem was that initially in the Terms of Reference and Inception Report 19 themes were indicated but that during the process seven more themes had to be explored and studied.

16.3.1.3 Content of Strategy Development

The content of strategies came forward in terms of issues and challenges to be addressed during the diagnostic search and baseline studies. The practitioners concluded on what the findings of what real problems were to be approached with effective strategies and measures for adaptation. In a parallel trajectory, the long term vision, mission and goals for adaptation were developed. It became clear in this way what the direction of strategies needed to be, aiming at goal achievement. The first orientation of issues, challenges and possible strategies and measures was discussed in the many workshops and delta ateliers all over the country which led to a rich volume of potential strategies and measures. It was necessary to structure this large amount, often differentiating in scale, in general occurrence or not, in specific subjects. So the three strategy categories were formed: National level, Hotspot level and thematic strategies. The large volume of potential strategies was funneled to realistic and as next step to preferred strategies in the three categories as further described in Ch. 9. Experts from the assistance team contributed from various disciplines, e.g., to the specific aspects of Flood Risks, e.g., in relation to standards for embankments or sea defenses. Fresh water supply demanded other technology in terms of water balance, demand and supply issues. Besides the water resources related disciplines, contribution from other disciplines was incorporated as, e.g., on coast and rivers, water supply and sanitation, agriculture, shipping, urban and spatial development, environment, ecology and biodiversity, economy, institutional, management and organization.

In relation to the findings of tracking strategies and plans, it was clear that many proposals from earlier studies and plans have not been implemented. As many of these proposals were containing still useful ideas and solutions, a number of them have been incorporated into the strategies. Although many projects were realized in the past, there is still a huge challenge to sufficiently protect and develop Bangladesh. Many strategies contain therefore no regret measures which can be executed on the short term and stay valid and useful on the longer term. Elaboration of long term strategies and measures into adaptation pathways proved to be more difficult. Probably because of the state of the Bangladesh delta which causes that many measures will stay effective for considerable time. Besides, it appears to be hard for involved institutions and experts to think about measures becoming less effective and then become replaced by more effective interventions. Also problematic was imagining tipping points or even discontinuities in developments like climate change which are supposed to indicate a possible shift in measures. Most difficult is that the government cannot or doesn't want to decide on long term adaptation pathways most probably because multi-sectoral governance isn't working like that (yet) in Bangladesh and certainly due to funding constraints and uncertainties from the various sources.

16.3.1.4 Institutions, Governance and Stakeholders

Government institutions play a dominant role. Institutional features and development is often related to the formation of a delta and its characteristics which requires certain measures to make the area habitable. Population form institutions over time and water resources management has emerged in different forms in various river deltas. A good practice example of this integrated approach to water institutions is the Delta Program in the Netherlands and Bangladesh. Through the Delta Program and Delta Plan, the governments not only signaled priority attached to water but it also instituted a holistic approach to delta management incorporating climate change. All aspects of delta management including planning, investments, operations and maintenance, and research are handled by the Delta Program and implemented in the framework of water management agencies. However, each delta is located in a different country with different culture, state form, government characteristics and planning and implementation traditions. Moreover, one needs to understand 'delta management' is not another connotation of 'water management. Delta management goes far and beyond which is holistic and comprehensive. Delta management relates to economic growth and development by promoting 'water-centric' economic development. This means that a strategic planning and transformative approach needs to be considered whether or not they can be organized, embedded in socio-political circumstances and achieved in terms of culture and knowledge.

Holistic or integrated delta analysis and planning and implementation are important characteristics of adaptation to climate change. Deltas are formed under specific river and sea interactions and circumstances while they have multiple ecosystem functions and human uses. The competing demands and values can create conflicts in terms of water and land ecosystems, spatially as well as within a specific area. Depending upon the source of demand for water, responsibility for managing that particular demand might lie with a specific agency of the government. Consequently, more often than not, multiple public agencies are involved in managing water internally. Without an integrated approach recognizing the role of external drivers as climate change, the risks for conflicting policies, competing water development programs and inefficient use of water are high.

In connection with the former point, national, regional and thematic strategies play a role in various forms.

Transboundary coordination of water cooperation among riparians. Many rivers originate in one country, travel through several other countries and finally drain out to the sea in another country. These rivers are a common resource for all riparians. More than half the world's population depends daily upon water resources shared by more than one country. Some 40% of the world's population lives in river and lake basins that comprise two or more countries and 90% live in countries that share basins. Yet adverse developments resulting from non-cooperative behavior of an upstream country could create havoc in terms of water shortages and or flooding for downstream countries. Rather than go to war and end up in a lose-lose situation, many delta countries sharing a common river have agreed to share the water as a common resource.

Vertical and horizontal coordination focus on institutions for internal water cooperation: Similar to the role of regional cooperation, the cooperation between various actors within a country dealing with water and other aspects of the delta (land use, environment, biodiversity) is an extremely important feature of successful delta management. In water alone, managing multiple uses, shortages and trade-offs on the demand side and multiple agencies providing and or managing the water resources on the supply side must be involved in making decisions that take into account the private interests of each group of users and the social interest of ensuring sustainable use of clean and safe water. While integrated water management takes care of supply side coordination, on the demand side cooperation among users is necessary to manage water shortages, conflicts and trade-offs. Internal water conflicts arise in both urban and rural areas, although in developing economies rural water conflicts are more serious in nature. This is partly because local voices are not well represented in national institutions dealing with rural water issues. Consequently, rural water investments and allocations often tend to bypass the priorities and interests of communities. Practices of well-coordinated internal water cooperation show that local water groups have emerged as important players in resolving water conflicts and trade-offs nationwide. More generally, active participation of user groups is an essential requirement for an effective delta management (Van Alphen et al., 2021).

Funding is of course a major point of attention given the required massive investments once the water sector is involved: government allocations over time with respect to continued adaptation; financing contributions by IFI' and DP's and private sector investments. Important is how all investments will be balanced between the strategies, national and regional as well as thematic per sector. The specific costs for operation and maintenance are extremely important with respect to a well maintained delta and upcoming adaptations. Besides, investment in the relevant delta institutions in terms of professional staff with required skills is needed together with investing in research, tools and knowledge generation. This especially is important for formation of development of water user institutions and principles as water taxes, cost recovery of beneficiary pay principle.

16.3.1.5 Developing Adaptation Capacities

The GoB decided to start an adaptation process with the intention of preparing a comprehensive delta plan BDP 2100, following a philosophy of 'adaptation by development'. For realizing this ambition against the background of increasing threats to the delta, international support has been mobilized, especially from the Netherlands. This led to a mixed team of practitioners, for BDP 2100 Formulation project (2014–2018) a team of 12 international experts and 42 Bangladeshi experts and about 5 support staff. Almost all staff was appointed on intermittent basis. For SIBDP (2018–2023), 4 international experts and 15 Bangladeshi with 4 support staff have been appointed, again most of them on intermittent basis. The experience and skill mix of the practitioners were very useful to produce the baseline studies and to perform the practices to deliver the BDP 2100.

The counterpart of the team was GED with a small number of dedicated officers and strong leadership by Project Director and the leading official Member (senior Secretary) of the Planning Commission, Ministry of Planning. At the start of the BDP 2100 Formulation project, there were some confusing shortcomings in terms of lacking external support from a spatial planning expert and an external team of economists. These start-up problems were settled with EKN and GED. The international team members brought in their experience in the Netherlands with the Dutch delta Program and ideas of adaptive delta management. Given the choices of GoB to develop its own delta plan with a broad multi-sectoral scope with the ambition of utilizing the delta Plan in becoming a middle income country, the practices of the experts needed to be adjusted. Not only was the delta significantly different from the Dutch delta, also the ambition and scope of the Bangladeshi delta plan was totally different. Close formal as well as informal interaction with government officials led to good understanding and collaboration. Mutual adjustment with informal communication made it possible for experts to coordinate their work and also communication between peers are the crucial activity which made progress possible. Numerous meetings, creative sessions and workshops took place also with stakeholders to come to the agreed results. In doing this, knowledge was shared and capacities on performing the studies, developing the content of scenarios and strategies and preparation of implementation were build. But, ambiguities, differences of opinion, frictions, sometimes power play are all part of the game. Leadership of professionals in the strategy process is of utmost importance in this whole process with complicated challenges and government and stakeholder setting. This means leadership in the strategy making praxis in close collaboration and understanding with the government institutions in order to know what they want and are heading for. Also the leadership in the government decision making and approval praxis is highly demanding, trying to fit the strategy process and its outputs in the government performance and get it finally approved at the highest level. SIBDP followed on this trajectory more or less in the same way, by working on five capacities to improve and transform institutions, knowledge, skills, funding and elaborate on adaptive programming and support transformations taking place.

16.3.1.6 Reflection on the Conceptual Model

After going through the whole process in Bangladesh during nine years, and living the strategy process (Gratton, 2000), the used concepts can be discussed and expanded while they may suggest a kind of repertoire for how to address adapting to climate change and dealing with other issues at the same time. So, in addition to the conceptual model presented in Chap. 2, the model can be expanded as follows in Fig. 16.119.

Overall the content of the diagnostic search and strategy process with its various technologies and pathways is considered of great importance in addition to strategy process design and dynamics. The same counts for the governance aspects such as steering, approval process and decision making. Very interesting are also the stakeholders roles in terms of (a) communicating expectations, (b) participation in problem identification, (c) participation in visioning and defining scenarios, (d) finding options for strategies and actions, (e) consultations and feedback, (f) influencing decision making, (g) localization of project actions.

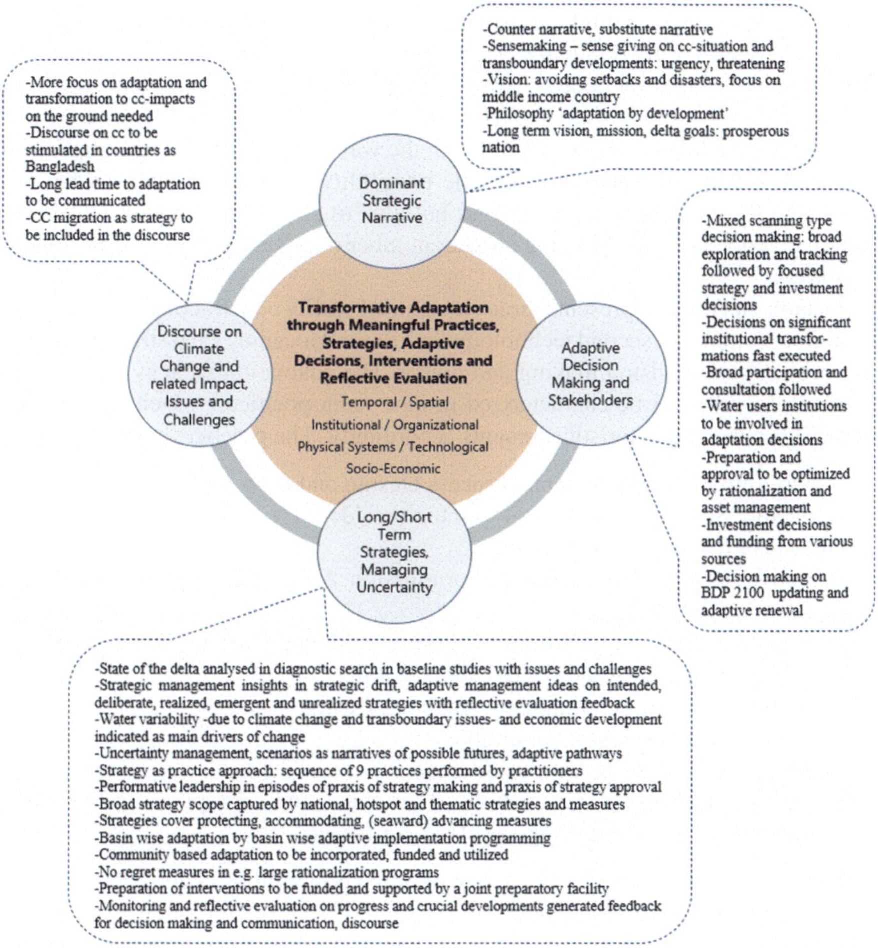

Fig. 16.1 Expanded conceptual model

16.3.2 Practices of Adaptation and Transformation

Thinking about the concepts and experiences per practice, reminded us of Noel Tichy's work on Managing Strategic Change. Although this is primarily focused on strategic change in organizations, the way of thinking can be applied to adapting to climate change and related issues at societal level in a transformative way. Tichy presents three generic dimensions which he sees as problem cycles over time, the Technical, Political and Cultural cycle which he confronts with the factors usually addressed in strategic change management.[2]

[2] Tichy, Managing Strategic Change (1983) e.g. pg. 13 and 119.

Inspired by this way of thinking we rearranged these concepts in relation to the nine practices in Annex 3. By doing this, a repertoire of possible performative ingredients opens up in the adaptation process, which practitioners can be aware of or can make a choice about in other situations. The Bangladesh delta is a highly complicated delta for which adequate strategy making in the way it was desired, is a complicated process. The table in Annex 3 offers the possibility not only to realize what has been covered by the BDP 2100 process but also to choose for another version of strategy making and, e.g., utilizing another number of practices or elements of these if required.

The table in Annex 3 presents three elements of reality construction, first: interaction and sensemaking, second technology (in its broad meaning) and strategy content aspects and third, decision making and power. The adaptation strategy process can as seen in Fig. 15.117 be characterized in nine main practices, which each can be elaborated in the characteristic elements accordingly. The practices are:

- Strategy Initiation, Agreements, Process Design and Start-Up.
- Diagnostic Search: Tracking of Strategies; Identifying Characteristics, Joint Diagnosis of Issues and Challenges.
- Interactive Scenario-based Strategy Formulation, Adaptation by Development and related Adaptive Measures.
- Developing a Dedicated Institutional Frame Work and Decision Tools.
- Investment Planning, Adaptive Decision Making
- Regional Programming, Managing Adaptive Design.
- Realization of Measures/ Interventions.
- Arranging Funding from Various Sources.
- Performing Reflective Evaluation, Updating and Adaptive Adjustment; Conducting Adaptation as Ongoing Transformation.

The results of this exercise offer action oriented possibilities to generate a process of adaptation and desired transformations.

In the former chapters it became clear, that application of the practices characterizes the whole BDP 2100 formulation process as well as the process of support to implementation with the SIBDP project. The strategies are practiced by practitioners in close collaboration with government officials and stakeholders in numerous sessions and formal meetings. How this has been done is shown by elaboration of the nine main practices. Much more can be said and studied about this to reflect on and as input for the way forward. The knowledge where the strategizing and practices are based on, was to some extent available from earlier planning exercises and support activities from development partners and financial institutions. Additional knowledge was created by conducting 26 Baseline Studies. Besides, a knowledge agenda for future research shows that knowledge gaps still exist. But the generated knowledge was highly appreciated by governments and stakeholders because of uncertainties about climate change but also about the strategy process and the possible measures and interventions. In this sense, the strategy process was experienced as 'a paradigm shift in Bangladesh planning'.

Adaptation to climate change is dealt with in a rather abstract way in many reports and documents. The contribution of the whole strategy exercise as presented before is to highlight adaptation as action and development process, requiring changes and transformation in many cases. It becomes clear, that strategizing is performative, in constituting its subjects and shaping its objects in terms of measures, projects and funding. Interesting is that strategizing can in this way be understood as aesthetic performance (Kornberger & Clegg, 2011) where power dwells in the representation of science driven facts and socio-economic and political values of safety, security and wellbeing. So, this type of strategy making is a socio-economic and, given the political background, values and decisions a socio-political practice utilizing practitioners' technology to generate insight and understanding, form stakeholders opinions on which choices and decisions of how to act in order to adapt can be made. In doing this, crucial factors and practices in the praxis of adaptation to climate change come forward, together with the role practitioners play in collaboration with related stakeholders.

References

Delta Program. (2009). Delta Program of the Netherlands with annexes. https://english.deltaprogramma.nl

Government of the Netherlands, Delta Commissioner. (2018). *Monitoren en signaleren voor adaptief deltamanagement. Alert op veranderende omstandigheden.* Retrieved from: www.Rijksoverheid.Nl/Deltaprogramma.

Gratton, L. (2000). *Living strategy. Putting people at the heart of corporate purpose.*

Kornberger, M., & Clegg, S. R. (2011). Strategy as performative practice: The case of Sydney 2030. *Strategic Organization, 9*(2), 136–162.

Syvitski, J. P. M., Kettner, A. J., Overeem, I., Hutton, E. W. H., Hannon, M. T. G., Brakenridge, R., Day, J., Vörösmarty, C., Saito, Y., Giosan, L., & Nicholls, R. J. (2009). Sinking deltas due to human activities. *Nature Geoscience, 2*, 681–686.

Tichy, N. (1983). *Managing strategic change. Technical, Political, Cultural Dynamics.*

Van Alphen, J., De Heer, J. M, & Minkman, E. (2021). Strategies for climate change adaptation: Lessons learnt from long-term planning in the Netherlands and Bangladesh. In *Water International* (Vol. 46(4), pp. 477–504).

Chapter 17
Conclusions and Way Forward

Abstract Wrapping up the findings and conclusions, also in light of the messages in Chap. 1. Indications for further empirical research and theoretical considerations will be given.

Keywords Contingencies · Uncertainties · Adaptation · Adaptive capacity · Scope · Strategy · Practice · Transformation · Technology · Governance · Organizational dynamics · Protecting · Accommodating · Advancing · Seaward advancing · Retreat · Migration · Way forward · Research · IFIs

17.1 Concluding Remarks

Coming back to initial messages at the beginning of this book, the following remarks can be made.

Adaptation to climate change has as described especially for Bangladesh and Vietnam indeed been focused on multi-sectoral challenges concerning water safety and security next to utilization of opportunities for socio-economic development. Although the Netherlands focused more on water safety and security in the early years of the Delta Program, later on spatial planning and agriculture as well as urban water and ecology got much more attention. In addition, adaptation to climate change concerns managing the contingencies and uncertainties in a strategy process and in related transformational changes. These have been embedded in the countries' governance, policies, planning and programming including a financing arrangement and budget allocation. To achieve this in depth understanding and attention for administrative systems and culture is required. Besides, design of strategies and measures fits by basin wise adaptation programming and adaptive design as well as rationalization of infrastructures in a well understood natural resources system, while where possible revitalizing these. There are still traditional preferences, ways of doing or ignoring things with respect to adaptive management; information sharing and multi-sectoral collaboration are also difficult to achieve. Hence, adaptation to climate change is,

J. M. de Heer and G. A. Choudhury, *Strategies for Adaptation to Climate Change in a Transformative Approach*, Springer Climate,
https://doi.org/10.1007/978-3-031-88884-7_17

although to a large extent techno-economic in terms of issues, strategies and solutions, to an even larger extent depending on willingness to adapt, on power play in and between organizations. This concerns in essence a cultural process in which interests, values and beliefs dominate sense making and sense giving in the interactions on uncertain climatic conditions and possible impacts, on opportunities and threats for socio-economic development and on dealing with risks and challenges. Besides, power and influence play an important role during and after the strategy process, starting, e.g., with the process design, choices during strategizing, role of stakeholders, results of Baseline Studies, text proposals and improvements on strategies and measures, use of adaptive pathways, selection of interventions, stakeholder claims, setting priorities, funding, multi-sectoral discussions and decision making. On all those items where choices and selection and of course decisions come forward, influences and exercise of formal or informal power play a significant role. This is why in Fig. 4 the praxis of strategy making has been separated from the praxis of strategy approval, while Whittington, 2006 presented his model with one praxis circle. Leadership of the strategy processes is next to managing the process and content, often compared to playing chess on many boards at the same time. Important is to prepare as good as possible especially together with insiders in the administrative culture and procedures.

In the chosen holistic approach, multiple involved aspects (or disciplines) and sectors, the knowledge and practices of these aspects and sectors can be utilized for adaptation and transformation and make this so called 'transformative technology management' of crucial importance. Innovation is being described as a creative process of finding new combinations; this is highly stimulated by a holistic approach where combinations (see, e.g., Sect. 16.2) offer chances for adaptation by development.

Methodology on adaptation to climate change impacts can be fruitfully derived from theories and practices of strategic management and change as well as from decision making theory and practices, of course in conjunction with the various scientific disciplines on climate change, civil engineering etc. on a range of complicated and challenging scientific and technical aspects. In the course of time from the start early 2014 till 2024, episodes with their own characteristic focus and activities have been identified and indicated with the main applied practices. These are labeled as (1) Strategy Initiation, Agreements, Process Design and Start-Up; (2) Diagnostic Search: Identifying Characteristics, Issues and Challenges; (3) Interactive Scenario-based Strategy Formulation, Adaptation by Development and related Adaptive Pathways and Measures; (4) Developing a Dedicated Institutional Frame Work and Tools; (5) Investment Planning, Decision Making and Adaptive Programming; (6) Managing Adaptive Design and Realization of Measures/ Interventions; (7) Arranging Funding from Various Sources; (8) Conducting Adaptation as Ongoing Transformation; (9) Performing Reflective Evaluation, Updating and Adaptive Adjustment. The indicated practices focus on the formulated subjects and can possibly be described better but this has proven to be a very useful way of strategy making. Usually, several phases or stages are indicated to structure a strategy process, as has been done in, e.g., Fig. 4. 2 (Sect.4.2.2). Because adaptation to climate change will not happen in one go and a process is in fact open ended, phases or stages are less useful. This is the reason for talking about a sequence of practices which can be followed by a selection of practices where needed. This is even more important when looking to execution of measures, which will not take place in an implementation phase but rather in parallel trajectories of separate programs and projects while at the same time other practices, e.g. for updating and renewal, may be performed as well.

Adaptation to climate change requires thorough stocktaking with diagnostic search and analysis of the area involved in relation to the expected climate change impact and other physical system and societal characteristics and developments. The baseline studies provide insight in baseline situation, state indicators and conditions. Based on that a diagnostic analysis shows what the issues and challenges in the delta and administrative systems are to be addressed. Of course there is a variation in scope and scale. The Bangladesh Delta Plan 2100 concerns a highly complicated delta in a multi-sectoral long term view of water safety, food security and socio-economic development. The Netherlands shows a smaller profiled Delta Program, mainly focused on water safety and fresh water with some attention for spatial development cs. The focus in the Vietnamese Mekong delta plan is on agriculture and water resources.

For managing uncertainty of how climatic conditions will be in future, a fruitful approach is to utilize interactive formulated external scenario's as narratives of possible futures. These can be based on main uncertain drivers with high potential impacts including climate change, e.g., transboundary and or economic developments. Strategies performing well under various scenarios are considered to be robust and useful. Together with the adoption of a long term vision and goals and organizing necessary transformations in the relevant systems, robust strategies and selected measures lead to a long term adaptive plan which can be elaborated into an investment plan. Important is to identify the by climate change impacted sectors and disciplines, to consider the per sector involved body of knowledge, tools, practices, solutions (sectoral technology) and to address the concerned issues and challenges with state of the art technology to contribute to transformative and adaptive management of the issues and challenges.

Adaptation is basically a process aimed at not only adjustment but on interventions and transformation to deal with life threatening contextual developments and related physical and social and economic issues and challenges, see, e.g., Fig. 17.1. This process consists of diagnosing of what is happening or could happen, accepting uncertainty due to a lack of knowledge, nonetheless preparing meaningful measures and choices and deciding on them by adaptive decision making. This is certainly not an easy task. The OECD statement (2011), that the water crises is in fact a governance crisis can, based on the former analysis, be confirmed. The best technical strategies with great and smart adaptive pathways and solutions can be designed, but if strategies do not lead to sufficient support and action or if the basic conditions of governance and involved organizations are weak, investment budgets are constrained and when yearly budgets are also too limited for crucial functions as maintenance, while transformations take considerable time, adaptation to climate

Fig. 17.1 Mighty river at danger level, limited time to adapt. Photo J.M. de Heer

change will be highly problematic for a long time. This may lead to serious consequences, the reason why the sense of urgency to take adaptive action on the ground needs to increase rapidly. The tendency to dive into the scientific aspects and in too many detailed technicalities of adaptation to climate change, its impacts and related procedures has to be better balanced, especially by the IFIs and other funding institutions, by paying much more attention to real contributions to measures, execution of adaptive programs and projects while at the same time supporting improvement and effectiveness of the complicated elements of governance, of financing and executive organizations.

So, important as way forward are swift actions to prepare adaptation and necessary transformation, perform the urgent tasks and measures. Constraints to adaptation have to be addressed. Not all adaptations concern a highly complicated delta in which many requirements and options exist for action. Substantial no regret adaptive interventions can take place in combination with capital maintenance and asset management, e.g., through execution of rationalization programs. In other cases some of the mentioned practices, measures and adequate interventions may already lead to sufficient adaptation to negative impacts. This means that planning and programming can be more limited or scaled to the situational issues and challenges, although the core elements and practices of adaptation may be more or less the same as presented before.

Adaptation of the physical system will in some places in the near future not be possible anymore, when sea level rise and salinity intrusion or water scarcity and drought already went too far. Strategies including retreat and migration are not considered here but require more attention because of societal implications and international sensitivities or tensions. The choice of Bangladesh on 'living with water' and to utilize adaptation and transformation in a mix of protecting, accommodating and seaward advancing as opportunities for development show an optimistic stance towards adaptation possibilities. Reflective evaluation in the course of BDP 2100 will learn how this will work out and whether or not this will be sufficient.

This publication explores the concepts and theoretical backgrounds as presented and makes use of the experiences with strategy making and execution as well as with governance and organizational dynamics. There is of course much more to study and say about the presented subjects and performed practices. What has been said about a delta plan, that it should be always science and knowledge driven, counts for every adaptation plan. In this way, shortcomings, gaps, improvements and new knowledge or innovations can be utilized in the course of time. It is therefore important to develop the relevant curricula and research programs in universities and knowledge institutions, to understand climate change impact in a holistic way but also to provide practical knowledge and applications, e.g., to executing agencies. Strengthening of adaptive capacities by science, technology and good governance is of crucial importances, e.g., to further develop and professionally perform each of the nine presented practices.

Finally, plans and pathways as well as interventions and reflection come forward in organizations, based on human actors making sense of issues and challenges with creative thinking and disputed choices on a desired perspective requiring determined joint action in which next generations, see Fig. 17.2, will play a crucial role.

Together utilizing knowledge and practices as good as possible, adaptation by development, that's what people are looking for. Dealing with their daily concerns and

Fig. 17.2 Interaction with curious young people during stakeholder meetings and study tour in the Haor area. Photo J.M. de Heer

worries about the future, making choices and working hard to achieve a prosperous nation and safe future. How technological advanced strategies, plans, interventions and transformations may be, they still are expressions of what people decide to do and so, of the politics of culture in a process of becoming a laboriously adapted world.

Reference

OECD. (2011). *Water governance in OECD countries: A multi-level approach.* OECD Publishing, Paris.

Annex 1: List of Major Water Policies, Acts and Programs

Policies/Programs	Responsible agency/Ministry
The IECO Master Plan (1964)	Bangladesh Water Development Board (BWDB)
The National Water Plan I (1986)	Master Plan Organization (MPO)
The National Water Plan II (1991)	Master Plan Organization (MPO)
Floodplain Management Pilot Project	Ministry of Water Resources
Flood Action Plan FAP (1990/95) with 11 main studies and 15 supporting Components:	Ministry of Water Resources
FAP 1: River Training Studies of the Brahmaputra River	Idem
FAP2: North West Regional Study	Idem
FAP 3: North Central Regional Flood Control/Drainage Std	Idem
FAP 4: Southwest Area Water Resources Management Project (1993)	Idem
FAP 5: South East Regional Study	Idem
FAP 6: North East Regional Study	Idem
FAP 7: Cyclone Protection Study	Idem
FAP 8: Dhaka Town Protection	Idem
FAP 9: Other Town Protection	Idem
FAP 10: Flood Forecasting and Warning Project	Idem
National Minor Irrigation Development Project (1996)	Ministry of Agriculture
National Policy for Safe Water Supply and Sanitation 1998	Ministry of Local Government, Rural Development and Co-operatives
Meghna Estuary Study, Master Plan (1998)	Bangladesh Water Development Board (BWDB)
National Water Policy (1999)	Ministry of Water Resources

(continued)

J. M. de Heer and G. A. Choudhury, *Strategies for Adaptation to Climate Change in a Transformative Approach*, Springer Climate, https://doi.org/10.1007/978-3-031-88884-7

(continued)

Policies/Programs	Responsible agency/Ministry
Guidelines for Participatory Water Management (2000)	Ministry of Water Resources (WARPO)
The National Water Management Plan (NWMP) (2001)	Water Resources Planning Organization (WARPO)
Action Plan for Solid Waste Management in 19 towns	Department of Public Health Engineering
District Development Plan of WARPO (2005)	Ministry of Water Resources (WARPO)
The Coastal Zone Policy (2005)	Ministry of Water Resources
District Perspective Plan for Development and Management of Water Resources (2005–06)	Local Government Engineering Department
The Master Plan for Haor Areas (2010)	Bangladesh Haor and Wetland Development Board
Dredging Plan 2012	Ministry of Shipping and Inland Water Transport
Sector Development Plan (SDP) for Water Supply and Sanitation Sector in Bangladesh (FY 2011–25)	Local Government Division, Ministry of Local Government, Rural Development and Cooperatives
The National Water Act (2013)	Ministry of Water Resources
Bangladesh Water Development Board Act (2013)	Ministry of Water Resources
Bangladesh Haor & Water Bodies Development Board Act (2014)	Ministry of Water Resources
Participatory Water Management Rules (2014)	Ministry of Water Resources

Source: BDP 2100 Project; compiled from government sources

Annex 2: Basic Delta Results Framework of BDP 2100

Basic Delta Result Framework of BDP 2100	
Goal 1: Ensure safety from floods and climate change related disasters (1) Risk free zones from natural disasters • Average flood affected area • Catastrophic flood affected area • Drought (extreme) free area • Storm surge affected area • Dry season salinity intrusion free area • Water logging free area (2) Population vulnerable to natural disasters • Flood vulnerable people • Cyclone vulnerable people • Erosion vulnerable people • Water logging vulnerable people (3) Physical Changes (4) Societal Developments	Goal 2: Ensure water security and efficiency of water usages (1) Dry season flow availability (Jan-May) (2) Dry season irrigation coverage (3) Irrigation water efficiency (4) Urban domestic water efficiency (5) Rural population with safe drinking water access (6) Rural population with safe sanitation (7) Surface water polluted by industrial wastes (8) Surface water sources polluted by other wastes (9) Physical Changes (10) Societal Developments
Goal 3: Ensure sustainable and integrated river systems and estuaries management (1) Erosion along major rivers • Area eroded along Jamuna • Area eroded along Ganges • Area eroded along Padma • Area eroded along Lower Meghna (2) Area of reclaimed lands (3) Physical Changes (4) Societal Developments	Goal 4: Conserve and preserve wetlands and ecosystems and promote their wise use (1) Permanent wetland with connectivity (2) Seasonal wetland with connectivity (3) Habitat protection • Area of perennial aquatic habitat • Area of seasonal aquatic habitat • Area of marine habitat (4) Harnessing of ecosystem services and goods (5) Physical Changes (6) Societal Developments

(continued)

J. M. de Heer and G. A. Choudhury, *Strategies for Adaptation to Climate Change in a Transformative Approach*, Springer Climate, https://doi.org/10.1007/978-3-031-88884-7

(continued)

Goal 5: Develop effective institutions and equitable governance for in-country and transboundary water resources management	Goal 6: Achieve optimal and integrated use of land and water resources
(1) Rural people with adequate capacity for WRM (2) Equitable share of water among users (3) Adequate monitoring mechanism (4) O&M budgeting (5) Physical Changes (6) Societal Developments	(1) Spatial zoning of integrated land and water uses (2) Spatial standardization of drainage density (3) Flood control, drainage and irrigation capacity • Area under irrigation schemes • Area under drainage schemes • Area under drainage and irrigation schemes • Area under FCD schemes • Area under FCDI schemes • Area under coastal FCD schemes • Area under coastal FCDI schemes (4) Sectoral use of water • Surface water used for irrigation • Groundwater used for irrigation • Domestic water use • Industrial water use (5) Navigation capacity • Wet season navigation course • Dry season navigation course (6) Physical Changes (7) Societal Developments

Annex 3: Adaptation and Transformation Practices

The Table in this Annex presents three elements of reality construction, first: interaction and sensemaking, second technology (in its broad meaning) and strategy content aspects and third, decision making and power. The practices are:

- Strategy Initiation, Agreements, Process Design and Start-Up
- Diagnostic Search: Identifying Characteristics, Issues and Challenges
- Interactive Scenario-based Strategy Formulation, Adaptation by Development and related Adaptive Pathways and Measures
- Developing a Dedicated Institutional Frame Work and Tools
- Investment Planning, Decision Making and Adaptive Programming
- Managing Adaptive Design and Realization of Measures/ Interventions
- Arranging Funding from Various Sources
- Conducting Adaptation as Ongoing Transformation
- Performing Reflective Evaluation, Updating and Adaptive Adjustment

J. M. de Heer and G. A. Choudhury, *Strategies for Adaptation to Climate Change in a Transformative Approach*, Springer Climate, https://doi.org/10.1007/978-3-031-88884-7

	Adaptation and transformation practices		
	Interaction and sensemaking	Technology and content	Decision making and power
Strategy initiation, agreements, process design and start-up	• General communication and awareness • Awareness of ongoing emission and pollution • Pressure on urgency to act • Focuses attention on important aspects as biodiversity, ecosystems -IPCC tries to convince people, governments; • Accessibility of reports is a challenge in itself for most people • Pressure groups claim attention; play power game, pressurize	• IPCC generates knowledge and awareness of strategic choices; • Discourse is important during strategizing, most people show awareness and involvement during participation and consultation;	• Awareness that response to climate change impacts will take a lot of time and financial resources; -IPCC frequent updates increases urgency to act; • Recent weather conditions and flooding demand structural measures
Diagnostic search: tracking of strategies and identifying characteristics, issues and challenges	• Guiding principles for initiation of strategy making elaborated in exploration of earlier policies and plans • Causes of uncertainty in relation to existing knowledge and modes of operation • Guiding principles and level of ambition for defining expected results and project design, setup and budget and time frame	• Identification of earlier solutions and statistics of implementation • Dominant narrative meaning government willingness to act • Meaning of the narrative shows values and willingness to formulate and implement strategies and interventions • Guiding on inspiration and ambition level • Acceptance of outcomes in terms of investments and system transformations	• Earlier policies and plans may still be valid and in use; • Earlier applied solutions may offer opportunities to further adapt or blocking adaptation • Narrative still guiding the actions • Monitoring progress in perspective of the in narrative communicated vision

(continued)

(continued)

	Adaptation and transformation practices		
	Interaction and sensemaking	Technology and content	Decision making and power
Interactive scenario-based strategy formulation, adaptation by development and related adaptive path-ways and measures	• Stakeholders experiences and expectations on issues and challenges; • Consultation sessions; Delta Ateliers; • Review and feedback on diagnostic results	• Stakeholders sense making of issues and solutions; • Jointly building scenarios; • Interactive strategy making, from potential to realistic to preferred; • Ideas about interventions and required transformations; • Consultation; • Review and feedback on strategy process and outcomes	Stakeholders involvement in local/ basin wise planning transformation and execution; • Participation in O&M activities; • Articulate needs or future expectations and vision in water user institutions; • Contribute in 'beneficiaries pay' for water and climate services arrangements
Developing a dedicated institutional frame work and tools	• Deciding to initiate climate change adaptation and study phase; • Recognizing possible needs for transformation, reform; • Tracking of earlier policies / plans; • Diagnosing institutional Readiness • Concluding about what works and what not; • Utilizing earlier valuable proposals;	• Developing long term vision, ambitions and goals, considering the scope of vision, goals; • Back casting vision and goals to present situation and expected issues and challenges; • Deciding to develop strategies to address the diagnosed issues and challenges; • Deciding on stakeholders involvement in various spheres and types of influence; • Deciding on transformations, approach and (legal) requirements; • Deciding on funding over time; Deciding on approval of the Delta Plan;	• Preparing Readiness to implementation by improving on 5 capacities: Administrative, Connective & Organization, Knowledge and Learning, Funding and Physical System Adaptive capacity; • Preparing intervention selection framework; • Arranging funding for project preparations and execution; • Preparing O&M planning, execution; • Considering progress and decision making on updating the plan and follow up;

(continued)

(continued)

	Adaptation and transformation practices		
	Interaction and sensemaking	Technology and content	Decision making and power
Investment planning, decision making and adaptive programming	• Basic characteristics and understanding, e.g., by conducting baseline studies of deltas or other situations are crucial for strategy making and effective implementation; • Tracking earlier and existing strategies and plans learns a lot about what already has been done and about success and failures; • An integrated approach brings in multiple disciplines and related technologies which lead to promising results when managed in coherence;	• Readiness and implementation build further on the diagnostic knowledge and insights; • Technology management is important to prepare projects in terms of adaptive design, economic and social as well as environmental assessments of interventions;	• Important awareness that findings of diagnostic search are not static, issues and challenges change with changing impacts; • For adaptation and updating purposes, diagnostic search and reflection remain crucial;

(continued)

(continued)

	Adaptation and transformation practices		
	Interaction and sensemaking	Technology and content	Decision making and power
Regional programming and managing adaptive design	• Understanding the situation by diagnostic search of the delta and of crucial developments; • Learning what has been done by tracking of strategies and plans; • Providing perspective by visioning and goal setting; • Back casting help clarify the adaptive pathways towards achieving goals and vision; • Scenarios providing narratives of possible futures as setting for strategy making; • Issues and challenges with related technologies form questions to be answered by strategies and interventions; • Institutional framework is crucial to enable approval, decision making and reforms; • Funding is also crucial given the high costs involved in interventions;	Readiness factors such as manpower, training; skills, practices, tools, budgets and equipment need to be early addressed given the growing work package coming out of the strategy making in addition to the usual ongoing work process; • Basin wise implementation offers advantage of logical planning entities where interventions and stakeholders are linked with each other • Adaptive design related to the strategies and adaptive pathways has to be explored and applied efficiently; • Technology management is of utmost importance to realize good value solutions and project results;	• Monitoring of progress in the strategy process can be related to the major developments and uncertainties to stay up to date; • Reflective evaluation, taking time out provides valuable insights and often new energy for follow up actions;

(continued)

(continued)

	Adaptation and transformation practices		
	Interaction and sensemaking	Technology and content	Decision making and power
Realization of measures/ Interventions	• Understanding of the intervention in terms of new structure or other provision or that it is in fact rationalization of existing infrastructure etc • Rationalization projects may require shorter preparation procedures • What does adaptive design mean in the situation • When is further adaptation expected	• Detailed exploration of the basin or area and preparations • Project preparation including feasibility study with check on BDP 2100 requirements and societal, economic and environmental impact assessments, detailed adaptive design • Project execution with selection of adequate contractor or other party • Localization, fitting in the situation involving stakeholders	• Decision making by executing organization to take up the intervention, setting man power and funding available • Decision on adaptive approach and design • Technology management to realize the intervention • Building in the monitoring and evaluation indicators

(continued)

(continued)

	Adaptation and transformation practices		
	Interaction and sensemaking	Technology and content	Decision making and power
Arranging funding from various sources	• Multi-sectoral studies and used technologies; • Thematic Baseline Studies; • Characteristics of the delta physical system in cross boundary context; • Characteristics of related fields of interest; • Diagnosing existing or expected issues and challenges Nationally, Hotspot wise and Thematic;	• Addressing uncertainty by building and considering scenarios; • Utilizing multi sectoral or thematic technologies in understanding issues and challenges in relation to strategy making; • Funneling strategies from potential, to realistic and choosing preferred strategies; • Checking robustness of strategies against scenarios; • Developing interventions in line with preferred strategies; • Developing knowledge agenda and management; • Making decision support system on selection of interventions and considering impacts; • Elaborating interventions in Project Concept Notes with estimates of costs and impacts; • Making Investment Plan	• Providing decision support system • Regional/ basin wise programming of investments; • Preparing result/ outcome based M&E system • Studying feasibility of projects; • Executing the projects; • Performing M&E; • Performing O&M tasks;

(continued)

(continued)

	Adaptation and transformation practices		
	Interaction and sensemaking	Technology and content	Decision making and power
Reflective evaluation and conducting adaptation as ongoing transformation	• Making sense of the applied interventions in relation to ongoing climate change and impacts • Increasingly shared feeling that business as usual isn't an option • Usual standards and working methods are questioned on effectiveness and sustainability • Signs of inertia, acting on it is subject of debate from denial to felt urgency • Argument that there is still lead time • Adaptation initially seen as possibility for no regret measures • Improvements become renewal, requiring transformation of thinking and acting • Measures develop depending on felt need • Serious events lead to adaptation measures doing things differently • Frequent reflection feeds debate	• Evaluating effectiveness of measures and technology • Looking after alternatives, innovations • Content of measures stays initially close to usual way of operating and maintaining • Frontrunner knowledge from academia and environmentalists, with some on adaptation; most on adjustment • Impacts of climate change require out of the box thinking under various degrees of uncertainty • Renewed scenarios indicate serious problems in future • Transformation of institutions, rules and regulations or proposals and funding • Gradual system transformation by maintenance and renewal planned • Technology utilized for innovative thinking, imagination and draft designs • Alternative, possibly significant system transformations applied	• Further discourse and debate on need for adaptation; political differences • Some policy makers come forward with in their view necessary measures • Changes in mindset or paradigm form as such significant transformation • Dominant narrative changes based on that • Resistance of some parties • Private sector moves gradually • Awareness and push by other parties • Adaptive decision making on required measures and additional investments • Significant interventions and transformations may be needed to deal with climate change impacts, leading to tough decision making and power play on significant transformations and their funding

(continued)

(continued)

	Adaptation and transformation practices		
	Interaction and sensemaking	Technology and content	Decision making and power
Performing reflective evaluation, updating and adaptive adjustment	• The status of the system, also in terms of maintenance can make use of state indicators as basis for a monitoring framework; • Reflection on the diagnostic search provides issues and challenges input for strategies of national scale, hot spot or thematic;	• Monitoring and reflective evaluation delivering insights in degree of goal and vision achievement; • Monitoring and reflection related to actual or expected climate change impacts brings forward new steering input and direction; • Updating of strategies and interventions will be required after 5 years;	• Providing instruments for monitoring, reflective evaluation and procedures for decision making • Defining ins and outs of the updating approach with respect to climate change and other crucial data and constraints', including new targets for achieving goals and vision; • Second order learning about how monitoring, reflective evaluation and updating are actually working for gaining the expected achievements

Glossary

Adaptive Actions Actions to adjust or transform a physical or institutional system, e.g., modification and or intensification, include or switch to a different action; or extended (add an action) at low cost or having small societal impact. Flexible actions do not result in lock-ins and have little influence on potential future options (have less path-dependencies).

Adaptation Pathway A sequence of effective measures to achieve a set of predefined goals under changing uncertain external conditions, such as climate change, socio-economic factors or other developments This is related to Sell-by year of an action or strategy: point in time, in a given external scenario, when an action or strategy will no longer effectively meet a set of predefined policy or strategic goals or standards.

Cumec a unit of flow equal to one cubic meter of water per second.

Delta A delta is a geo-morphological area, largely defined by its low-lying surface form and location in the landscape and coastal area that forms at the mouth of a river. Deltas form from deposition of sediment carried by a river as the flow discharges into the sea. Over long periods, this deposition builds a dynamic and characteristic geographic, ecological and social pattern of the delta and its features.

Delta Atelier Delta Ateliers are a social design type of one or two-day workshops in which the issues at stake are analyzed and discussed, ideas and views are generated and presented while interests of stakeholders are identified and linked with the discussed issues and are elaborated in visualized and supported results. These results have a certain status as building blocks in the strategy formulation process. The Delta Ateliers form an interactive strategy making concept which has been frequently used for important issues to the BDP 2100 formulation process

Holistic 'Holistic' refers to viewing and understanding a system and its properties as they happen as whole, not as a collection of separate elements. This includes the view that systems cannot be fully understood solely in terms of their component elements. A delta forms an interesting application of holism; as a whole it includes, e.g., social, biological, physical, chemical, and economic aspects in a given area.

J. M. de Heer and G. A. Choudhury, *Strategies for Adaptation to Climate Change in a Transformative Approach*, Springer Climate, https://doi.org/10.1007/978-3-031-88884-7

Holistic planning This is a type of planning to address problems of a system in a comprehensive and coherent way, viewing the system in relation with its interdependent factors and actors. Multiple policy domains may be involved accordingly, requiring coherent governance and budget allocation.

Integrated approach Combination of relevant elements in a comprehending approach to understand more of the total system.

Learning cycle A learning cycle is a concept of how people learn from experience, steps in the cycle are: 1. Doing something, having an experience; 2. Reflecting on the experience; 3. Concluding from the experience, developing a suitable approach; 4. Planning the next steps, to apply or test this approach. Adaptation is often compared with learning cycles.

Lock-in Situation in which earlier investments determines future actions with the choosing solution. Deviating actions can only be implemented against high costs or high societal impact because of the earlier choices and investments.

Measure or action Individual intervention, often shaped as project, which may be infrastructure but also institutional, legal, economic, knowledge / capacity development and may be at specific spatial, sector- or national-general level. Measures can be part of one strategy but can also fit in multiple strategies and programs.

No regret and win–win actions No-regret actions are useful and cost-effective on the short term and under a range of future conditions and do not involve hard trade-offs with other policy objectives. Win–win actions contribute to a central objective while also having other, e.g., social, economic and environmental, policy impacts and benefits

Path-dependency Extent to which a policy action (in a pathway) is limited by actions implemented in the past or by actions planned anterior in the pathway.

Plan A coherent framework of analysis of issues and challenges against well-chosen vision and goals or targets (expected results), as well as proposed set of measures, approaches, methods and resources following a time schedule.

Practice activities that people do mostly with a certain purpose, e.g. with respect to strategy making, those activities performed to develop or craft a strategy, using knowledge and experience but also underlying influences of, e.g., informal discussions, what is happening in a team, in organizations or dynamics in the broader context.

Practitioner Professional who has knowledge, qualifications and experience to perform certain from their (universities or knowledge) institutions derived professional practices. These actors form an important link in the chain of academia, government institutions, private sector and the public.

Praxis the relevant setting of strategy making where practitioners craft their strategic actions and besides, all actors and factors which influences that.

Robust actions Actions that result in acceptable indicator values under a wide variety of futures. Strategies and measures are considered as robust when performing well in a range of scenarios.

Scenario Coherent descriptions, narratives, of alternative hypothetical futures that reflect different perspectives on past, present and future developments, which can serve as a basis for action. In this publication, scenarios are presented as external scenarios that describe external developments that cannot be influenced and are thus policy-free.

Sector A sub-division within the government, focused on a policy domain with its discourse, legal arrangements and policy and planning instruments.

Sectoral planning Planning to address problems within a sector or policy domain, e.g., by a ministry mandated to act on it.

Signposts Information that should be tracked in order to determine whether implementation of action or reassessment of the plan is needed.

Strategy A from Delta Vision point of view coherent combination of measures that contributes to reaching the Delta Goals and achieve the vision over time.

Technology the application of scientific or otherwise professional or experiential knowledge for practical purposes. In a holistic approach or multi-sectoral plan, a broad body of knowledge comes available and needs to be managed to utilize it for strategizing and execution of measures and transformation. This is also called transformative technology management.

Tipping Point Threshold conditions under which an action or strategy will no longer effectively meet a set of predefined policy or strategic goals or standards

Transformation A process of meaningful change, in a sometimes dramatic and marked episode of change in, e.g., organization, physical form, nature or appearance.

Transformation management Making sense of the way an organization or other entity is performing in relation to opportunities or challenges, developing ideas or vision of how it could be and defining direction or proactive adjustments and the subsequent management of the moves to achieve the new vision.

References

Aggarwal, P. K., Kumar S. N., Pathak H. (2010). *Impacts of climate change on growth and yield of rice and wheat in the Upper Ganga Basin.* WWF Report.

Ahmad, Q. K., Biswas, A. K., Rangachari, R., & Sainju, M. M. (Eds.). (2001). *Ganges-Brahmaputra-Meghna region. A framework for sustainable development.* UPL.

Ahmed, M., Choudhury, G. A., & De Heer, J. M. (2023). Bangladesh Bodwip: Ekush Shotoker Bybosthapona. In *Bangla on the Bangladesh delta and delta plan 2100.* Prothoma Prokashoni.

Ahmed, M., & Suphachalasai, S. (2014). *Assessing the costs of climate change and adaptation in South Asia.* Asian Development Bank.

Ahmed, Y., Choudhury, G. A., & Ahmed, M. S. (2017). Strategy formulation and adaptation pathways generation for sustainable development of western floodplain of Ganges. *Journal of Water Resource and Protection, 9*, 663–691.

Akter, R., & Roy, J. K. (2017). The impacts of non-performing loan on profitability: An empirical study on banking sector of Dhaka stock exchange. *International Journal of Economics and Finance, 9*(3), 126–132.

Alam, M., Alam, M. M., Curray, J. R., Chowdhury, M. L. R., & Gani, M. R. (2003). An overview of the sedimentary geology of the Bengal basin in relation to the regional tectonic framework and basin-fill history. *Sedimentary Geology, 155*, 179–208.

Alam, S. (2015). *The economy in transition: Challenging issues.*

Alam, S., De Heer, J. M., & Choudhury, G. A. (Eds.). (2018). *Bangladesh delta plan 2100: Baseline studies.* https://plancomm.gov.bd/

Alam, S., De Heer, J. M., & Choudhury, G. A. Eds. (2018b). *Bangladesh Delta Plan 2100: Baseline Studies on Water Resources Management. Volume 1.* General Economics Division, Planning Commission, Government of Bangladesh. https://plancomm.gov.bd/

Alam, S., De Heer, J. M., & Choudhury, G. A. (Eds.). (2018c). *Bangladesh Delta Plan 2100: Baseline Studies on Disaster and Environment Management. Volume 2.* General Economics Division, Planning Commission, Government of Bangladesh. https://plancomm.gov.bd/

Alam, S., De Heer, J.M., & Choudhury, G.A. (Eds.). (2018d). *Bangladesh Delta Plan 2100: Baseline Studies on Land Use and Infrastructure Development. Volume 3.* General Economics Division, Planning Commission, Government of Bangladesh. https://plancomm.gov.bd/

Alam, S., De Heer, J. M., & Choudhury, G. A. (Eds.). (2018e), *Bangladesh Delta Plan 2100: Baseline Studies on Agriculture and Food Security. Volume 4.* General Economics Division, Planning Commission, Government of Bangladesh. https://plancomm.gov.bd/

Alam, S., De Heer, J. M., & Choudhury, G. A. (Eds.). (2018f), *Bangladesh Delta Plan 2100: Baseline Studies on Socio Economic Aspects of Bangladesh Delta. Volume 5.* General Economics Division, Planning Commission, Government of Bangladesh. https://plancomm.gov.bd/

J. M. de Heer and G. A. Choudhury, *Strategies for Adaptation to Climate Change in a Transformative Approach*, Springer Climate, https://doi.org/10.1007/978-3-031-88884-7

Alam, S., De Heer, J. M., & Choudhury, G. A. (Eds.). (2018g). *Baseline Studies on Governance and Institutional Development. Volume 6.* General Economics Division, Planning Commission, Government of Bangladesh. https://plancomm.gov.bd/

Ali, M. Z., Carlile, G., & Giasuddin, M. (2020). Impact of global climate change on livestock health: Bangladesh perspective. *Open Veterinary Journal, 10*, 178–188. https://doi.org/10.4314/ovj.v10i2.7

Anwar, M. S., Rahman, K., Bhuiyan, M. A. E., & Saha, R. (2022). Assessment of sea level and morphological changes along the eastern coast of Bangladesh. *Journal of Marine Science and Engineering, 10*(4), 527.

Aquastat, FAO. (2011). *FAO's information system on water and agriculture.* FAO, Food and Agriculture Organization of the United Nations.

Arif, H. (2022). Government to launch Tk 249cr project to facilitate irrigation in Rajshahi-Rangpur division. *The Business Post.*

Asian Development Bank. (2021). *Sovereign blue bonds.* Publication 731026.

Aziz, M. S. B., Hasan, N. A., Mondol, M. M. R., Alam, M. M., & Haque, M. M. (2021). Decline in fish species diversity due to climatic and anthropogenic factors in Hakaluki Haor, an ecologically critical wetland in northeast Bangladesh. *Heliyon, 7*(1), e05861.

BADC. (2021). *BADC Annual Report.*

Barḳat, A., Khan, S. H., Majumder, S., Badiuzzaman, M., Sabina, N., Ahamed, K., Abdullah, Md. (2015). *Local governance and decentralization in Bangladesh. Politics and Economics.* Pathak Shamabesh, Dhaka.

Bhatt, G. D. (2001). Knowledge management in organizations: Examining the interaction between technologies, techniques, and people. *Journal of Knowledge Management, 5*, 68–75.

Bangladesh Bureau of Statistics (BBS). (2022a). *Household Income and Expenditure Survey (HIES).*

Bangladesh Bureau of Statistics (BBS). (2022b). *Population and Housing Census (Census).*

Barua, P., Rahman, S. H., Barua, S., & Rahman, I. M. (2020). Climate change vulnerability and responses of fisherfolk communities in the South-Eastern coast of Bangladesh. *Water Conservation and Management, 4*(1), 20–31.

Barwick, M.A. (2011). *Checklist to assess organizational readiness.*

BBS. (2010, 2022). *Statistical Year Book of Bangladesh, Bureau of Bangladesh Statistics, Government of People's Republic of Bangladesh.*

BDP 2100 Knowledge Portal: http://bdp2100kp.gov.bd

Brammer, H. (1996). *The geography of the soils of Bangladesh.*

Brammer, H. (2014). *Climate change, sea-level rise and development in Bangladesh.* UPL.

Burton, I., Dickinson T. and Howard Y. (2008). Upscaling adaptation studies to inform policy at the global level. The Integrated Assessment Journal Bridging Sciences & Policy Vol. 8, Iss. 2 (2008), Pp. 25–37.

BWDB. (2021). *BWDB Annual Report.*

BWDB-UNDP. (1982). *Groundwater Survey: The hydrological conditions of Bangladesh.* UNDP Technical Report DP/UN/BGD-74-009/1

Caesar, J., & Janes, T. (2018). Regional climate change over South Asia. (2018) In R. J. Nicholls, C. W. Hutton, W. N. Adger, S. E. Hanson, M. Munsur Rahman, M. Salehin. (Eds.). *Ecosystem services for well-being in deltas* (pp. 207–221).

Carbon Brief. (2018). *Mapped: How every part of the world has warmed—and could continue to.* Infographics-multiple authors. URL: https://www.carbonbrief.org/mapped-how-every-part-of-the-world-has-warmed-and-could-continue-to-warm

Carter, C., Clegg, S. R., & Kornberger, M. (2008). *A very short, fairly interesting and reasonably cheap book about studying strategy.* Sage Publications.

CEGIS and Department of Environment. (2011). *Final report on programs containing measure to facilitate adaption to climate change of the second national communication project of Bangladesh.* Bangladesh.

CEGIS. (2021). *Technical note: Impacts of projected sea level rise on coastal embankments.* Bangladesh.

Centre for Effective Services. (2022). *An introductory guide to implementation*. https://thecenterforimplementation.com/toolbox/tag/Readiness+and+context
Chakraborty, J., Grineski, S. E., & Collins, T. W. (2019). Hurricane Harvey and people with disabilities: Disproportionate exposure to flooding in Houston, Texas. *Social Science and Medicine, V226*, 176–181.
Child, J. (1972). Organizational structure, environment and performance: The role of strategic choice. *Sociology, 6*, 2–22.
Choudhury, G. A., Terwisscha van Scheltinga, C., Van den Bergh, D., Chowdhury, F., De Heer, J. M., Hossain, M., & Karim, Z. (2012). *Preparations for the Bangladesh Delta Plan*. Alterra Report 2300. ISSN 1566-7197.
Climate Analytics. (2018). *Sea level legacy: 20cm more rise by 2300 for each 5-year delay in peaking emissions*, https://climateanalytics.org
Clingendael Institute, Van Genderen, R., & Rood. J. (2011). *Water Diplomacy Report*.
Crimmins, A., Balbus, J., & Gamble, J.L. et al. (2016). *The impacts of climate change on human health in the United States: a scientific assessment*. U.S. Global Change Research Program.
CSIRO. (2014). *Bangladesh integrated water resources assessment: Final report*.
Davenport, T. H., & Prusak, L. (2000). *Working knowledge: How organizations manage what they know* (pp. 1–7). Harvard Business School Press.
Davis, K. F., Bhattachan, A., D'Odorico, P., & Suweis, S. (2018). A universal model for predicting human migration under climate change: Examining future sea level rise in Bangladesh. *Environmental Research Letters, 13*(6), 064030.
Dewulf, A., & Termeer, C. (2015). Governing the future? The potential of adaptive delta management to contribute to governance capabilities for dealing with the wicked problem of climate change adaptation. *Journal of Water and Climate Change, 6*(4), 759–771.
De Geus, A. (1988, Mar–Apr 2–6). Planning as learning. *Harvard Business Review*.
De Heer, J. M. (1991). *Strategie en verandering in organisaties onder druk (Strategy and Change in Organizations Under Pressure)*. (PhD Dissertation). Leiden University. VUGA.
De Heer, J. M., Van Marrewijk, A., & Veenswijk, M. (2009a). Making sense of complexity in cross cultural collaboration. In *Unravelling organizational dynamics*. Prentice Hall/Pearson Education.
De Heer, J. M. (2009b). *Inaugural address for the chair 'Cultural Change and Strategy'*. VU University Amsterdam.
De Heer, J.M., Choudhury, G.A. et al. (2014). *Inception report Bangladesh delta plan 2100 formulation project*.
De Heer, J. M., Veenswijk, M., Merkus, S. (2015). *De bestuurlijke cultuur van besluitvorming in de Nederlandse infrastructurele sector*.
De Heer, J. M., Choudhury G. A., Islam, R. (2016). *Study Report on Governance*, BanDuDeltAS.
De Heer, J. M., & Aartsen, M. (2019, Mar). Bangladesh prepares itself for a climate resilient future. *Water Governance, 03/2019*.
De Heer, J. M., Minkman, E., & Van Alphen, J. (2021). Adaptation strategies and water governance in the Netherlands and Bangladesh. In: *China Water Risk (CWR)*, 22–11–2021.
De Heer, J. M. (2022). The Bangladesh Delta Plan 2100 (Chap. 12). In: de Wilde, K. (Ed.). *Moving Coastlines. Emergence and use of land in the Ganges–Brahmaputra–Meghna Estuary*. UPL.
De Heer, J. M., Oliemans W., & Choudhury G. A. (2022). Meaning of BDP 2100 for the coastal zone: Coastal development in BDP 2100 perspective (Chap. 13). In: de Wilde, K. (Ed.) *Moving Coastlines. Emergence and use of land in the Ganges-Brahmaputra-Meghna Estuary*. UPL.
De Heer, J. M. (2023, Mar). Update to: Bangladesh prepares itself for a climate resilient future. *Water Governance, 02/2023*.
De Heer, J. M., Choudhury, G. A., Oliemans, W., Ahmed, F., Tahsin, A., Hossain, J., Slager, K., Islam, T., Ara Parveen, I., Sadiq, S., Haque, Z., Rashid, A., Schasfoort, F., Van de Vat, M., Rahman, M.A., Bani, A., & Ahmed, S. (2023a). *Implementation program for the north-west region (Hurasagar Basin)*. SIBDP.

De Heer, J.M., Choudhury, G.A., Hossain, Md A., Ahmed, S., Haque Khan, Z., Hossain, Md J., Islam, T., Ara Pervin, I., Chandra Mridha, G., Alam, R., Kanti Magumdar, T., Uddin, Md S., Salam Sikder, Md A., & Tahsin, A. (2023b). *North-East (Haor) basin management plan: Implementation program*. SIBDP.

De Heer, J.M., Choudhury, G.A., Ahmed, F., & Ahmed, S. et al. (2023c). *Implementation program for the urban areas*. SIBDP.

De Heer, J.M., Choudhury, G.A., Islam, R., Hossain, Md A., & Ahmed, S. et al. (2023d). *Chattogram hill tracts implementation program*. SIBDP.

De Heer, J.M., Choudhury, G.A., Oliemans, W., Ahmed, F., Rahman, Md M., Sadik, Md S., Haque, Z., Rashid, A., Rahman, Md A., Banik, A., Hossain, Md J., Islam, T., Ara Pervin, I., Begum, M., Slager, K., Schasfoort, F., Tahsin, A., Hossain, Md A., & Ahmed, S. et al. (2023e). *Extended coastal zone basin management plan*. SIBDP.

De Heer, J.M., Choudhury, G.A., Akter, J., Tahsin, A., Haque Sarker, M., Hossain, M.A., & Ahmed, S. et al. (2023f). *Major river systems and estuaries: Implementation Program*. SIBDP

De Heer, J. M., Choudhury, G. A., Khan, F. A., Monowar-ul Haq, M., Huq Mou, B., Ahmed, S. *Monitoring and evaluation report*. SIBDP.

De Ruijter, P., & Alkema, H. (2014). *Scenario based strategy. Navigate the future*.

De Wit, B., & Meyer, R. (2010). *Strategy: Process, content*. Cengage Learning.

DPHE. (2001). *Groundwater studies for Arsenic contamination in Bangladesh*. Final Report, Rapid Investigation Phase, Department of Public Health Engineering, Government of Bangladesh.

DPHE-BGS. (2001). *Arsenic contamination of groundwater in Bangladesh*. Department of Public Health Engineering, Government of Bangladesh and British Geological Survey, rapid investigation phase, Final Report.

Duncan, N. E., & Wack, P. (1994). Scenarios designed to improve decision making. *Planning Review*, *22*(4).

Embassy of the Kingdom of the Netherlands. (2013). *Project Document. The Formulation of the Bangladesh Delta Plan 2100*.

Etzioni, A. (1967). Mixed-scanning: A "third" approach to decision-making. *Public Administration Review*, *27*(5), 385–392.

Etzioni, A. (1986, Jan–Feb). Mixed scanning revisited. *Public Administration Review*, *46*(1), 8–14.

Etzioni, A. (1989). Humble decision making. *Harvard Business Review*, *67*(4), 122–126.

Ezaz, G. T., Zhang, K., Li, X., Shalehy, M. H., & Hossain, M. A. (2022). Spatiotemporal changes of precipitation extremes in Bangladesh during 1987–2017 and their connections with climate changes, climate oscillations, and monsoon dynamics. *Global and Planetary Change, 208*, 103712.

FAO. (2011). *Aquastat*.

FFWC, BWDB: Flood Forecasting and Warning Centre, Bangladesh Water Development Board. www.ffwc.gov.bd

Fisher, F., & Forrester, J. (Eds.). (1993). *The argumentative turn in policy analysis and planning*. Durham.

Forastiere, F. (2010). Climate change and health: A challenge for epidemiology and public health. *International Journal of Public Health, 55*, 83–84.

Gioia, D. A., & Chittipeddi, K. (1991). Sensemaking and sensegiving in strategic change initiation. *Strategic Management Journal, 12*(6), 433–448.

Global Water Partnership. (2023). IWRM and Tools etc. https://www.gwp.org/

Government of Bangladesh (2013). Water Act 2013.

Government of Bangladesh. (2020). *Making vision 2041 a reality. Bangladesh perspective plan 2021–2041*.

Government of the Netherlands, Delta Commissioner. (2018). Monitoren en signaleren voor adaptief deltamanagement. Alert op veranderende omstandigheden. Retrieved from: www.Rijksoverheid.Nl/Deltaprogramma.

Government of the Netherlands, Delta Commissioner. (2018). https://english.deltaprogramma.nl/delta-programme/what-is-the-delta-programme/adaptive-delta-management

Gratton, L. (2000). Living Strategy. Putting people at the heart of corporate purpose.
GSB. (1990). *Geological Survey of Bangladesh.*
Gulich, E., & Quasthoff, U. M. (1986). Story-telling in conversation: cognitive and interactive aspects. *Poetics.*
Haasnoot, M. (2013). *Anticipating change: Sustainable water policy pathways for an uncertain future.*
Haasnoot, M., Diermanse, F., Kwadijk, J., De Winter, R., & Winter, G. (2019). *Strategieën voor adaptatie aan hoge en versnelde zeespiegelstijging. Een verkenning.* Deltares report 11203724-004.
Habiba, U., Abedin, M. A., Shaw, R., & Hassan, A. W. R. (2014). Salinity-induced livelihood stress in coastal region of Bangladesh. In *Water insecurity: A social dilemma* (Vol. 13, pp. 139–165). Emerald Group Publishing Limited.
Hodgson, G., Ali, R., Turner, J., Ahmed, M., Dawes, W., Masud, M. S., Hossain, M. J., Alam, S., Islam, M. M., & Saha, G. K. (2014). Bangladesh integrated water resources assessment supplementary report. *Water Table Trends and Associated Vertical Water Balance in Bangladesh.* CSIRO.
Hossain, P. R., Amjath-Babu, T. S., Krupnik, T. J., Braun, M., Mohammed, E. Y., & Phillips, M. (2021). Developing climate information services for aquaculture in Bangladesh: A decision framework for managing temperature and rainfall variability-induced risks. *Frontiers in Sustainable Food Systems, 5*, 677069.
Hussain, S.G. (2011). Assessing impacts of climate change on cereal production and food security in Bangladesh. In *Climate change and food security in South Asia* (pp. 459–476).
IPCC. (1990). *Strategies for Adaptation to Sea Level Rise.* Report of the Coastal Zone Management. Subgroup, IPCC Response Strategies Working Group.
IPCC. (2022). Climate Change 2022: Impacts, Adaptation and Vulnerability. In Pörtner, H.-O., Roberts, D. C., Tignor, M., Poloczanska, E. S., Mintenbeck, K., Alegría, A., Craig, M., Langsdorf, S., Löschke, S., Möller, V., Okem, A., Rama, B. (Eds.). *Contribution of working group II to the sixth assessment report of the intergovernmental panel on climate change* (pp. 3056). Cambridge University Press. https://doi.org/10.1017/9781009325844s
IPCC. (2023). *Climate change 2023: Synthesis report. Contribution of working groups i, ii and iii to the sixth assessment report of the intergovernmental panel on climate change.* In Lee, H., & Romero, J. (eds.).
International Institute for Sustainable Development. Global Environment Facility. (2020). *Credit-enhancement-instruments.*
International Institute for Sustainable Development. Global Environment Facility, GEF. (2023). Funding. https://www.thegef.org/who-we-are/funding.
International Institute for Sustainable Development. Global Environment Facility, GEF. (2019). Adaptation Fund. Retrieved from https://climatefundsupdate.org/the-funds/adaptation-fund
Islam, M. K., Rahaman, M., & Ahmed, Z. (2018). *Blue Economy of Bangladesh: Opportunities and challenges for sustainable development.*
Jakariya, M., & Islam, M. R. (Eds.). (2021). *Climate Change in Bangladesh. A cross disciplinary framework.*
Jarzabkowski, P. (2004). Strategy as practice: Recursiveness, adaptation and strategic practices-in-use. *Organization Studies, 25*(4).
Johnson, G., & Scholes, K. (2002). *Exploring corporate strategy* (6th ed.).
Kautto-Koivula, K. (1999). *Can societies learn from global organization*? EGEDL99 Conference. http://www.uta.fi/EGEDL/outline/kauttokoivula.htms
Kautto-Koivula, K. (1999). *Main operations for societal knowledge management.*
Khan, S. I., Islam, A. A., & Haque, M. I. (2008). *Political culture, political parties and the democratic transition in Bangladesh.* UPL.
Kingdon, J. W. (1995). *Agendas, alternatives, and public policies* (2nd ed.).

Kirby, M., Mainuddin, M., Palash, W., Qadir, E., & Shah-Newaz, S. M. (2014). *Bangladesh Integrated Water Resources Assessment supplementary report: Approximate regional water balances.*

Kornberger, M., & Clegg, S. R. (2011). Strategy as performative practice: The case of Sydney 2030. *Strategic Organization, 9*(2), 136–162.

Kwakkel, J. H., Haasnoot, M., & Walker, W. E. (2015). Developing dynamic adaptive policy pathways: A computer-assisted approach for developing adaptive strategies for a deeply uncertain world. *Climate Change, 132*, 373–386.

Leary, N., Adejuwon, J., Barros, V., Burton, L., & Lasco, R. (2007). *Climate change and adaptation.* Earthscan.

Maroon, M. (2023). *From poverty to progress: Understanding humanity's greatest achievement.*

Marsooli, R., Lin, N., Emanuel, K., Feng, K. (2019). Climate change exacerbates hurricane flood hazards along US Atlantic and Gulf Coasts in spatially varying patterns. *Nature Communications, 10*, 3785.

Malhi, G. S., Kaur, M., & Kaushik, P. (2021). Impact of climate change on agriculture and its mitigation strategies: A review. *Sustainability, 13*(3), 1318.

Mengel, M., Nauels, A., Rogeli, J & Schleussner, C. F. (2018). Committed sea-level rise under the Paris agreement and the legacy of delayed mitigation action. *Nature Communications, 9*(601).

Merkus, S., De Heer, J. M., & Veenswijk M. (2014). Decision-making as performative struggle: Strategic political-executive practices influencing the actualization of an infrastructural development. *Journal of Organizational Ethnography.* ISSN: 2046-6749

Merkus, S. (2014). *Struggling for meaning in the Dutch infrastructure sector*. VU University.

Middendorp, T. (2022). *De klimaat generaal. Bouwen aan weerbaarheid. Podium, Amsterdam.*

Ministry of Environment, Forest, and Climate Change. (2022). *National adaptation plan of Bangladesh (2023–2050).*

Ministry of Fisheries, Department of Fisheries. (2021). *Yearbook of fisheries statistics of Bangladesh.*

Ministry of Planning, Five Year Plans. https://plancomm.gov.bd/

Ministry of Planning, General Economics Division, Bangladesh Planning Commission. (2018). *Bangladesh delta plan (Bangladesh in the 21st Century), Volume 1: Strategy; Volume 2: Investment Plan.* https://plancomm.gov.bd/

Ministry of Planning, General Economics Division, Bangladesh Planning Commission (2015). *Seventh five year plan FY2016–FY2020 accelerating growth, empowering citizens.*

Ministry of Planning, General Economics Division, Bangladesh Planning Commission. (2012). *Perspective plan of Bangladesh 2010–2021.*

Ministry of Planning, General Economics Division, Bangladesh Planning Commission. (2011). *Sixth five year plan FY2011-FY2015 accelerating growth and reducing poverty.*

Ministry of Planning, GED, BDP 2100 Formulation Project. (2016). *Sixty years of water resources development in Bangladesh: Lessons Learnt.*

Ministry of Planning, General Economics Division, Bangladesh Planning Commission. (2021). *Promoting sustainable blue economy in Bangladesh through sustainable blue bond.*

Ministry of Planning, General Economics Division, Bangladesh Planning Commission. (2023). *BDP 2100 Knowledge portal.* http://bdp2100kp.gov.bd

Ministry of Science and Information and Communication Technology. (2028). *National ICT Policy.*

Ministry of Water Resources Bangladesh. (1999). *National Water Policy.*

Ministry of Water Resources Bangladesh. (2000). *Guidelines for participatory water management.*

Ministry of Water Resources Bangladesh. (2001). *National Water Management Plan.*

Ministry of Water Resources Bangladesh. (2005). *Coastal Zone Policy.*

Ministry of Water Resources, Bangladesh, Haor and Wetland Development Board (2012). Master Plan of Haor Area.

Ministry of Water Resources. (2015). *Rule on participatory water management.*

Mintzberg, H., & Waters, J. A. (1985). Of strategies, deliberate and emergent. *Strategic Management Journal, 6*(3).

Mirza, M. M. Q. (1997). Hydrological changes in the Ganges system in Bangladesh in the post-Farakka period. *Hydrological Sciences Journal, 42*(5), 613–631.

Morgan, G. (1997). *Images of organization* (2nd Ed.).

MPO. (1987). *Technical Report No 5. The Groundwater Resources and its Availability for Development.* Ministry of Irrigation, Water Development and Flood Control.

Muhith, A. M. A. (2007). *Agenda for Good Governance.* From lawlessness and corruption to a caring and prosperous democracy.

Mujeri, N., & Mujeri, M. K. (2022). South-South cooperation: Addressing development finance in Bangladesh. In: *The financial express.*

Naslund, L., & Pemer, F. (2012). The appropriated language: Dominant stories as a source of organizational inertia. *Human Relations, 65*(1), 89–110.

Netherlands Development Cooperation. (1993). *Flood Action Plan Bangladesh: A study of the debate on flood control in Bangladesh.*

Netherlands Institute of International Relations Clingendael. (2011). *Clingendael annual report.* Amsterdam.

Nonaka, I., & Toyama, R. (2015). The knowledge-creating theory revisited: Knowledge creation as a synthesizing process. In: *The essentials of knowledge management* (pp. 95–110).

OECD. (2011). *Water governance in OECD countries: A multi-level approach, OECD Studies on Water.* Paris: OECD Publishing

OECD. (2015). *Centre for entrepreneurship, SMEs, regions and cities.* Principles on Water Governance.

OECD. (2024). Principles on water governance. https://www.oecd.org/en/topics/sub-issues/water-governance/the-oecd-principles-on-water-governance-and-implementation-strategy.html

Ophra, S. J., Begum, S., Islam, R., & Islam, M. (2018). Assessment of bank erosion and channel shifting of Padma River in Bangladesh using RS and GIS techniques. *Spatial Information Research, 26*(6), 599–605.

Pal, L. A. (2014). *Beyond policy analysis—Public issue management in turbulent times* (5th ed.).

Pangare, G., & Nishat, B. (2014). Perspectives on Hydro diplomacy In: *Hydro-Diplomacy: Sharing water across borders.* Academic Foundation.

Patz, J. A., Stull, V. J., & Limaye, V. S. (2020) A low-carbon future could improve global health and achieve economic benefits. *JAMA 2.* https://doi.org/10.1001/jama.2020.1313

Pettigrew, A. M. (1992). The character and significance of strategy process research. *Strategic Management Review, 13*(2).

Quinn, J. B., Mintzberg, H., & James R. M. (1988). *The strategy process. Concepts, contexts, and cases.*

Rabbi, H. R., & Rahman, A. (2017). Ship breaking and recycling industry of Bangladesh; issues and challenges. *Procedia Engineering, 194*, 254–259.

Rahman, M. L., Shahjahan, M., & Ahmed, N. (2021). Tilapia farming in Bangladesh: Adaptation to climate change. *Sustainability, 13*(14), 7657.

Rahman, K. S., Islam, Z., Navera, U. K., & Ludwig, F. (2019). A critical review of the Ganges water sharing arrangement. *Water Policy, 21*(2), 259–276.

Rahman, M. M., & Bhattacharya, A. K. (2014). Saline Water Intrusion in Coastal Aquifer: A case study from Bangladesh. *Journal of Engineering, 4*(1), 7–13.

Rashid, M. H. A., Islam, A. B. M. J., Shirazy, B. J., & Shahidullah, S. M. (2017). Cropping systems and land use pattern in Rajshahi region. *Bangladesh Rice Journal, 21*(2), 237–254.

Rashid, S., & Rahman, R. (Eds.). (2010). *Water resource development in Bangladesh historical documents.* UPL.

Ritchie, H. (2024). *Not the end of the world. How we can be the first generation to build a sustainable planet.* ISBN13:9781784745011.

Saito, Y. (2005). Uplift of the Himalayas and Tibetan Plateau and large river deltas in Asia: Their characteristics and Holocene evolution. *The Journal of the Geological Society of Japan, 111*(11), 717–724.

Schipper, E., Ayers, J., Reid, H., Huq, S., & Rahman, A. (2020). *Community-based adaptation to climate change.*

Schwenker, B., & Wulf, T. (2013). *Scenario-Based strategic planning.*

Singh, O. P., Khan, T. M. A., & Rahman, M. S. (2001). Has the frequency of intense tropical cyclones increased in the north Indian Ocean? *Current Science*, 575–580.

Shamsudduha, M., Taylor, R. G., Ahmed, K. M., & Zahid, A. (2011). The impact of intensive groundwater abstraction on recharge to a shallow regional aquifer system: evidence from Bangladesh. *Hydrogeology Journal, 19*(4), 901–916. https://doi.org/10.1007/s10040-011-0723-4

Syvitski, J. P. M., Kettner, A. J., Overeem, I., Hutton, E. W. H., Hannon, M. T., Brakenridge, G. R., Day, J., Vörösmarty, C., Saito, Y., Giosan, L., & Nicholls, R. J. (2009). Sinking deltas due to human activities. *Nature Geoscience* 2, 681–686.

Shaw, R., Luo, Y., Cheong, T. S., Abdul Halim, S., Chaturvedi, S., Hashizume, M., Insarov, G. E., Ishikawa, Y., Jafari, M., Kitoh, A., Pulhin, J., Singh, C., Vasant, K., & Zhang, Z. (2022). Asia. In: Pörtner, H.-O., Roberts, D.C., Tignor, M., Poloczanska, E. S., Mintenbeck, K., Alegría, A., Craig, M., Langsdorf, S., Löschke, S., Möller, V., Okem, A., Rama, B. (Eds.) *Climate Change 2022: Impacts, adaptation and vulnerability. Contribution of working group II to the sixth assessment report of the intergovernmental panel on climate change.* Cambridge University Press.

SMRC. (2003). *The vulnerability assessment of the SAARC coastal region due to sea level rise: Bangladesh case.* SAARC Meteorological Research Centre.

Taylor, C. (2013). *The discourses of climate change.*

Tichy, N. M. (1983). *Managing strategic change, technical, political, cultural dynamics.*

Thompson, P. M., Sultana, P., & Islam, N. (2003). Lessons from community-based management of floodplain fisheries in Bangladesh. *Journal of Environmental Management, 69*(3), 307–321.

Timmermans, J., Haasnoot, M., Kwakkel, J., & Rutten, M. (2015). *Adaptive delta management: Roots and branches.* IAHR 2015.

UNCCD, www.unccd.int/land-and-life etc.

UNFCCC. (2012). *Climate Displacement in Bangladesh.*

UNFCCC. (2015). *Paris Agreement.*

UNFCCC COP28. (2023). United Nations Climate Change Conference or Conference of the Parties of the UNFCCC.

UNFCCC. (1987). *Montreal protocol on substances that deplete the ozone layer.*

UNFCCC. (1997). *Kyoto Protocol.*

Varma, H.K., Dhingra, A. S., & Raghu Rama Swamy, D. T. V. (2012). *Exploring public–private partnership in the irrigation and drainage sector in India.* Asian Development Bank.

Van Alphen, J., De Heer, J. and Minkman, E. (2021). Strategies for climate change adaptation: Lessons learnt from long-term planning in the Netherlands and Bangladesh *Water International*, 46(4), 477–504.

Van der Heijden, K. (1996). *Scenarios. The art of strategic conversation.* Wiley.

Wack, P. (1985, Sept–Oct). Scenarios: Uncharted waters ahead. *Harvard Business Review*, 73–89.

Weick, K. (1995). *Sensemaking in organizations.* Sage Publications.

Weiner, B. J. (2009). A theory of organizational readiness for change. *Implementation Science, 4*, 67.

Werners, S. E., Wise, R. M., Butler, J. R. A., Totin, E., & Vincent, K. (2021). Adaptation pathways: A review of approaches and a learning framework. *Environmental Science and Policy, 116*, 266–275.

Whittington, R. (2006). Completing the practice turn in strategy research. *Organization Studies, 27*(5), 614–635.

Whittington, R. (2017). Greatness takes practice: on practice theory's relevance to "great strategy". *Strategy Science, 3*(1), 343–351.

Wiig, K. (1995). *Knowledge management methods.*

World Bank Group. (2021). *Climate risk country profile: Bangladesh.*

World Bank Group. (2023). *Open data readiness assessment (ODRA).*

Zahid, A. (2015). *Groundwater management aspects in Bangladesh.*

Zahid, A., Qumrul Hassan, M., Imes, J. L., & Clark, D. W. (2009). Hydraulic characterization of aquifer(s) and pump test data analysis of deep aquifer in the affected Meghna river floodplain of Bangladesh. *Environmental Research Journal, 3*(2/3).

Zevenbergen, C., Van Alphen, J., Shah Alam Kahn, M., Terwisscha van Scheltinga, C., Veerbeek, W. (2018). Adaptive delta management a comparison between the Netherlands and Bangladesh delta program. *International Journal of River Basin Management, 16*(3), 1–7.

Index

J. M. de Heer and G. A. Choudhury, *Strategies for Adaptation to Climate Change in a Transformative Approach*, Springer Climate, https://doi.org/10.1007/978-3-031-88884-7

C

K

P

R

W

The manufacturer's authorised representative in the EU is Springer Nature Customer Service Centre GmbH, Europaplatz 3, 69115 Heidelberg, Germany. If you have any concerns regarding our products, please contact ProductSafety@springernature.com

Printed and bound by CPI Group (UK) Ltd, Croydon, CR0 4YY
01/10/2026
02225971-0016